T0139779

Mathematics of Planet Earth

Volume 8

Springer's Mathematics of Planet Earth collection provides a variety of well-written books of a variety of levels and styles, highlighting the fundamental role played by mathematics in a huge range of planetary contexts on a global scale. Climate, ecology, sustainability, public health, diseases and epidemics, management of resources and risk analysis are important elements. The mathematical sciences play a key role in these and many other processes relevant to Planet Earth, both as a fundamental discipline and as a key component of cross-disciplinary research. This creates the need, both in education and research, for books that are introductory to and abreast of these developments.

Springer's MoPE series will provide a variety of such books, including monographs, textbooks, contributed volumes and briefs suitable for users of mathematics, mathematicians doing research in related applications, and students interested in how mathematics interacts with the world around us. The series welcomes submissions on any topic of current relevance to the international Mathematics of Planet Earth effort, and particularly encourages surveys, tutorials and shorter communications in a lively tutorial style, offering a clear exposition of broad appeal.

More information about this series at http://www.springer.com/series/13771

Nedjeljka Žagar · Joseph Tribbia
Editors

Modal View of Atmospheric Variability

Applications of Normal-Mode Function Decomposition in Weather and Climate Research

Editors
Nedjeljka Žagar
Meteorological Institute
Universität Hamburg
Hamburg, Germany

Joseph Tribbia
National Center for Atmospheric Research
Boulder, CO, USA

ISSN 2524-4264 ISSN 2524-4272 (electronic)
Mathematics of Planet Earth
ISBN 978-3-030-60965-8 ISBN 978-3-030-60963-4 (eBook)
https://doi.org/10.1007/978-3-030-60963-4

Mathematics Subject Classification: 35-XX, 65-XX, 76Nxx, 86-XX

This Springer imprint is published by the registered company Springer Nature Switzerland AG
The registered company address is: Gewerbestrasse 11, 6330 Cham, Switzerland

Preface

This book reviews the theory and applications of normal mode functions in weather and climate dynamics and prediction. A major work on the formulation of the Hough normal mode functions and the construction of software packages for their computation took place at the National Center for Atmospheric Research in Boulder, Colorado, in 1970-1980. In particular, Akira Kasahara's work on normal mode functions, which is documented in Chapter 1 of this book, provided guidance and inspiration for many follow-on studies including the work by other authors contributing to this volume.

The history of numerical weather prediction is closely related to the application of the normal mode functions in data assimilation and initialization. In Chapter 2, an overview of nonlinear normal-mode initialization is presented including the pros and cons of its implementation in operational numerical weather prediction models. This is followed by Chapter 3 describing the application of normal mode functions in data assimilation, from the early attempts in optimal interpolation and three-dimensional variational data assimilation to recent efforts focusing on tropical aspects and the diagnosis of analysis uncertainties, forecast error covariances and predictability. In Chapters 4 to 6, the vast work of Hiroshi Tanaka and collaborators on a number of problems involving theoretical and numerical aspects of atmospheric dynamics and prediction is presented, including applications to barotropic and baroclinic instability, spatio-temporal variability and energy transfers.

The collaboration among the authors of various chapters of the book during the past decade resulted in MODES, a software package for the Hough normal mode function decomposition of global three-dimensional analyses and numerical model outputs.

As the results of research by other groups using MODES or equivalent diagnostic software packages emerge, we hope that this volume will find an audience among the next generation of atmosphere and ocean scientists who will revisit the old and find new applications of the Hough normal-mode functions.

This book is dedicated to the author of its leading chapter, Dr. Akira Kasahara who over the years has been a colleague, a mentor and a friend. It is his gentle

personality and lifelong research of normal modes that kept us on the track towards this volume.
Nedjeljka Žagar, Universität Hamburg, Germany
Joe Tribbia, National Center for Atmospheric Research, Colorado, USA
Hiroshi L. Tanaka, University of Tsukuba, Japan

Hamburg, Boulder and Tsukuba
August 2020

Acknowledgements

The work presented in this volume was often a result of a collaborative effort with our colleagues and students, and we gratefully acknowledge the contributions of everyone who contributed to the presented material. In particular, we are indebted to Paul N. Swarztrauber, who developed the code for the computation of the Hough harmonics that was used in most of studies employing the normal-mode functions and was built into MODES software package. The development of MODES and its application for variability analysis during 2012-2016 was supported by the European Research Council (ERC), Grant Agreement 280153.

We are grateful to Sergiy Vasylkevych and Nils Gustafsson for reviewing chapters 1 and 3, respectively, and to John Boyd and anonymous reviewers for their comments on the book.

It is also a pleasure to thank our Springer editor Martin Peters and his assistant Leonie Kunz for their support and guidance during the process of making this volume.

The following organisations and individuals provided permissions to reproduce the figures: European Centre for Medium Range Weather Forecasts (ECMWF), Society for Industrial and Applied Mathematics (SIAM), American Meteorological Society (AMS), American Geophysical Union (AGU), Dr. David Parrish, Tellus, Taylor & Francis, Wiley and Springer.

Contents

Chapter 1
3D Normal Mode Functions (NMFs) of a Global Baroclinic Atmospheric Model

Akira Kasahara

Abstract Characteristics of the atmospheric motions can be learned from the 3D normal modes of adiabatic, inviscid and linearized equations with respect to resting states. One popular atmospheric model consists of the hydrostatic baroclinic primitive equations (PEs) together with the thermodynamic and continuity equations. This chapter describes how to construct the 3D normal modes of the hydrostatic PE model on a sphere to aid the development of computer software of the normal mode functions.

1.1 Historical Perspective

Basic properties of the atmospheric motions can be investigated from solutions of the system of inviscid and adiabatic atmospheric prediction equations, describing small amplitude motions (called perturbations) superimposed on the hydrostatic basic state of no motion. For example, Monin and Obukhov (1959) examined such solutions of a compressible rotating system on the tangent f-plane. They showed that the solutions consist of essentially two different kinds of oscillations: One is the acoustic waves which arise from the compressibility of fluid. The other is the inertio-gravity waves which are due to the restitutive forces of the gravity and rotation of the Earth. The inertio-gravity (IG) waves play an essential role in the medium to small-scale flows such as frontal motions, severe storms, and tropical cyclones. The IG motions are also important for the large-scale flows particularly in the tropics (Žagar et al., 2009a,b). In contrast, the importance of acoustic (or sound) waves is limited only in very fast motions which play little role in slow weather phenomena.

The majority of weather events are governed by another kind of large-scale wave motion that is now referred to as the Rossby wave which was rediscovered in the late 1930-40s by Carl-Gustaf Rossby (1898-1957) and his collaborators through examination of upper and synoptic weather charts of the Northern Hemisphere. Platzman (1968) presents a detailed history of Rossby wave as well as its physics. Quoting his personal conversation with Jerome Namias, a noted climatologist, Platzman writes

N. Žagar and J. Tribbia (eds.), *Modal View of Atmospheric Variability*,
Mathematics of Planet Earth 8, https://doi.org/10.1007/978-3-030-60963-4_1

that "the discovery of Rossby wave is the important event in the period between the time of formulation of the Polar Front Theory and the inception of numerical weather prediction with the barotropic model." The Rossby wave is governed essentially by a restitutive wave mechanism associated with the meridional variation of the Coriolis parameter. The ingenuity of Rossby was the invention of so-called "beta plane" on the Cartesian geometry and his analysis on the property of planetary-scale motions based on the conservation of the vertical component of the absolute barotropic vorticity (Rossby et al., 1939). Bernhard Haurwitz (1905-1986) who had worked on the oscillations of the global atmosphere (Haurwitz, 1940) realized the connection between the Rossby wave and the oscillations of the second class found in the 19th century by Margules (1892) and independently by Hough (1898). It turns out that the oscillations of the first class, which they found, correspond to the IG waves mentioned earlier. We will come back to the discussion on the first and second class oscillations later.

Research on the oscillations of the atmosphere has a relatively long history. According to Taylor (1936), the subject was first treated by Laplace who showed that in an atmosphere of uniform temperature the oscillations are identical with those of an ocean of uniform depth. G. I. Taylor examined oscillations of a compressible baroclinic atmosphere and showed that the 3D system can be split into two separate structure equations in the horizontal and vertical which are connected through a parameter, referred to as the equivalent depth (or height). This practice has been extensively used in the theory of atmospheric tides (e.g. Wilkes, 1949; Siebert, 1961; Chapman and Lindzen, 1970).

In the early days of operational numerical forecasting over the Northern Hemisphere, quasi-geostrophic models over-predicted by a large margin the movement of planetary-scale (Rossby) waves of, say the longitudinal wavenumbers 1 through 3 (Bolin, 1955; Cressman, 1958). The reason for this phenomenon was that, even though the planetary-scale waves are quasi-geostrophic, the prediction models constructed under the assumption of small horizontal divergence relative to the vertical component of vorticity are too restrictive to describe global-scale weather changes. A satisfactory solution to this problem is to use the hydrostatic primitive equations (HPE) of motion which imposes no constraint on the role of horizontal divergence. In fact, all atmospheric general circulation models developed during the late 1950 to 1960s adopted the HPE (Washington and Kasahara, 2010). With an increased computing capability, operational forecasts organizations had started to use the HPE models over the Northern Hemisphere in the late 1960s (Shuman and Hovermale, 1968).

One complication from using the HPE model for weather forecasting is the need of special procedure to properly balance the mass and wind fields, called initialization, to set-up initial conditions. Because the HPE model is capable of handling fast propagating inertio-gravity motions, the lack of balance between the mass and wind fields in the initial conditions can yield large-amplitude IG motions immediately after time integration and mask the large-scale quasi-geostrophic flows (Hinkelmann, 1951). This was experienced during the late 1960s at the operational numerical forecasting centers and various methods were tried to balance the initial conditions. The problem of initialization turned out to be a rather difficult one and efforts to arrive at a satisfactory solution lingered well into the 1990s.

One of the early efforts to the practice of initialization, which is relevant to the theme of this article, was the work of Thomas Flattery. He was at the Department of Meteorology of University of Chicago and wrote a Ph.D. thesis under Prof. George Platzman entitled "Hough functions", which he published only in the form of technical report (Flattery, 1967). Using the Hough functions, he developed so-called the "Flattery objective analysis", which was used during the 1970s at the National Meteorological Center (now National Center for Environmental Prediction), NOAA in Washington, D.C. (Flattery, 1970). The Hough functions are eigenfunctions of the Laplace's tidal equations. The essence of the Flattery's analysis is as follows: The 3D hemispherical (or global) distributions of geopotential height and two horizontal wind components are expressed by a series involving the sum of products of Hough functions in the horizontal and empirical orthogonal functions (determined from observations for mass and horizontal wind components) in the vertical. These expansion functions were used to fit observed data and the analysis was derived by a least-square method. There are two problems in this procedure. One is that the analyzed mass and wind fields are not in strict balance due to the use of empirical functions in the vertical instead of normal mode functions. The other is that Flattery used only the Hough functions of the barotropic mode, as we explain later, so that the initial conditions contain practically no divergence, which is unrealistic.

In anticipation of the major improvement in weather forecasts with the use of expanded observations collected during the Global Weather Watch conducted worldwide in 1978-79 under the Global Atmospheric Research Program (GARP), the multivariate statistical interpolation (MSI) method was begun to be developed as an alternative to the Flattery analysis. Details of objective analysis methods, including the MSI, are described in an excellent textbook by Daley (1991). It turned out that replacing the Flattery analysis by the MSI analysis yielded an unexpected outcome. The MSI method indeed gave a better fit of observational data to the analysis than the Flattery method, but the dynamical balance between the mass and wind fields was much worse than the Flattery analysis. Thus, this experience gave a strong motivation to tackle the problem of initialization.

A satisfactory solution to this problem was proposed by Machenhauer (1977) and independently by Baer (1977), Baer and Tribbia (1977) and Tribbia (1979). When the initial wind filed is divergence free, just like the case of Flattery analysis, the motions of the first kind (IG modes) show up due to nonlinear effects in the HPE model, which become noise during the time integration. Therefore, it is better to have the initial wind field with an appropriate amount of horizontal divergence in such a way that the time tendency of horizontal divergence vanishes. To achieve this goal in the PE model we can work with the prognostic spectral equations of motions written in normal mode space. By setting the time tendencies of the spectral coefficient equations to be zero, we get the resulting diagnostic equations which are nonlinear. These nonlinear diagnostic equations are solved iteratively. For this reason, this balance procedure is referred to as nonlinear initialization in contrast to the linear initialization such as done by Flattery.

Because the construction of spectral prognostic equations for global motions involves the use of Hough functions and the solutions of nonlinear normal mode initialization (NNMI) are obtained in spectral domain, it was rather difficult to understand the physics of NNMI. Leith (1980) investigated the NNMI using a PE model on the mid-latitude beta plane in which the horizontal eigenfunctions can be expressed in terms of harmonic functions rather than the Hough functions in the global case. Leith introduced the concept of manifold to describe a collection of slow (low-frequency) rotational modes and fast (high-frequency) IG modes and represented the combination of slow and fast eigenvectors in the phase space on the orthogonal plane in graphical form called manifold diagram. Application of the geometric manifold diagrams facilitated a visual understanding of the movement of eigenvectors in the phase space connected, for example, in the process of obtaining solutions of variational NNMI (Tribbia, 1982). Daley (1991) used the manifold diagrams extensively to discuss the conditions for convergence of iteration to solve the NNMI.

It is not our intention to describe the history of initialization in this Chapter, except for the role of 3D normal mode functions (NMFs) played in the initialization for global numerical weather prediction. Independently from the application of NMFs to objective analysis by Flattery (1970), Dickinson and Williamson (1972) addressed the use of NMF representation to aid understanding of the features of various wave motions in the global atmosphere and constructed the NMFs of a two-layer primitive equation model by a finite-difference method. Later, Williamson and Dickinson (1976) developed the NMFs for a grid-point version of the NCAR general circulation model (GCM) and diagnosed GCM model simulation outputs in terms of the NMFs. With the advent of NNMI, operational forecasting centers started to develop the 3D NMFs for their data initialization. Temperton and Williamson (1981) and Williamson and Temperton (1981) constructed the 3D NMFs for the multi-level grid-point global model of the European Center for Medium-Range Weather Forecasts and applied it to the NNMI. For the Canadian operational spectral forecasting model, Daley (1979) formulated NMFs using the spectral method in the horizontal and the finite-element method in the vertical.

This chapter describes the formulation of 3D NMFs for the global primitive equation model based on the work of Kasahara and Puri (1981). In Section 2, we described the hydrostatic primitive equation (HPE) model which is extensively used for numerical weather prediction and climate simulation. In Section 3, we present the linearized HPE for the basic state at rest and derive the two systems of horizontal and vertical structure equations by means of the separation of variables. In Section 4 and 5, we discuss the properties of the eigensolutions of the vertical structure equation (VSE) and the numerical solutions of VSE, respectively. In Section 6 and 7, we discuss the properties of the horizontal structure equations (HSEs) and the numerical method of solutions of the HSEs, In Section 8, we demonstrate that the NMF expansion permits the partition of total energy into the kinetic and available potential energy of each vertical mode separately. In Section 9, we present the expansion of 3D global data in terms of the NMFs and in Section 10, we develop Hough spectral primitive equation model. Conclusions are stated in Section 11.

1.2 Atmospheric Model and Basic Equations

One of the popular atmospheric models is the hydrostatic primitive equation model that consists of the horizontal momentum equations together with the continuity and thermodynamic equations. The hydrostatic primitive equations are derived from the original Euler equations with two major assumptions, "shallowness" and "traditional approximations" in addition to a minor assumption that the geopotential of the Earth is a sphere. The derivation of the HPE prediction model can be seen for example in White et al. (2005). While the consequence of these two major approximations is under investigation, the HPE prediction model is believed to be suitable to describe large- to medium-scale atmospheric motions within the height of 100 km above the ground.

We adopt spherical coordinates (λ, φ, t) in the horizontal denoting, respectively, longitude, latitude, and time. In the vertical direction, we use the σ–coordinate (Phillips 1971) defined by

$$\sigma = p/p_s \quad , \tag{1.1}$$

where p and p_s denote the pressure and surface pressure, respectively. We assume that vertical velocity vanishes at the surface, $p = p_s$, and the model top, $p = 0$, for the mass conservation so that

$$\dot{\sigma} = \frac{d\sigma}{dt} = 0 \;\; \text{at} \;\; \sigma = 0 \;\; \text{and} \;\; \sigma = 1 \; , \tag{1.2}$$

where $\dot{\sigma}$ is referred to as the vertical σ velocity and d/dt denotes the total derivative as defined later.

Atmospheric prediction equations without dissipation and forcing for this coordinate system are presented by Kasahara (1974), which is referred to as K74, so that we summarize here only pertinent equations for the derivation of normal mode functions. The horizontal equations are given by (see (6.12) of K74)

$$\frac{du}{dt} - \left(2\Omega \sin\varphi + \frac{u \tan\varphi}{a}\right) v = \frac{-1}{a\cos\varphi}\left(g\frac{\partial z}{\partial \lambda} + RT\frac{\partial}{\partial \lambda}(\ln p_s)\right) , \tag{1.3}$$

$$\frac{dv}{dt} + \left(2\Omega \sin\varphi + \frac{u \tan\varphi}{a}\right) u = \frac{-1}{a}\left(g\frac{\partial z}{\partial \varphi} + RT\frac{\partial}{\partial \varphi}(\ln p_s)\right) , \tag{1.4}$$

where (u, v) denote the longitudinal and meridional velocity components, (a, g, Ω) are the radius, gravity and angular rotation rate of the Earth which are assumed as constant, z stands for geopotential height, (T, R) are the temperature and gas constant of air. The total derivative is defined by

$$\frac{d}{dt} = \frac{\partial}{\partial t} + \mathbf{V}\cdot\nabla + \dot{\sigma}\frac{\partial}{\partial \sigma} , \tag{1.5}$$

with the horizontal velocity $\mathbf{V} = (u, v)$ and

$$\mathbf{V} \cdot \nabla = \frac{u}{a \cos \varphi} \frac{\partial}{\partial \lambda} + \frac{v}{a} \frac{\partial}{\partial \varphi} . \tag{1.6}$$

The model atmosphere is assumed to be in hydrostatic equilibrium which is expressed by [see (6.13) of K74]

$$g \frac{\partial z}{\partial \sigma} = - \frac{RT}{\sigma} . \tag{1.7}$$

By integrating the above with respect to σ from the Earth's surface, $\sigma = 1$, where the elevation is denoted by $z = H(\lambda, \varphi)$, we obtain

$$gz = gH + R \int_{\sigma}^{1} \frac{T}{\sigma} d\sigma . \tag{1.8}$$

The mass continuity equation is expressed by [see (6.15) of K74]

$$\frac{\partial p_s}{\partial t} + \nabla \cdot (p_s \mathbf{V}) + p_s \frac{\partial \dot{\sigma}}{\partial \sigma} = 0 . \tag{1.9}$$

By integrating (1.9) with respect to σ from 0 to 1 and using the boundary condition (1.2), we get

$$\frac{\partial}{\partial t}(\ln p_s) = - \nabla \cdot \tilde{\mathbf{V}} - \tilde{\mathbf{V}} \cdot \nabla(\ln p_s), \quad \text{with} \quad \tilde{\mathbf{V}} = \int_0^1 \mathbf{V} \mathrm{d}\sigma . \tag{1.10}$$

Using the above equation, (1.9) can be expressed as

$$\frac{\partial \dot{\sigma}}{\partial \sigma} + \nabla \cdot \mathbf{V} + \mathbf{V} \cdot \nabla(\ln p_s) - \nabla \cdot \tilde{\mathbf{V}} - \tilde{\mathbf{V}} \cdot \nabla(\ln p_s) = 0 . \tag{1.11}$$

By integrating the above with respect to σ and using the boundary condition (1.2), we get

$$\dot{\sigma} = - \frac{1}{p_s} \int_0^{\sigma} \nabla \cdot (p_s \mathbf{V}) \, d\sigma + \frac{\sigma}{p_s} \nabla \cdot (p_s \tilde{\mathbf{V}}) . \tag{1.12}$$

The thermodynamic equation without external heating is given by [see (6.23) of K74]

$$\frac{\mathrm{d}T}{\mathrm{d}t} = \frac{RT}{c_p \, p} \frac{\mathrm{d}p}{\mathrm{d}t} , \tag{1.13}$$

where c_p denotes the specific heat at constant pressure.

By taking the total time derivative of the logarithm of (1.1), we get

$$\frac{1}{p} \frac{\mathrm{d}p}{\mathrm{d}t} = \frac{1}{\sigma} \frac{\mathrm{d}\sigma}{\mathrm{d}t} + \frac{1}{p_s} \frac{\mathrm{d}p_s}{\mathrm{d}t} = \frac{\dot{\sigma}}{\sigma} + \frac{1}{p_s} \frac{\partial p_s}{\partial t} + V \cdot \nabla(\ln p_s) . \tag{1.14}$$

The last two terms of the above expression come from the definition of $\mathrm{d}/\mathrm{d}t$ from (1.5) with the fact that p_s does not depend on σ. Thus, using (1.10) we can write (1.13) as

$$\frac{\mathrm{d}T}{\mathrm{d}t} = \frac{RT}{c_p}\left[\frac{\dot{\sigma}}{\sigma} - \nabla\cdot\tilde{\mathbf{V}} + (\mathbf{V}-\tilde{\mathbf{V}})\cdot\nabla(\ln p_s)\right]. \tag{1.15}$$

We now have a complete system of atmospheric prediction equations (without any additional terms for frictional and diabatic effects) when the initial conditions are given for horizontal wind components u and v, temperature T (or geopotential height z) at σ levels in the vertical, and the surface pressure p_s together with the boundary conditions (1.2) over a sphere. As supplementary initial conditions the geopotential height z (or the temperature T) and the vertical σ-velocity are obtained by the diagnostic equations (1.8) (or Eq. (1.7) and (1.12), respectively.

Now, we are ready to advance the state of atmosphere at $t = 0$ to $t = \Delta t$ (time increment) in the following way: 1. The velocity components (u, v) can be advanced by using (1.3) and (1.4) with the definition of $\mathrm{d}/\mathrm{d}t$ by (1.5) and (1.6).

2. The surface pressure p_s can be advanced by using (1.9).

3. The geopotential height z (or the temperature T) and the vertical σ-velocity are calculated diagnostically in the same way as it is done initially.

We repeat the process involving steps 1 to 3 to advance the atmospheric state to $t = n\Delta t$, where n is an integer.

The operational numerical weather prediction using the primitive equation system began during the 1960s. For more detail see Shuman and Hovermale (1968) and Robert et al. (1972).

1.3 Linearized Basic Equations and Separation of the Variables

The atmosphere is a vibrating system and has natural modes of oscillations just like a musical instrument. We can learn a great deal about the properties of motions by examining the atmosphere as a vibrating system. Although the atmospheric equations are nonlinear, they can be linearized if we are interested in small-amplitude motions as perturbations around the atmosphere at rest with no external forcing and heating. Solutions of such a system with appropriate boundary conditions are referred to as normal modes.

In this section we will derive the linearized version of the primitive equations and show how we obtain two separate systems of equations in the horizontal and vertical directions. For this purpose we introduce two new variables P and W defined by

$$P = gz + RT_0(\sigma)\ln p_s \quad \text{and} \quad W = \dot{\sigma} - \sigma\,[\nabla\cdot\tilde{\mathbf{V}} + \tilde{\mathbf{V}}\cdot\nabla(\ln p_s)]\,. \tag{1.16}$$

Using variable P, we rewrite the horizontal equations of motion (1.3) and (1.4) as

$$\frac{\partial u}{\partial t} - 2\Omega\sin\varphi\, v + \frac{1}{a\cos\varphi}\frac{\partial P}{\partial\lambda} = C_1, \tag{1.17}$$

$$\frac{\partial v}{\partial t} + 2\Omega \sin\varphi\, u + \frac{1}{a}\frac{\partial P}{\partial \varphi} = C_2, \tag{1.18}$$

where

$$C_1 = -V \cdot \nabla u - \dot{\sigma}\frac{\partial u}{\partial \sigma} - \frac{RT'}{a\cos\varphi}\frac{\partial(\ln p_s)}{\partial \lambda} + \frac{u\, v\tan\varphi}{a}, \tag{1.19}$$

$$C_2 = -V \cdot \nabla v - \dot{\sigma}\frac{\partial v}{\partial \sigma} - \frac{RT'}{a}\frac{\partial(\ln p_s)}{\partial \varphi} - \frac{u^2\tan\varphi}{a}, \tag{1.20}$$

and T' is the deviation of temperature T from the time-independent mean temperature $T_0(\sigma)$ so that

$$T' = T - T_0(\sigma). \tag{1.21}$$

Using variable W, we rewrite the continuity equation (1.11) as

$$\frac{\partial W}{\partial \sigma} + \nabla \cdot \mathbf{V} = -\mathbf{V} \cdot \nabla(\ln p_s). \tag{1.22}$$

We now need to express the thermodynamic equation (1.15) using variables P and W. By differentiating the definition of P in (1.16) with respect to σ and using the hydrostatic equation (1.7), we obtain

$$\frac{\partial P}{\partial \sigma} = -\frac{RT}{\sigma} + R\frac{dT_0}{d\sigma}\ln(p_s). \tag{1.23}$$

By differentiating the above equation with respect to t and using (1.15), (1.10), and (1.16), we get, after some manipulation,

$$\frac{\sigma}{R\Gamma_0}\frac{\partial}{\partial t}\left(\frac{\partial P}{\partial \sigma}\right) + W = C_3, \tag{1.24}$$

where

$$C_3 = \frac{1}{\Gamma_0}\left(\mathbf{V} \cdot \nabla T' - \frac{RT'}{c_p\sigma}W + \dot{\sigma}\frac{\partial T'}{\partial \sigma} - \frac{RT'}{c_p}\mathbf{V} \cdot \nabla(\ln p_s)\right) \tag{1.25}$$

and

$$\Gamma_0 = \frac{RT_0}{c_p\,\sigma} - \frac{dT_0}{d\sigma} \tag{1.26}$$

is a measure of the static stability of the basic state as a function of σ. We assume that $\Gamma_0 \neq 0$ for physical reality.

The boundary conditions for the original system are given by (1.2). We need to rewrite them for the new system using P and W. At the model top, we have

$$W = 0 \ \text{ at } \ \sigma = 0, \tag{1.27}$$

which is obvious. At the Earth's surface, $\sigma = 1$, and we have from (1.16) that

$$P_s = gH + R(T_0)_s \ln p_s \quad \text{and} \quad W_s = -[\nabla \cdot \tilde{\mathbf{V}} + \tilde{\mathbf{V}} \cdot \nabla(\ln p_s)]\,. \tag{1.28}$$

The subscript s for P and W indicates their values at the Earth's surface where its elevation is denoted by $H(\lambda, \varphi)$. Using (1.10), we get from (1.28) that

$$\frac{\partial P_s}{\partial t} - R(T_0)_s W_s = 0 \quad \text{at} \quad \sigma = 1\,, \tag{1.29}$$

where T_0 is assumed time independent.

We now consider motions of small amplitude superimposed on the basic state at rest with temperature T_0 which depends only on σ. We denote perturbation variables by a prime. The right-hand sides of (1.17), (1.18), (1.22), and (1.24) represent nonlinear terms. Therefore, by neglecting those right-hand side terms, we obtain the following system of linearized equations:

$$\frac{\partial u'}{\partial t} - 2\Omega \sin\varphi\, v' = -\frac{1}{a\cos\varphi}\frac{\partial P'}{\partial \lambda}\,, \tag{1.30}$$

$$\frac{\partial v'}{\partial t} + 2\Omega \sin\varphi\, u' = -\frac{1}{a}\frac{\partial P'}{\partial \varphi}\,, \tag{1.31}$$

$$\frac{\partial W'}{\partial \sigma} + \nabla \cdot V' = 0\,, \tag{1.32}$$

$$\frac{\sigma}{R\Gamma_0}\frac{\partial}{\partial t}\left(\frac{\partial P'}{\partial \sigma}\right) + W' = 0\,. \tag{1.33}$$

By eliminating W' between (1.32) and (1.33), we get

$$\frac{\partial}{\partial t}\left[\frac{\partial}{\partial \sigma}\left(\frac{\sigma}{R\Gamma_0}\frac{\partial P'}{\partial \sigma}\right)\right] - \nabla \cdot \mathbf{V}' = 0\,. \tag{1.34}$$

Equations (1.30), (1.31), and (1.34) constitute the system of equations for u', v', and P' that we are going to solve with the boundary conditions. The simplest form of boundary conditions consistent with (1.27), (1.29) and (1.33) is obtained as follows. At $\sigma = 1$, we get from (1.33) that

$$\frac{1}{R\Gamma_0}\frac{\partial}{\partial t}\left(\frac{\partial P'}{\partial \sigma}\right) + W'_s = 0\,, \tag{1.35}$$

where the subscript s for W' denotes its surface value. By eliminating W'_s in the above equation using (1.29), we can get the lower boundary condition containing only the variable P' as

$$\frac{\partial P'}{\partial \sigma} + \frac{\Gamma_0}{T_0}P' = 0 \quad \text{at} \quad \sigma = 1\,. \tag{1.36}$$

The upper boundary condition can be derived as

$$\sigma \frac{\partial P'}{\partial \sigma} = 0 \quad \text{at} \quad \sigma = 0 \tag{1.37}$$

from (1.27) and (1.33).

We now demonstrate that this system can be split into two groups of equations in the horizontal and vertical directions by the method of separation of variables. Let us assume that

$$(u', v', P') = (\hat{u}, \hat{v}, g\hat{h})\, G(\sigma), \tag{1.38}$$

where $(\hat{u}, \hat{v}, \hat{h})$ are functions of (λ, φ, t). By substituting (1.38) into (1.34), we get

$$\frac{1}{G}\frac{d}{d\sigma}\left(\frac{\sigma\, g}{R\Gamma_0}\frac{dG}{d\sigma}\right) = \nabla \cdot \hat{V}\left(\frac{\partial \hat{h}}{\partial t}\right)^{-1} = \frac{-1}{D}. \tag{1.39}$$

Note that the left-hand side term depends only on σ, while the middle term is a function of (λ, φ, t). Therefore, each term can be equal to $-1/D$, in which D is a separation constant, having the dimension of length.

By substituting (1.38) into (1.30), (1.31), and using (1.39), we can express the time-dependent horizontal system of equations as

$$\frac{\partial \hat{u}}{\partial t} - 2\Omega \sin\varphi\, \hat{v} = -\frac{g}{a \cos\varphi}\frac{\partial \hat{h}}{\partial \lambda}, \tag{1.40}$$

$$\frac{\partial \hat{v}}{\partial t} + 2\Omega \sin\varphi\, \hat{u} = -\frac{g}{a}\frac{\partial \hat{h}}{\partial \varphi}, \tag{1.41}$$

$$\frac{\partial \hat{h}}{\partial t} + D\nabla \cdot \hat{\mathbf{V}} = 0, \tag{1.42}$$

and the diagnostic vertical structure equation for $G(\sigma)$ as

$$\frac{d}{d\sigma}\left(\frac{\sigma g}{R\Gamma_0}\frac{dG}{d\sigma}\right) + \frac{1}{D}G = 0. \tag{1.43}$$

The top and bottom boundary conditions (1.36) and (1.37) are expressed by

$$\frac{dG}{d\sigma} + \frac{\Gamma_0}{T_0}G = 0 \quad \text{at} \quad \sigma = 1, \tag{1.44}$$

$$\sigma \frac{dG}{d\sigma} = 0 \quad \text{at} \quad \sigma = \sigma_T, \tag{1.45}$$

where σ_T is the model top.

We essentially followed the method of variable separation used by Taylor (1936) who referred to D as equivalent height. We will discuss in Sections 1.4 and 1.5 the properties of the vertical structure equation (1.43) and how to determine the value of D and the functional form of $G(\sigma)$, respectively.

1.4 Eigensolutions of the Vertical Structure Equation (VSE)

The VSE (1.43) with the boundary conditions (1.44) and (1.45) constitutes a Sturm-Liouville problem (Courant and Hilbert, 1953; Hildebrand, 1958). We present in this section some properties of its solutions. First, we rewrite (1.43) in a dimensionless form as

$$\frac{d}{d\sigma}\left(\frac{\sigma}{S}\frac{dG}{d\sigma}\right) + \mu G = 0, \tag{1.46}$$

where

$$S = \frac{R\Gamma_0}{gH_*} \quad \text{and} \quad \mu = \frac{H_*}{D}, \tag{1.47}$$

in which H_* is a scaling constant with the dimension of height and μ is the dimensionless inverse of equivalent height scaled by H_*.

We also rewrite the boundary conditions (1.44) and (1.45) as

$$\frac{dG}{d\sigma} + rG = 0, \quad \text{where} \quad r = \frac{\Gamma_0}{T_0} \quad \text{evaluated at} \quad \sigma = 1, \tag{1.48}$$

$$\sigma\frac{dG}{d\sigma} = 0 \quad \text{at the model top} \quad \sigma = \sigma_T\ . \tag{1.49}$$

Nontrivial solutions of (1.46) under boundary conditions (1.48) and (1.49) exist only for certain values of the parameter μ. The permissible values of μ are referred to as eigenvalues and the associated solutions of $G(\sigma)$ are called eigenfunctions, also known as characteristic or structure functions.

First we show that the values of μ must be positive. By multiplying (1.46) by G, integrating the resulting equation from σ= 0 to 1, and using the following identity,

$$G\frac{d}{d\sigma}\left(\frac{\sigma}{S}\frac{dG}{d\sigma}\right) = \frac{d}{d\sigma}\left(G\frac{\sigma}{S}\frac{dG}{d\sigma}\right) - \frac{\sigma}{S}\left(\frac{dG}{d\sigma}\right)^2 \tag{1.50}$$

and boundary conditions (1.48) and (1.49), we get

$$\int_0^1 \frac{\sigma}{S}\left(\frac{dG}{d\sigma}\right)^2 d\sigma + \left|\frac{r}{S}G^2\right|_{\sigma=1} = \mu\int_0^1 G^2 d\sigma\,. \tag{1.51}$$

Since the terms on the left-hand side are positive, it follows that the values of μ are positive. Namely, the equivalent height D is positive.

Next, we demonstrate that the eigenfunctions $G(\sigma)$ are orthogonal. Let us write (1.46) for one of the eigenfunctions as G_i corresponding to μ_i:

$$\frac{d}{d\sigma}\left(\frac{\sigma}{S}\frac{dG_i}{d\sigma}\right) + \mu_i G_i = 0\,. \tag{1.52}$$

Similarly, we can write (1.46) as

$$\frac{d}{d\sigma}\left(\frac{\sigma}{S}\frac{dG_j}{d\sigma}\right) + \mu_j G_j = 0, \tag{1.53}$$

for eigenfunction G_j corresponding to another eigenvalue μ_j. By multiplying (1.52) by G_j and (1.53) by G_i and subtracting the resultant equations from each other, we get

$$(\mu_j - \mu_i) G_i G_j = G_j \frac{\mathrm{d}}{\mathrm{d}\sigma}\left(\frac{\sigma}{S}\frac{dG_i}{d\sigma}\right) - G_i \frac{\mathrm{d}}{\mathrm{d}\sigma}\left(\frac{\sigma}{S}\frac{dG_j}{d\sigma}\right) = \frac{\mathrm{d}}{\mathrm{d}\sigma}\left[\frac{\sigma}{S}\left(G_j \frac{\mathrm{d}G_i}{d\sigma} - G_i \frac{dG_j}{\mathrm{d}\sigma}\right)\right]. \tag{1.54}$$

By integrating (1.54) with respect to σ from 0 to 1 and using boundary conditions (1.48) and (1.49), we obtain

$$(\mu_j - \mu_i) \int_0^1 G_i\, G_j\, \mathrm{d}\sigma = 0\ . \tag{1.55}$$

Since we assumed that $\mu_j \neq \mu_i$, the above result shows that the functions G_i and G_j are orthogonal. Combining the normalization of eigenfunctions, we can state its orthogonality as

$$\int_0^1 G_i(\sigma)\, G_j(\sigma)\, \mathrm{d}\sigma = \delta_{ij}\ , \tag{1.56}$$

where $\delta_{ij} = 1$ if $i = j$ and zero otherwise.

Solutions of the VSE in the σ-system were examined earlier by Jacobs and Wiin-Nielsen (1966), Simons (1968), and Wiin-Nielsen (1971a,b) in connection with quasi-geostrophic modeling. They assumed a constant basic-state temperature lapse rate in geopotential height and obtained analytical solutions. Their studies followed much earlier investigations of the VSE with geopotential height as the vertical coordinate in connection with the study of atmospheric tides. Since Taylor (1936) demonstrated that a global compressible atmospheric model can be split into the two systems of horizontal and vertical structure equations connected with a separation constant, many investigators undertook the study of atmospheric tides as a problem of determining the solutions of these two systems. In particular, considerable efforts were made to calculate values of the equivalent height from solutions of the VSE under various basic state temperature profiles and upper boundary conditions. An early history on the theory of atmospheric tides is found in Wilkes (1949) and Siebert (1961). Later development on the classical theory of atmospheric tides is described by Chapman and Lindzen (1970). However, there is one important difference in the approaches to studying atmospheric tides and calculating global normal mode functions. We will discuss what that difference is in Section 1.6.

It is instructive to calculate analytically the eigenvalues and associated eigenfunctions of the VSE (1.46) with boundary conditions (1.48) and (1.49) by assuming a specific profile of the basic state temperature $T_0(\sigma)$ or the stratification parameter $\Gamma_0(\sigma)$. As one example of illustrating the eigensolutions of the VSE, we assume the following basic state temperature profile $T_0(\sigma)$,

$$T_0(\sigma) = (T_s - T_\infty)\exp\left(\frac{R}{c_p}\ln\sigma\right) + T_\infty, \tag{1.57}$$

where T_s and T_∞ denote the temperatures at the surface and at infinity ($\sigma = 0$), respectively.

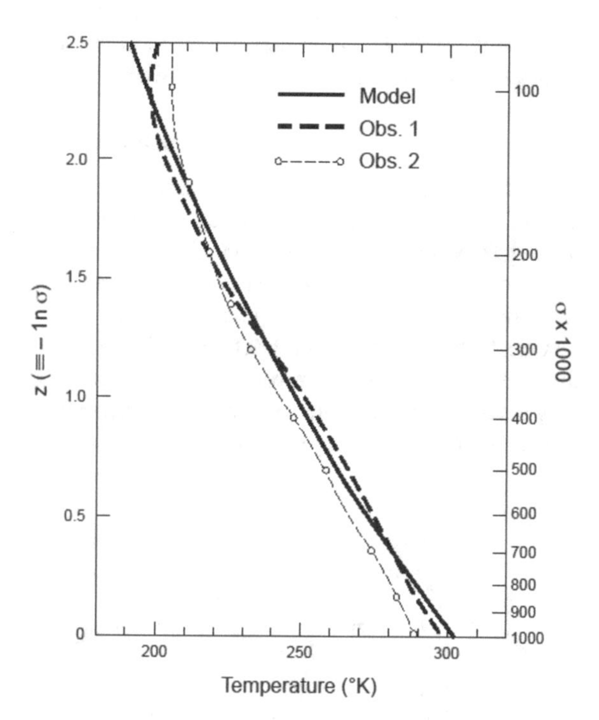

Fig. 1.1 Profile of T_0 as a function of $Z = -\ln\sigma$. The scale of σ multiplied by 1000 is shown on the right axis. Obs. 1 denotes a mean tropical temperature distribution (Jordan, 1958) and Obs. 2 a global mean temperature distribution during the FGGE (Tanaka, 1985). From Kasahara and Tanaka (1989), Fig. 1. ©American Meteorological Society. Used with permission.

Figure 1.1 shows the profile of T_0 as a function of dimensionless vertical scale $Z(= -\ln\sigma)$ on the left. The scale for σ multiplied by 1000 is shown on the right. Obs. 1 denotes a mean tropical temperature distribution (Jordan, 1958) and Obs. 2 a global mean temperature profile during FGGE (Tanaka, 1985). The solid line for model is calculated with the values of T_s= 302.53 K and T_∞= 83.265 K. Fulton and Schubert (1980) adopted the same form of $T_0(\sigma)$ and showed that, because T_∞ is finite, the geopotential height of the basic state calculated from the hydrostatic equation diverges, i.e. the value approaches infinity at $\sigma = 0$. This creates a problem in that the spectrum of the Sturm-Liouville operator defined by (1.46)-(1.49) may not be discrete as explained later. Therefore, the top of the model is placed at a finite level of Z_T $(= -\ln\sigma_T)$ for analytical calculations.

For the basic state temperature profile of the form (1.57), the stability parameter Γ_0 as defined by (1.26) and its dimensionless form S as defined in (1.47) are expressed as

$$\Gamma_0 = \frac{1}{\sigma}\left(\frac{T_\infty R}{c_p}\right) \text{ and } S = \frac{1}{\sigma}\left(\frac{T_\infty R^2}{c_p g H_*}\right) . \tag{1.58}$$

The analytical solutions of the VSE corresponding to the above form of Γ_0 or S were obtained by Gavrilin (1965), Fulton and Schubert (1980) , Staniforth et al. (1985), Kasahara and Tanaka (1989), and Terasaki and Tanaka (2007). Figure 1.2 shows the profiles of eigenfunctions G_m for the first twelve vertical modes as functions of $Z(= -\ln\sigma)$ with the top at $Z_T = 2.5$ (Kasahara and Tanaka, 1989). The numbers beside the profiles indicate mode index m. The corresponding values of the equivalent height D_m in meters are listed at the top of the figure. Observe that the values of equivalent height D_m as an inverse of the eigenvalue μ_m form a discrete spectrum

$$D_1 > D_2 > D_3 > ... > D_m > 0 , \tag{1.59}$$

where the integer subscript m can be chosen as large as one wishes to calculate. The case of $m = 1$ is the lowest mode referred to as the external mode corresponding to the largest equivalent height D_1 and its eigenfunction G_1 has no zero-crossing point in its profile. The remaining cases of $m \geq 2$ are referred to as the internal modes and G_m have $m - 1$ zero-crossing (nodal) points.

One feature of the eigenfunction profiles in Fig. 1.2 which should be noted is that the vertical scale of zero-crossing points for higher mode index m becomes very small, if the G_m were plotted in the linear scale of σ. Moreover, the amplitude of G_m becomes very large toward the model top. Figure 2 in Terasaki and Tanaka

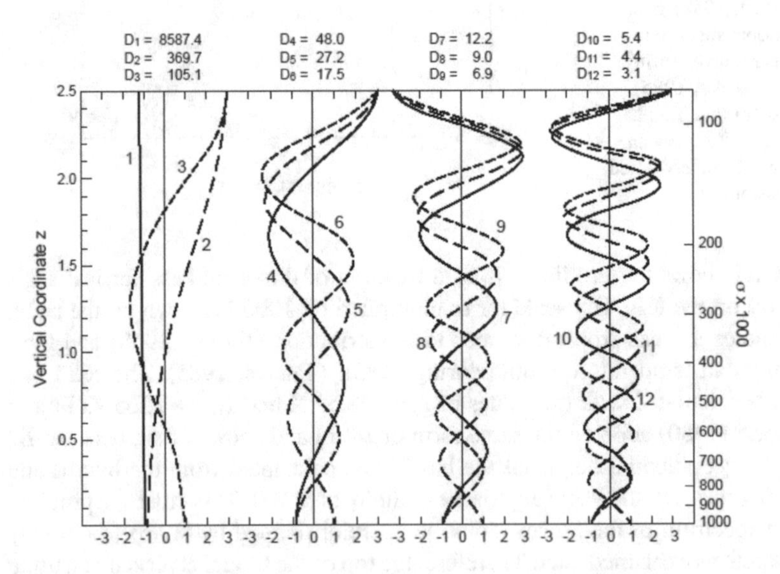

Fig. 1.2 Profiles of eigenfunctions G_m for the first twelve vertical modes as functions of Z. Numerals besides the profiles indicate modal index m. The values of equivalent height D_m are listed at the top. From Kasahara and Tanaka (1989), Fig. 2. ©American Meteorological Society. Used with permission.

(2007) shows a much more dramatic increase of the amplitudes with height, when the model top is placed much higher at Z_T approximately 7 instead of 2.5. Staniforth et al. (1985) noted that the basic state atmosphere is unbounded, if Γ_0 is inversely proportional to σ. Namely, the amplitude of G_m diverges as σ approaches to zero. In fact, they investigated the solutions for a more general form of $\Gamma_0(\sigma) = A\sigma^{1-\alpha}$, where A and α are positive values. For $1 < \alpha \leq 2$, they obtained discrete solutions in terms of the Bessel functions with respect to σ which do not diverge for $\sigma \to 0$.

The existence of well-behaved eigenfunctions in σ was investigated by Cohn and Dee (1989) who posed two basic questions concerning the eigensolutions of VSE for a more general profile of $\Gamma_0(\sigma)$ or $T_0(\sigma)$. First, under what condition the spectrum of eigenvalue of VSE becomes discrete or continuous? Second, what boundary condition is appropriate at the top? They showed that the spectrum is discrete provided the basic atmosphere is bounded, a condition which is also consistent with the approximations required to derive PE. Actually it is impractical to deal with the continuous spectrum of eigensolutions for the expansion of atmospheric data given at discrete σ-levels. They show that the spectrum is discrete and the vertical velocity vanishes as $\sigma \to 0$, provided the basic atmosphere is bounded. The basic state atmosphere is said to be bounded when the geopotential height for a specific basic temperature profile is finite. A necessary condition for the bounded atmosphere is expressed as

$$T_0(\sigma) \to 0 \text{ for } \sigma \to 0 . \tag{1.60}$$

Analytical solutions are instructive, but they are too idealized for a practical use, because the profile $\Gamma_0(\sigma) = A\sigma^{1-\alpha}$ is hardly representative of the atmosphere which also embraces the stratosphere and beyond. As an example of the temperature distribution of real atmosphere, we show in Fig. 1.3 the profile of $T_0(z)$ of the U.S. standard atmosphere (NAS, 1962) as a function of geometric height. Figure 1.4 shows the profile of Brunt-Väisälä frequency (BVF) in units of sec^{-1} which is another kind

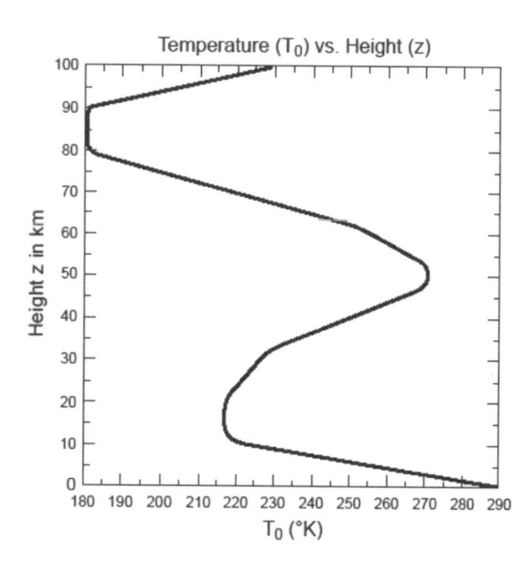

Fig. 1.3 Profile of $T_0(z)$ in degrees K of the U.S. standard atmosphere as a function of geometric height in km.

of stratification parameter related to Γ_0. The BVF is defined by

$$BVF = \left[\frac{g}{T_0}\left(\frac{g}{c_p} + \frac{dT_0}{dz}\right)\right]^{1/2} . \tag{1.61}$$

Because the vertical derivative of the temperature is discontinuous, we had to smooth the graph of BVF. To obtain the solutions of VSE for real data analysis, it is clear that we must resort to a numerical method.

Before presenting the numerical method to obtain the eigensolutions of VSE in the next section, we should note the completeness of eigenfunctions as expected from the solutions of a Sturm-Liouville problem. Namely, we can represent any well-behaved continuous function $F(\sigma)$ between $\sigma = 0$ and 1 by a linear combination of G_m as

$$F(\sigma) = \sum_{m=1}^{M} C_m G_m \sigma \quad \text{in the limit of} \quad M \to \infty , \tag{1.62}$$

where C_m is the expansion coefficient which can be obtained from the inverse transform

$$C_m = \int_0^1 F(\sigma)\, G_m \, d\sigma \tag{1.63}$$

by virtue of the orthogonality condition (1.62). In reality we only use a finite number of modes to represent $F(\sigma)$.

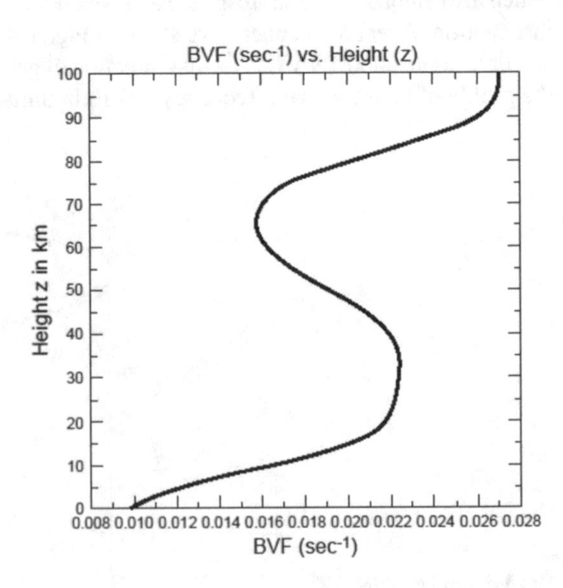

Fig. 1.4 Profile of Brunt-Väisälä frequency (BVF) in units of $\sec^{-1}$ calculated from $T_0(z)$ of the U.S. standard atmosphere.

1.5 Numerical Solutions of the VSE

Since the vertical profile of $\Gamma_0(\sigma)$ in the atmosphere varies considerably as seen from Fig. 1.4, we must resort to a numerical method to obtain the eigensolutions of the VSE. There are three basic numerical formulations to solve the VSE. Staniforth and Daley (1977) proposed a finite-element formulation for vertical discretization in the sigma system which was later used by Daley (1979) to solve the VSE for non-linear normal mode initialization. Sasaki and Chang (1985) applied a finite-element method for solution of the VSE. Kasahara (1984) and Castanheira et al. (1999) solved the VSE using a spectral (Galerkin) method with the Legendre polynomials as basis functions. Kasahara and Puri (1981, referred here to KP81) adopted a finite-difference formulation to solve the VSE. Although some authors attempted to evaluate the performance of various numerical schemes by comparing their results with analytical solutions for relatively simple profiles of $\Gamma_0(\sigma)$, we do not yet have a clear indication as to the best numerical scheme for a realistic profile of $\Gamma_0(\sigma)$. In this section we, therefore, adopt a finite-difference scheme following KP81 to solve the VSE (1.46) with (1.47) and boundary conditions (1.48) and (1.49).

Figure 1.5 illustrates discrete σ levels in solid lines and layer increment $\Delta\sigma_j$, in which the subscript j denotes integer index for the levels where the eigenfunction $G(j)$ is calculated. Index j ranges from 1 to the maximum number of layers J. Half-integers represent mid-levels, denoted by dashed lines, of two consecutive σ levels. The additional σ level corresponding to $j = J + 1$ is created to handle the bottom boundary condition (1.48).

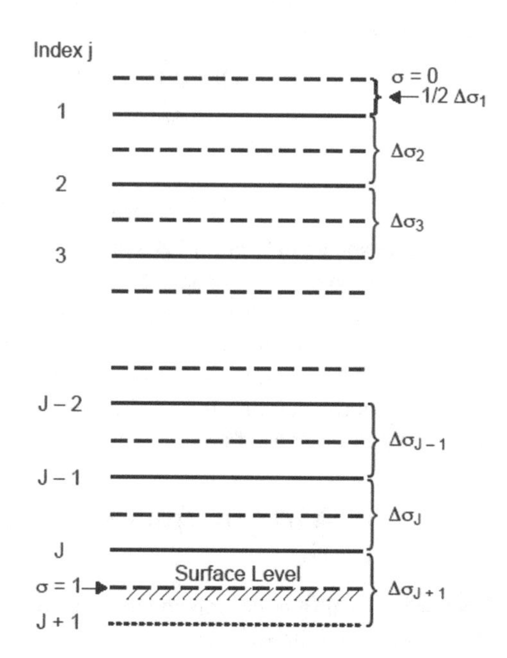

Fig. 1.5 Discrete representation of variables in the vertical. Solid lines show sigma levels where the eigenfunctions are calculated. From Kasahara and Puri (1981), Fig. 1. ©American Meteorological Society. Used with permission.

The difference form at level $j = 1$ is given by

$$\frac{2}{(\Delta\sigma_1 + \Delta\sigma_2)} \left[\left(\frac{\sigma}{S}\right)_{1\frac{1}{2}} \frac{G(2) - G(1)}{\Delta\sigma_2} - 0 \right] + \mu\, G(1) = 0 . \tag{1.64}$$

The upper boundary condition at $\sigma = 0$ is indicated by zero as the second term in the brackets.

The difference form at level $2 \leq j \leq J - 1$ is given by

$$\frac{2}{(\Delta\sigma_j + \Delta\sigma_{j+1})} \left[\begin{array}{l} \left(\frac{\sigma}{S}\right)_{j+\frac{1}{2}} \frac{G(j+1)-G(j)}{\Delta\sigma_{j+1}} \\ -\left(\frac{\sigma}{S}\right)_{j-\frac{1}{2}} \frac{G(j)-G(j-1)}{\Delta\sigma_j} \end{array} \right] + \mu G(j) = 0 . \tag{1.65}$$

The difference form at level J is given by

$$\frac{2}{(\Delta\sigma_J + \Delta\sigma_{J+1})} \left[\begin{array}{l} \left(\frac{\sigma}{S}\right)_s \frac{G(J+1)-G(J)}{\Delta\sigma_{J+1}} \\ -\left(\frac{\sigma}{S}\right)_{J-\frac{1}{2}} \frac{G(J)-G(J-1)}{\Delta\sigma_J} \end{array} \right] + \mu G(J) = 0 . \tag{1.66}$$

The subscript s for (σ/S) in the brackets indicates the value at the surface, namely at $\sigma = 1$. The value of $G(J + 1)$ in the brackets can be evaluated from the boundary condition (1.48) which is expressed by the following difference form,

$$\frac{G(J+1) - G(J)}{\Delta\sigma_{J+1}} = -0.5\, r_s [G(J+1) + G(J)] \tag{1.67}$$

Thus, the unknown $G(J + 1)$ can be expressed in terms of $G(J)$. Once it is done, we can rewrite (1.67) as

$$\frac{G(J+1) - G(J)}{\Delta\sigma_{J+1}} = -\frac{b}{\Delta\sigma_{J+1}} G(J) , \tag{1.68}$$

where

$$b = r_s\, \Delta\sigma_{J+1} \left(1 + 0.5\, r_s\, \Delta\sigma_{J+1}\right)^{-1} . \tag{1.69}$$

We have shown in Section 1.4 that the eigenfunctions $G(\sigma)$ are orthogonal. When the eigenfunctions are evaluated at the discrete levels, we can apply the same principle of proof to show the discrete eigenfunctions are orthogonal, but the proof requires a slight modification as described in the following. Namely, we can show from the system of difference equations (1.64)-(1.68) that two eigenfunctions $G_l(j)$ and $G_m(j)$ at level σ_j corresponding to two different eigenvalues μ_l and μ_m, respectively, are orthogonal in the sense that

$$\sum_{j=1}^{J} G_l(j)\, G_m(j)\, 0.5(\Delta\sigma_j + \Delta\sigma_{j+1}) = 0 \quad for \quad l \neq m . \tag{1.70}$$

This orthogonality condition suggests the proper definition of the vertical structure function in the finite-difference framework should be

$$\Phi(j) = G(j)\left[0.5\left(\Delta\sigma_j + \Delta\sigma_{j+1}\right)\right]^{1/2}, \tag{1.71}$$

where the vertical weighting is used to insure orthonormality. Note, also, that the $\Phi(j)$ eigenvectors do not satisfy the vertical boundary conditions and the $G(j)$ eigenfunctions must be used to produce proper field values.

With this definition of $\Phi(j)$, system (1.64)-(1.66) can be rewritten in the matrix form

$$\mathbf{M\Phi} = \mu\mathbf{\Phi}, \tag{1.72}$$

where $\mathbf{\Phi}$ is the eigenvector

$$\mathbf{\Phi} = [\Phi(1) \quad \Phi(2) \quad \ldots \quad \Phi(J-1) \quad \Phi(J)]^T \tag{1.73}$$

and $\mathbf{M}$ is the tri-diagonal matrix as

$$\left|\begin{array}{cccccccccc} \frac{C_1}{d_1{}^2} & \frac{-C_1}{d_1 d_2} & 0 & 0 & . & . & . & 0 & . & 0 \\ \frac{-C_1}{d_1 d_2} & \frac{C_1+C_2}{d_2{}^2} & \frac{-C_2}{d_2 d_3} & 0 & . & . & . & 0 & . & 0 \\ & & \frac{-C_{j-1}}{d_{j-1} d_j} & \frac{C_{j-1}+C_j}{d_j{}^2} & \frac{-C_j}{d_j d_{j+1}} & & & & & \\ 0 & . & 0 & \frac{-C_{J-2}}{d_{J-2} d_{J-1}} & \frac{C_{J-2}+C_{J-1}}{d_{J-1}{}^2} & \frac{-C_{J-1}}{d_{J-1} d_J} & & & & \\ 0 & . & 0 & 0 & \frac{-C_{J-1}}{d_{J-1} d_J} & \frac{C_{J-1}+C_J}{d_J{}^2} & & & & \end{array}\right|,$$

in which

$$C_j = \left(\frac{\sigma}{S}\right)_{j+\frac{1}{2}} (\Delta\sigma_{j+1})^{-1}, \qquad j = 1, 2, \ldots, J-1 \quad , \tag{1.74}$$

$$C_J = \frac{r_s}{S_s}(1 + 0.5\, r_s\, \Delta\sigma_{J+1})^{-1} = \frac{gH_*}{R\,(T_0)_s}(1 + 0.5\, r_s\, \Delta\sigma_{J+1})^{-1} \quad , \tag{1.75}$$

and

$$d_j = [0.5\,(\Delta\sigma_j + \Delta\sigma_{j+1})]^{1/2}, \; j = 1, 2, \ldots, J\ . \tag{1.76}$$

Since the matrix $\mathbf{M}$ is symmetric and of order J, we obtain J eigenvalues of μ and J eigenvectors of $\mathbf{\Phi}$ which are real. By combining with the normalization of eigenvectors, the orthogonality of $\mathbf{\Phi}$ is expressed by

$$\sum_{j=1}^{J} \Phi_l(j)\,\Phi_m(j) = \delta_{lm}\ . \tag{1.77}$$

We can calculate the equivalent height D_m from the eigenvalue μ_m using its definition (1.47). Then we get a sequence of D_m as shown by (1.59).

1.6 Properties of the Horizontal Structure Equations (HSEs)

We now discuss the properties of the horizontal structure equations (HSEs) given by (1.40)-(1.42). In order to let this section to stand by itself, we write the HSEs again in the form

$$\frac{\partial u}{\partial t} - 2\Omega \sin\varphi\, v = -\frac{g}{a\cos\varphi}\frac{\partial h}{\partial \lambda}, \tag{1.78}$$

$$\frac{\partial v}{\partial t} + 2\Omega \sin\varphi\, u = -\frac{g}{a}\frac{\partial h}{\partial \varphi}, \tag{1.79}$$

$$\frac{\partial h}{\partial t} + \frac{D}{a\cos\varphi}\left[\frac{\partial u}{\partial \lambda} + \frac{\partial}{\partial \varphi}(v\cos\varphi)\right] = 0\,. \tag{1.80}$$

These are identical in form to linear shallow-water equations in the basic state of no motion and constant depth D and are referred originally to as Laplace's tidal equations (cf. Lamb 1932) without tide-generating force.

After multiplying (1.78) - (1.80) by uD, vD and g, respectively, adding the resulting three equations, and integrating the sum over the entire sphere, we get

$$\frac{\partial}{\partial t}\int \frac{1}{2}\left[D(u^2+v^2)+gh^2\right] a^2\cos\varphi d\varphi d\lambda = 0\,. \tag{1.81}$$

This shows that the sum of the total kinetic and potential energies is invariant with respect to time.

Although the solutions of HSEs have been investigated since Margules (1892) and Hough (1898) in the late 19th century, our full understanding of the solutions did not come until the 1960s when high-speed electronic computers became available.

1.6.1 Case of Forced Oscillations – Tidal Problem

Historically the solutions of the HSEs were investigated in connection with the theory of atmospheric tides as reviewed for example by Siebert (1961) who referred to the eigenfunctions of the HSEs to as Hough functions, because the computing scheme developed by Hough (1898) had been used for calculating the eigenfunctions of HSEs up to the mid 1960s. For tidal problems, we know the periods of oscillations, such as diurnal and semi-diurnal, which are generated by gravitational forces of the moon and the sun and by thermal forces due to the absorption and emission of heat connected with periodic incoming solar radiation. Therefore, for forced problems with a known frequency of oscillation what has to be determined from (1.78) - (1.80) is the value of equivalent height D. Once D is obtained, then the VSEs similar in form to (1.46), expressed using geometric height as the vertical coordinate and with some forcing terms added, is solved to get the profile of structure function G. They usually use an open upper boundary condition instead of assuming the vertical motion to vanished at the model top. In contrast, for a construction of the normal mode functions for representation of atmospheric circulations, the equivalent height D must be determined first as the eigenvalue of VSE (1.46). Then the corresponding frequencies of free oscillations are determined as the eigenvalue of HSEs (1.46)-(1.49) for a specific value of D.

Since we have demonstrated in Section 1.4 that the equivalent height D as the eigenvalue of the VSE (1.46) is positive, we feel obliged to mention that in the case of tidal problems of known frequency the equivalent height D as the eigenvalue of the HSEs can become negative. In fact, there is an intriguing history behind the discovery of negative equivalent heights as alluded to by Platzman (1996).

It has been well known that the semidiurnal (12 hr) surface pressure oscillation over the globe is very regular and can easily be detected Wilkes (1949). The mechanism of semidiurnal tide became clear by the turn of the 1960 as the excitation of large-scale atmospheric oscillations by the direct absorption of incoming solar radiation by water vapor in the troposphere (Siebert, 1961) and atmospheric ozone in the stratosphere (Butler and Small, 1963). In contrast, the diurnal (24 hr) surface pressure oscillation has been known to be rather irregular and obscure and its geographical distribution became only clear after Haurwitz (1965) made an extensive study. The question was raised then why the diurnal tide is hard to detect compared with the semidiurnal tide since the excitation mechanism due to the absorption of insolation by water vapor and ozone should be more favorable as diurnal forcing. The answer to this question became clear when Lindzen (1966) and Kato (1966), independently realized that the HSEs (1.78) - (1.80) can yield negative eigenvalues of equivalent height D and the corresponding Hough functions are necessary for the calculation of diurnal tide. Moreover, they found that the solutions of the VSE corresponding to negative D are not oscillatory (harmonic), but locally trapped (exponential) at the level of forcing. In the case of semidiurnal tide, the values of D are all positive. Therefore, disturbances generated by forcing can propagate vertically. In contrast, in the case of diurnal tide the Hough modes corresponding to negative D play major role. Hence, disturbances generated in the stratosphere stay mostly in situ and do not propagate to the surface. In fact, the diurnal tide is dominant in the upper atmosphere (McLandress, 2002). In a nutshell, a traditional way of thinking have assumed that the value of D must be positive, perhaps, from an analogy of the HSEs as the shallow-water equations. A comprehensive review on the atmospheric tides and their mechanisms based on the modern dynamical theory is given by Chapman and Lindzen (1970).

The solutions of the HSEs corresponding to negative D were noted by various investigators independently during the mid 1960s. Flattery (1967) and Longuet-Higgins (1968) calculated the eigensolutions of the HSEs for positive and negative values of D. Their calculations show that the negative eigenvalues of D appear when the frequency of oscillation is less than 2Ω or the period of oscillation becomes greater than 12 hours. We wish to discuss this matter further.

First, we rewrite HSEs (1.78) - (1.80) in a dimensionless form by the following transformation:

$$\begin{pmatrix} u \\ v \\ h \end{pmatrix} = \begin{pmatrix} \tilde{u}\, U_* \\ \tilde{v}\, U_* \\ \tilde{h}\, H_* \end{pmatrix} \exp[i\,(k\lambda - 2\Omega \upsilon t)] \quad , \tag{1.82}$$

where U_* and H_* denote constants for scaling horizontal speed and vertical length, respectively. Also, k denotes the longitudinal wavenumber and υ dimensionless frequency, both of which are specified. The equivalent height D becomes the eigenvalue in the case of forced oscillations.

$$\upsilon\,\tilde{u} \; - \; i\,\tilde{v}\,\sin\varphi \; - \; \tilde{C}_*\,\frac{k}{\cos\varphi}\,\tilde{h} = 0, \tag{1.83}$$

$$\upsilon\,\tilde{v} \; - \; \tilde{u}\,\sin\varphi \; - \; \tilde{C}_*\,\frac{d\tilde{h}}{d\varphi} = 0, \tag{1.84}$$

$$\upsilon\,\tilde{h} + \frac{\tilde{D}}{\cos\varphi}\left[-k\,\tilde{u} + \frac{d}{d\varphi}(i\,\tilde{v}\,\cos\varphi) \right] = 0, \tag{1.85}$$

where

$$\tilde{C}_* = \frac{gH_*}{2a\Omega U_*} \quad \text{and} \quad \tilde{D} = \frac{DU_*}{H_* 2a\Omega} \; . \tag{1.86}$$

By eliminating $\tilde{u}$ and $\tilde{v}$ from (1.83)–(1.85), we obtain the following equation for $\tilde{h}$ which is a familiar form of Laplace's tidal equation

$$\Xi(\tilde{h}) + \varepsilon\,\tilde{h} = 0, \tag{1.87}$$

in which

$$\Xi(\tilde{h}) \; = \; \frac{d}{d\mu}\left[\left(\frac{1-\mu^2}{\upsilon^2-\mu^2}\right)\frac{d\,\tilde{h}}{d\mu}\right] + \frac{1}{\upsilon^2-\mu^2}\left[\frac{k(\upsilon^2+\mu^2)}{\upsilon(\upsilon^2-\mu^2)} - \frac{k^2}{1-\mu^2}\right]\tilde{h}, \tag{1.88}$$

where

$$\mu \; = \; \sin\varphi \quad \text{and} \quad d\mu \; = \; \cos\varphi\, d\varphi, \tag{1.89}$$

$$\varepsilon \; = \; \left(\tilde{D}\,\tilde{C}_*\right)^{-1} \; = \frac{4\,a^2\Omega^2}{gD} \; . \tag{1.90}$$

The quantity ε is referred to as the Lamb's parameter (Longuet-Higgins, 1968).

For a specified value of the frequency υ, the eigenvalue ε and the corresponding eigenfunctions are determined under the following boundary conditions:

$$\tilde{h}(\mu) = 0 \quad \text{for} \quad \mu = -1 \text{ at } \varphi = -90^o \quad \text{and} \quad \mu = 1 \text{ at } \varphi = +90^o \; . \tag{1.91}$$

We now show that $\tilde{h}(\mu)$ is a set of orthogonal functions. For this purpose we construct the following integral identity from (1.87) based on two different eigenfunctions $\tilde{h}_l$ and $\tilde{h}_m$, corresponding to two different eigenvalues ε_l and ε_m, respectively:

$$\int_{-1}^{1} [\, \tilde{h}_l(\tilde{h}_m) - \tilde{h}_m(\tilde{h}_l)\,]\, d\mu \;= (\varepsilon_l - \varepsilon_m) \int_{-1}^{1} \tilde{h}_m \, \tilde{h}_l \, \mathrm{d}\mu \,. \tag{1.92}$$

By using (1.88) we perform the integration of the left-hand side of (1.92). This yields

$$\left[\tilde{h}_l \left(\frac{1-\mu^2}{\upsilon^2-\mu^2}\right) \frac{d\,\tilde{h}_m}{d\mu}\right]_{-1}^{1} - \left[\tilde{h}_m \left(\frac{1-\mu^2}{\upsilon^2-\mu^2}\right) \frac{d\,\tilde{h}_l}{d\mu}\right]_{-1}^{1} = (\varepsilon_l - \varepsilon_m) \int_{-1}^{1} \tilde{h}_m \, \tilde{h}_l \, \mathrm{d}\mu \,. \tag{1.93}$$

Flattery (1967) and Holl (1970) showed that, in the range of $-1 < \mu < 1$, $\mu^2 = \upsilon^2$ is not a singularity. Therefore, the left-hand side of (1.93) vanishes due to the boundary conditions (1.91) and we get

$$\int_{-1}^{1} \tilde{h}_m \, \tilde{h}_l \, \mathrm{d}\mu \;=\; 0 \;\text{ for }\; l \neq m \,. \tag{1.94}$$

This orthogonality condition of the Hough functions for forced oscillations is discussed by Kato (1966) and Flattery (1967).

Now, for deriving criterion for the existence of negative D, we follow Dikii (1965) who expressed the Laplace tidal equation in the following alternative form

$$\left[\left(1-\mu^2\right)\frac{\mathrm{d}}{\mathrm{d}\mu} - \frac{k\mu}{\upsilon}\right] \left[\frac{\left(1-\mu^2\right)}{\left(\upsilon^2-\mu^2\right)}\frac{\mathrm{d}}{\mathrm{d}\mu} + \frac{k\mu}{\upsilon\left(\upsilon^2-\mu^2\right)}\right] \tilde{h} = \left[\frac{k^2}{\upsilon^2} - \left(1-\mu^2\right)\varepsilon\right] \tilde{h} \,. \tag{1.95}$$

Note that the derivation of (1.95) is more straightforward from (1.83)-(1.85) than transforming it from (1.87). Dikii, then, introduces an auxiliary function $\tilde{\xi}(\mu)$ which is defined as

$$\left[\left(1-\mu^2\right)\frac{\mathrm{d}}{\mathrm{d}\mu} + \frac{k\mu}{\upsilon}\right] \tilde{h} = \left(\upsilon^2-\mu^2\right)\tilde{\xi} \,. \tag{1.96}$$

By substituting (1.96) into (1.95), we get

$$\left[\left(1-\mu^2\right)\frac{\mathrm{d}}{\mathrm{d}\mu} - \frac{k\mu}{\upsilon}\right] \tilde{\xi} = \left[\frac{k^2}{\upsilon^2} - \left(1-\mu^2\right)\varepsilon\right] \tilde{h} \,. \tag{1.97}$$

We now multiply (1.96) and (1.97) by $\tilde{\xi}$ and $\tilde{h}$, respectively, add the two resulting equations, and get

$$(1-\mu^2)\frac{\mathrm{d}}{\mathrm{d}\mu}(\tilde{\xi}\,\tilde{h}\,) \;=\; \frac{k^2}{\upsilon^2}\tilde{h}^2 + (\upsilon^2-\mu^2)\,\tilde{\xi}^2 - \varepsilon\,(1-\mu^2)\,\tilde{h}^2 \,. \tag{1.98}$$

After dividing the above equation by $(1-\mu^2)$, we integrate the result with respect to μ from -1 to $+1$ and use the boundary conditions (1.91). The result is

$$\int_{-1}^{1} \frac{k^2}{\upsilon^2(1-\mu^2)}\tilde{h}^2 \mathrm{d}\mu + \int_{-1}^{1} \left(\frac{\upsilon^2-\mu^2}{1-\mu^2}\right)\tilde{\xi}^2 \mathrm{d}\mu = \varepsilon \int_{-1}^{1} \tilde{h}^2 \,\mathrm{d}\mu \,. \tag{1.99}$$

From this relation, it is clear that ε is real and positive if $\upsilon > 1$, namely for periods smaller than 12 hrs. If $\upsilon < 1$ for larger periods, ε can become negative, i.e. $D < 0$. Similar argument for the condition permitting negative D is presented by Holl (1970) with a discussion of the completeness of the orthogonal eigenfunctions of the Laplace tidal equation.

1.6.2 Case of Free Oscillations – Normal Mode Problem

We now consider the case of calculating the normal modes of free oscillations in which both the positive D or ε and zonal wavenumber k are specified. In this case, the frequency of oscillation υ becomes the eigenvalue. By introducing the dimensionless variables

$$\tilde{u} = \frac{u}{\sqrt{g\,D}}, \quad \tilde{v} = \frac{v}{\sqrt{g\,D}}, \quad \tilde{h} = \frac{h}{D}, \quad \tilde{t} = 2\Omega t, \tag{1.100}$$

we can rewrite (1.78)-(1.80) as

$$\frac{\partial}{\partial \tilde{t}}\mathbf{W} + \mathbf{L}\mathbf{W} = 0, \tag{1.101}$$

where $\mathbf{W}$ denotes the vector dependent variable

$$\mathbf{W} = (\tilde{u}, \tilde{v}, \tilde{h})^T \,, \tag{1.102}$$

and $\mathbf{L}$ is the linear differential matrix operator

$$\mathbf{L} = \begin{bmatrix} 0 & -\sin(\varphi) & \frac{\gamma}{\cos(\varphi)}\frac{\partial}{\partial\lambda} \\ \sin(\varphi) & 0 & \gamma\frac{\partial}{\partial\varphi} \\ \frac{\gamma}{\cos(\varphi)}\frac{\partial}{\partial\lambda} & \frac{\gamma}{\cos(\varphi)}\frac{\partial}{\partial\varphi}[\cos(\varphi)(.)] & 0 \end{bmatrix}, \tag{1.103}$$

in which γ is a dimensionless parameter defined by

$$\gamma = \frac{\sqrt{gD}}{2a\Omega} = \frac{1}{\sqrt{\varepsilon}}, \tag{1.104}$$

as the inverse of the square-root of ε, Lamb's parameter given by (1.90).

Since (1.101) is a linear system, the solution $\mathbf{W}$ can be expressed as a linear combination of functions that have the form

$$\mathbf{W}(\lambda, \varphi, \tilde{t}) = \mathbf{H}_n^k(\lambda, \varphi)\exp(-i\upsilon_n^k \tilde{t}), \tag{1.105}$$

where $\mathbf{H}_n^k(\lambda, \varphi)$ represents the horizontal structure functions with zonal wavenumber k and meridional index n. The corresponding dimensionless frequency υ_n^k also depends on k and n.

Now, we define the global inner product as

$$< \mathbf{W}_l, \mathbf{W}_m^* >= \frac{1}{2\pi} \int_0^{2\pi} \int_{-1}^{1} (\tilde{u}_l\, \tilde{u}_m^* + \tilde{v}_l\, \tilde{v}_m^* + \tilde{h}_l\, \tilde{h}_m^*)\, \mathrm{d}\mu\, \mathrm{d}\lambda, \tag{1.106}$$

where $\mu = \sin\varphi$ and the asterisk denotes the complex conjugate. Subscript l refers to a particular mode corresponding to a zonal wavenumber k_l and a meridional index n_l. Subscript m is another modal index.

Then, we show that the linear operator (1.103) has the following property:

$$< \mathbf{W}_l, \mathbf{L}\mathbf{W}_m^* > + < \mathbf{L}\mathbf{W}_l, \mathbf{W}_m^* >= 0\,. \tag{1.107}$$

This can be verified by forming relevant inner products, integrating them globally, and using Green's theorem (Platzman, 1972).

By substituting (1.105) into (1.107) and noting that $\mathbf{H}_l$ is the eigenfunction of $\mathbf{L}$ such that

$$\mathbf{L}\mathbf{H}_l = i\, \upsilon_l \mathbf{H}_l\,, \tag{1.108}$$

we obtain

$$(\upsilon_l - \upsilon_m^*) < \mathbf{H}_l, \mathbf{H}_m^* >= 0\,, \tag{1.109}$$

where the minus sign comes from the complex conjugate.

We can consider the following two cases:

1. The case of $l = m$. Because $< \mathbf{H}_l, \mathbf{H}_m^* >$ becomes proportional to the total energy of the linearized system that must not vanish, it follows that $\upsilon_l = \upsilon_m^*$; therefore, the υ_l must be real and we can drop the asterisk from the notation of eigenfrequency.
2. The case of $l \neq m$. Because $\upsilon_l \neq \upsilon_m$, we must have $< \mathbf{H}_l, \mathbf{H}_m^* >$, meaning that $\mathbf{H}_l$ corresponding to υ_l must be orthogonal to the $\mathbf{H}_m$ associated with υ_m which is different from υ_l.

Since the magnitude of $\mathbf{H}_l$ is arbitrary, we normalize $\mathbf{H}_l$ so that

$$\frac{1}{2\pi} \int_0^{2\pi} \int_{-1}^{1} \mathbf{H}_l \cdot \mathbf{H}_m^* \mathrm{d}\mu \mathrm{d}\lambda = \delta_{lm}\,, \tag{1.110}$$

where the right-hand side is unity if $l = m$, and zero otherwise.

Because the oscillations in longitudinal direction are harmonic, we can express the Hough harmonic function $\mathbf{H}_n^k$ as

$$\mathbf{H}_n^k\,(\lambda, \varphi) = \mathbf{\Theta}_n^k(\varphi) \exp(ik\lambda)\ . \tag{1.111}$$

The meridional dependence is the Hough vector function

$$\mathbf{\Theta}_n^k(\varphi) = \begin{bmatrix} U_n^k(\varphi) \\ iV_n^k(\varphi) \\ Z_n^k(\varphi) \end{bmatrix}, \tag{1.112}$$

which has three components: zonal velocity U, meridional velocity V, and geopotential height Z, all having zonal wavenumber k and meridional index n. The factor i $(= \sqrt{-1})$ in front of V is introduced to account for the phase shift of $\pi/2$ with respect to U and Z [1].

By substituting (1.111) with (1.112) into (1.110), we find

$$\int_{-1}^{1} \mathbf{\Theta}_l \cdot \mathbf{\Theta}_m^* \, d\mu = \int_{-1}^{1} (U_l U_m^* + V_l V_m^* + Z_l Z_m^*) \, d\mu = \delta_{lm} . \tag{1.113}$$

This is the orthogonality condition for $\mathbf{\Theta}_l$ associated with frequency υ_l.

1.7 Solutions of the HSEs

Margules (Max 1892) seems to be the first who obtained the numerical solutions of HSEs. He published three articles during 1892-1893 which are now reproduced as a technical report. His articles are rather difficult to read due to the use of unfamiliar notation and the application of harmonic functions as the basis to construct the solutions. Nevertheless, what is surprising is that he identified the presence of two kinds of the normal modes known as the first and second class of free oscillations and exhibited the eigenfunctions which are qualitatively in agreement with those we know of today. Few years later without noticing the Margules work, Hough (1898) adopted the associated Legendre functions to solve the HSEs. He also investigated asymptotic solutions for small value of Lamb's parameter ε and identified the two kinds of normal modes as done by Margules. The solution algorithm that Hough developed became a standard to follow by many investigators noted earlier and we will do so too.

[1] The algorithm for solving the horizontal structure equations (HSEs) developed by Kasahara (1976) had the minus sign in front of the imaginary unit (i) in the meridional component of the Hough vector function in Eq. (1.112). The minus sign was used to ensure that the meridional wind was positive for a northward flow, as is the usual meteorological convention. However, Kasahara's computer codes are no longer available. The convention used throughout this book is consistent with the software package for the HSEs developed by Swarztrauber and Kasahara (1985) which is used by the book authors and in the NMF MODES software. This is the reason that the meridional profiles of U, V, Z, shown in Kasahara (1976) as Figures 4, 5, 6, 7, and 8 for zonal wavenumbers 1 to 3 have the sign of V reversed compared with those derived from the Swarztrauber and Kasahara (1985).

1.7.1 General Algorithm

We rewrite the HSEs (1.78)–(1.80) by introducing the stream function ψ and the velocity potential χ,

$$u = \frac{1}{a\cos\varphi}\frac{\partial\chi}{\partial\lambda} - \frac{\partial\psi}{a\,\partial\varphi}, \quad (1.114)$$

$$v = \frac{\partial\chi}{a\,\partial\varphi} + \frac{1}{a\cos\varphi}\frac{\partial\psi}{\partial\lambda}, \quad (1.115)$$

and

$$\nabla^2\chi = \frac{1}{a\cos\varphi}\left[\frac{\partial u}{\partial\lambda} + \frac{\partial}{\partial\varphi}(v\cos\varphi)\right], \quad (1.116)$$

$$\nabla^2\psi = \frac{1}{a\cos\varphi}\left[\frac{\partial v}{\partial\lambda} - \frac{\partial}{\partial\varphi}(u\cos\varphi)\right], \quad (1.117)$$

where ∇^2 denotes the horizontal Laplacian operator:

$$\nabla^2 = \frac{1}{a^2\cos\varphi}\left[\frac{1}{\cos\varphi}\frac{\partial^2}{\partial\lambda^2} + \frac{\partial}{\partial\varphi}\left(\cos\varphi\frac{\partial}{\partial\varphi}\right)\right]. \quad (1.118)$$

The two Laplacians in (1.116-1.117) represent, respectively, the horizontal divergence and the vertical component of vorticity.

Then, we assume that the dependent variables are proportional to harmonic functions in longitude λ with zonal wavenumber k and in time with the dimensionless frequency υ scaled by 2Ω. Together with this assumption, we also introduced the dimensionless velocity potential $\tilde{\chi}$, stream function $\tilde{\psi}$, and height $\tilde{Z}$ which correspond to their original variables through

$$\left(\frac{2\Omega}{gD}\chi,\ \frac{2\Omega}{gD}\psi,\ \frac{h}{D}\right) = \left(\tilde{\chi},\ \tilde{\psi},\ \tilde{Z}\right)\exp[i(k\lambda - \upsilon 2\Omega t)]. \quad (1.119)$$

Note that $\tilde{\chi}$, $\tilde{\psi}$, and $\tilde{Z}$ depend on latitude φ only.

After the above operations, the HSEs (1.78)–(1.80) are transformed to

$$(\upsilon\nabla_k^2 - k)(i\tilde{\chi}) + (\mu\nabla_k^2 + L)\tilde{\psi} = \nabla_k^2\tilde{Z}, \quad (1.120)$$

$$(\upsilon\nabla_k^2 - k)\tilde{\psi} + (\mu\nabla_k^2 + L)(i\tilde{\chi}) = 0, \quad (1.121)$$

$$\upsilon\tilde{Z} = -\frac{1}{\varepsilon}\nabla_k^2(i\tilde{\chi}), \quad (1.122)$$

where

$$\mu = \sin\varphi, \quad (1.123)$$

$$\nabla_k^2 = \frac{\mathrm{d}}{\mathrm{d}\mu}\left[\left(1-\mu^2\right)\frac{d}{d\mu}\right] - \frac{k^2}{1-\mu^2}, \quad (1.124)$$

$$L = (1 - \mu^2)\frac{\mathrm{d}}{\mathrm{d}\mu}, \tag{1.125}$$

and ε is the Lamb's parameter defined by (1.90).

Now, we express $\tilde{\chi}$, $\tilde{\psi}$, and $\tilde{Z}$ in the series of

$$\begin{pmatrix} \tilde{\chi} \\ \tilde{\psi} \\ \tilde{Z} \end{pmatrix} = \sum_{n=k}^{\infty} \begin{pmatrix} iA_n^k \\ B_n^k \\ C_n^k \end{pmatrix} P_n^k(\mu) \quad , \tag{1.126}$$

where $i\,(= \sqrt{-1})$ in front of A_n^k is introduced to account for the phase difference between $\tilde{\chi}$ and $\tilde{\psi}$, and $P_n^k(\mu)$ is the associated Legendre function of order n and rank k with its normalization condition

$$\int_{-1}^{1} P_n^k(\mu)\,P_{n'}^k(\mu)\,\mathrm{d}\mu = \begin{cases} 0, & n \neq n' \\ \frac{2}{2n+1}\,\frac{(n+k)\,!}{(n-k)\,!}, & n = n' \end{cases} \quad . \tag{1.127}$$

We list some formulas of functional operations which are useful:

$$\nabla_k^2\, P_n^k \;=\; -n\,(n\,+1)\,P_n^k \;\; for \;\; n > k < 0, \tag{1.128}$$

$$\mu\, P_n^k \;=\; \frac{n+k}{2n+1} P_{n-1}^k + \frac{n-k+1}{2n+1} P_{n+1}^k, \tag{1.129}$$

$$LP_n^k \;=\; \frac{(n+1)(n+k)}{2n+1}\, P_{n-1}^k \;-\; \frac{n\,(n-k+1)}{2n+1} P_{n+1}^k, \tag{1.130}$$

$$\mu\,\nabla_k^2\,P_n^k \;=\; \frac{-n\,(n+1)(n+k)}{2n+1} P_{n-1}^k \;-\; \frac{n\,(n+1)(n-k+1)}{2n+1}\, P_{n+1}^k\,. \tag{1.131}$$

Therefore, we have

$$(\mu\,\nabla_k^2 + L)P_n^k = -\frac{(n-1)(n+1)(n+k)}{2n+1}\, P_{n-1}^k - \frac{n\,(n+2)(n-k+1)}{2n+1}\, P_{n+1}^k\,. \tag{1.132}$$

By substituting the series representation of (1.126) into (1.120)–(1.122), using the formulas (1.128)–(1.131) and (1.132), and collecting the coefficients of P_n^k, we find

$$[\upsilon\,n\,(n+1)+k]A_n^k \;-\; \frac{n\,(n+2)(n+k+1)}{2n+3}\, B_{n+1}^k - \frac{(n-1)(n+1)(n-k)}{2n-1}\, B_{n-1}^k \;=\; -n\,(n+1)\,C_n^k, \tag{1.133}$$

$$-\,[\upsilon\,n\,(n+1)+k]\,B_n^k \;+\; \frac{n\,(n+2)(n+k+1)}{2n+3}\, A_{n+1}^k + \frac{(n-1)(n+1)(n-k)}{2n-1}\, A_{n-1}^k = 0, \tag{1.134}$$

$$\upsilon\, C_n^k \;=\; -\frac{n\,(n+1)}{\varepsilon}\, A_n^k\,. \tag{1.135}$$

To simplify writing of above three equations, we introduce the parametric notation as

$$K_n \;=\; \frac{-k}{n\,(n+1)}\,, \qquad r_n \;=\; \frac{-n\,(n+1)}{\varepsilon}\,, \tag{1.136}$$

$$p_n \;=\; \frac{(n+1)(n+k)}{n\,(2n+1)}\,, \qquad q_n \;=\; \frac{n\,(n-k+1)}{(n+1)(2n+1)}\,, \tag{1.137}$$

for $n = k, k+1, k+2, \ldots$

Using the definitions (1.136) and (1.137), equations (1.133)–(1.135) can be written in the form:

$$(-\upsilon \,+\, K_n)\, A_n^k \;+\; p_{n+1}\, B_{n+1}^k \;+\; q_{n-1}\, B_{n-1}^k \;-\; C_n^k \;=\; 0\,, \tag{1.138}$$

$$(-\upsilon \,+\, K_n)\, B_n^k \;+\; p_{n+1}\, A_{n+1}^k \;+\; q_{n-1}\, A_{n-1}^k \;=\; 0\,, \tag{1.139}$$

$$-\upsilon\, C_n^k \;+\; r_n\, A_n^k \;=\; 0\,. \tag{1.140}$$

By eliminating C_n^k between (1.138) and (1.140), we can get two equations for A_n^k and B_n^k. This is the traditional way of solving the HSEs. However, the drawback is that the resulting equations are quadratic in ε, whereas (1.138)–(1.140) can be solved by standard linear algebra methods.

1.7.1.1 Symmetric Modes

Equations (1.138)–(1.140) contain two independent systems. One consists of A_n^k and C_n^k for $n = k,\ k+2,\ \ldots$. and B_n^k for $n = k+1,\ k+3,\ \ldots$.. As seen from (1.140), in this case the velocity potential $\tilde{\chi}$ and the height $\tilde{Z}$ are symmetric with respect to the equator and the stream function $\tilde{\psi}$ is antisymmetric.

Let $\mathbf{X}$ be the column vector,

$$\mathbf{X} = [\, A_k^k,\ B_{k+1}^k,\ C_k^k,\ A_{k+2}^k,\ B_{k+3}^k,\ C_{k+2}^k,\ \ldots]^T\,, \tag{1.141}$$

and $\mathbf{A}$ be the matrix

$$\mathbf{A} = \begin{bmatrix} K_k & p_{k+1} & -1 & 0 & 0 & 0 & 0 & . \\ q_k & K_{k+1} & 0 & p_{k+2} & 0 & 0 & 0 & . \\ r_k & 0 & 0 & 0 & 0 & 0 & 0 & . \\ 0 & q_{k+1} & 0 & K_{k+2} & p_{k+3} & -1 & 0 & . \\ 0 & 0 & 0 & q_{k+2} & K_{k+3} & 0 & p_{k+4} & . \\ 0 & 0 & 0 & r_{k+2} & 0 & 0 & 0 & . \\ . & . & . & . & . & . & . & . \end{bmatrix}. \tag{1.142}$$

Then the frequency υ of the symmetric system can be obtained as the eigenvalue and the expansion coefficient vector $\mathbf{X}$ as the eigenvector of the matrix problem in the form

$$\mathbf{AX} = \upsilon \mathbf{X} . \tag{1.143}$$

Once the vector $\mathbf{X}$ is obtained, the latitudinal profiles of $\tilde{\chi}$, $\tilde{\psi}$, and $\tilde{Z}$ are calculated through (1.126).

1.7.1.2 Antisymmetric Modes

The other system consists of A_n^k and C_n^k for $n = k+1, k+3, ...$ and B_n^k for $n = k, k+2, ...$ In this case velocity potential $\tilde{\chi}$ and height $\tilde{Z}$ are antisymmetric with respect to the equator and the stream function $\tilde{\psi}$ is symmetric.

Let $\mathbf{Y}$ be the column vector,

$$\mathbf{Y} = [\, B_k^k,\ A_{k+1}^k,\ C_{k+1}^k,\ B_{k+2}^k,\ A_{k+3}^k,\ C_{k+3}^k,\ \ldots]^T , \tag{1.144}$$

and $\mathbf{B}$ be the matrix

$$\mathbf{B} = \begin{bmatrix} K_k & p_{k+1} & 0 & 0 & 0 & 0 & 0 & . \\ q_k & K_{k+1} & -1 & p_{k+2} & 0 & 0 & 0 & . \\ 0 & r_{k+1} & 0 & 0 & 0 & 0 & 0 & . \\ 0 & q_{k+1} & 0 & K_{k+2} & p_{k+3} & 0 & 0 & . \\ 0 & 0 & 0 & q_{k+2} & K_{k+3} & -1 & p_{k+4} & . \\ 0 & 0 & 0 & 0 & r_{k+3} & 0 & 0 & . \\ . & . & . & . & . & . & . & . \end{bmatrix} . \tag{1.145}$$

Then the frequency υ of the antisymmetric system can be obtained as the eigenvalue and the expansion coefficient vector $\mathbf{Y}$ as the eigenvector by solving the matrix problem,

$$\mathbf{BY} = \upsilon \mathbf{Y} . \tag{1.146}$$

Once the vector $\mathbf{Y}$ is obtained, the latitudinal profiles of $\tilde{\chi}$, $\tilde{\psi}$, and $\tilde{Z}$ are calculated through (1.126).

The general algorithm presented in this subsection was used by Kasahara (1976) to calculate the frequency υ and the coefficient vectors $\mathbf{X}$ and $\mathbf{Y}$ for positive value of ε or D. Since this algorithm does not assume the positivity of ε, it seems that the same algorithm should work for negative value of ε, but it has not been verified. Detailed discussions on the properties of solutions for negative values of ε, including the asymptotic behavior are presented by Longuet-Higgins (1968).

Since our objective is to construct the 3D normal mode functions with positive ε or D, we will not purse further the solutions of HSEs for negative D for forced oscillations. Therefore, from this point we only deal with the normal mode solutions corresponding to positive D.

Once the solutions of $\tilde{\chi}$, $\tilde{\psi}$, and $\tilde{Z}$ are obtained through (1.126), we can calculate the velocity components u and v. First, we introduce the dimensionless velocity components $\tilde{U}$ and $\tilde{V}$ which are related to u and v as

$$\begin{bmatrix} u/\sqrt{gD} \\ iv/\sqrt{gD} \\ h/D \end{bmatrix} = \begin{bmatrix} \tilde{U}(\varphi) \\ \tilde{V}(\varphi) \\ \tilde{Z}(\varphi) \end{bmatrix} \exp[i\,(k\lambda - \upsilon\, 2\Omega\, t) \quad . \tag{1.147}$$

The factor i ($= \sqrt{-1}$) in front of v is introduced to account for the phase shift of $\pi/2$ with respect to u and h. We added $\tilde{Z}$ to complete the dimensionless dependent variables.

Using (1.114–1.115) and (1.119), we get

$$\tilde{U} = \frac{1}{\sqrt{\varepsilon}\cos\varphi}(i\,k\,\tilde{\chi} - L\tilde{\psi}), \tag{1.148}$$

$$\tilde{V} = \frac{i}{\sqrt{\varepsilon}\,\cos\varphi}\left(i\,k\,\tilde{\psi} + L\tilde{\chi}\right) . \tag{1.149}$$

By substituting into (1.148) and (1.149) the series of $\tilde{\chi}$ and $\tilde{\psi}$ as given by (1.126), and using operator (1.128–1.131), we obtain the representation of $\tilde{U}$ and $\tilde{V}$ in terms of the series of A_n^k and B_n^k.

The magnitude of eigenvectors **X** and **Y** are arbitrary and must be decided by the normalization of eigenvectors. We invoke here the condition of normalization based on the conservation of integrated total energy derived from (1.81) over the sphere so that

$$\int \frac{1}{2}\left[D\,(u^2 + v^2) + gh^2\right] a^2 \cos\varphi\, \mathrm{d}\varphi\, \mathrm{d}\lambda = I \tag{1.150}$$

is invariant for each horizontal mode corresponding to each eigenfrequency.

By using the expressions (1.147)-(1.149) at $t = 0$, we can show that

$$I = \frac{1}{2}\, gD^2 a^2 \pi \int_{-1}^{1} [\tilde{Z}^2 - \varepsilon^{-1}(\tilde{\chi}\nabla_k^2\,\tilde{\chi} + \tilde{\psi}\nabla_k^2\,\psi)]\, \mathrm{d}\mu\,. \tag{1.151}$$

Then, using the formula (1.128) and the normalization (1.127) of P_n^k, we obtain

$$\sum_{n=k}^{\infty}\left\{(C_n^k)^2 + \frac{n\,(n+1)}{\varepsilon}[(A_n^k)^2 + (B_n^k)^2]\right\} \times \frac{2}{2n+1}, \frac{(n+k)\,!}{(n-k)\,!} = K, \tag{1.152}$$

where $K = 2I\,(gD^2\,a^2\pi)^{-1}$.

Similarly, we can express (1.150) using $\tilde{U}$, $\tilde{V}$, and $\tilde{Z}$, defined by (1.147) for $t = 0$ and get

$$\int_{-1}^{1} (\tilde{U}^2 + \tilde{V}^2 + \tilde{Z}^2)\, \mathrm{d}\mu = K\,. \tag{1.153}$$

Thus, we choose $K = 1$ in (1.152) to normalize the eigenvectors and check the accuracy of normalization using (1.153) as well as the orthogonality in the sense of (1.113).

1.7.2 Species of the Normal Modes

In this subsection, we present approximate analytical solutions for small value of positive ε following Hough (1898) and Longuet-Higgins (1968). A simplified system of (1.138)–(1.140) is obtained by eliminating C_n^k in the form

$$(-\upsilon + K_n - \frac{r_n}{\upsilon})\, A_n^k \;+\; p_{n+1}\, B_{n+1}^k + q_{n-1}\, B_{n-1}^k \;=\; 0, \tag{1.154}$$

$$(-\upsilon + K_n)\, B_n^k \;+\; p_{n+1}\, A_{n+1}^k + q_{n-1}\, A_{n-1}^k \;=\; 0\,. \tag{1.155}$$

The symmetric modes are represented by A_n^k for $n = k,\ k+2, ..$ and B_n^k for $n = k+1,\ k+3, \ldots$.

Let us introduce vector $\mathbf{X}_c$ and matrix $\mathbf{A}_c$ in the form

$$X_c = [A_k^k,\ B_{k+1}^k\, A_{k+2}^k,\ B_{k+3}^k,\ .\ .\ .]^T\,,$$

$$\mathbf{A}_c = \begin{bmatrix} (-\upsilon + K_k - \frac{r_k}{\upsilon}) & p_{k+1} & 0 & 0 & . \\ q_k & (-\upsilon + K_{k+1}) & p_{k+2} & 0 & . \\ 0 & q_{k+1} & (-\upsilon + K_{k+2} - \frac{r_{k+2}}{\upsilon}) & p_{k+3} & . \\ 0 & 0 & q_{k+2} & (-\upsilon + K_{k+3}) & . \\ . & . & . & . & . \end{bmatrix}.$$

Then, we can express (1.154)-(1.155) as $\mathbf{A}_c\mathbf{X}_c = 0$. For small value of ε, as we see from the definition of r_k in (1.136), we find that the diagonal terms in $\mathbf{A}_c$ dominate. Therefore, we get approximately the product

$$\prod_n \left(\upsilon - K_n + \frac{r_n}{\upsilon}\right)(\upsilon - K_{n+1}) \approx 0\,, \tag{1.156}$$

for $n = k, k+2, k+4, \ldots$. From this product we expect that there are two different kinds of oscillations.

1.7.2.1 Oscillations of the First Kind

By equating the expression in the first parentheses of (1.156) to zero, recalling the definitions of K_n and r_n in (1.136), and assuming that $\upsilon \neq 0$, we obtain

$$\upsilon^2 + \frac{k\upsilon}{n\,(n+1)} - \frac{n\,(n+1)}{\varepsilon} = 0, \tag{1.157}$$

for $n = k,\ k+2,\ k+4,\ \ldots$. This gives

$$\upsilon = \frac{-k}{2n\,(n+1)} \pm \left[\frac{k^2}{4n^2\,(n+1)^2} + \frac{n\,(n+1)}{\varepsilon}\right]^{1/2}. \tag{1.158}$$

Two solutions of (1.158) represent the frequencies of inertio-gravity waves that propagate eastward and westward and are referred to as the oscillations of the first kind (or class) by Margules (1892) and Hough (1898). The eigenfunctions corresponding to the first kind are approximated by the coefficients A_n^k so that the waves are dominated by irrotational motions as

$$\tilde{\chi} \propto A_n^k P_n^k(\sin\varphi)\exp[i(k\lambda - \upsilon\tilde{t})] \ \text{ and } \ \tilde{\psi} \approx 0. \qquad (1.159)$$

For a small value of ε, i.e. large D, (1.158) is further reduced to the frequency of gravity waves as

$$\nu = 2\Omega\upsilon = \pm\frac{\sqrt{n(n+1)gD}}{a}. \qquad (1.160)$$

We can repeat the same procedure of getting approximate solutions of (1.154)–(1.154) for the antisymmetric modes. It turns out that the same approximate solutions of (1.158) is obtained for $n = k+1,\ k+3,\ ..$ Therefore, the approximate solutions are valid for both symmetric and antisymmetric modes.

1.7.2.2 Oscillations of the Second Kind

The other kind of oscillations is obtained by equating the expression in the second parentheses of (1.156) to be zero. We then obtain the frequency

$$\nu = 2\Omega\upsilon = \frac{-2\Omega k}{n(n+1)}, \qquad (1.161)$$

for $n = k, k+2, k+4, \ldots$. We can repeat the same procedure for the antisymmetric modes of (1.154)–(1.155) and get the same form of frequency as (1.161). This type of westward propagating waves is called oscillations of the second kind which are rotational in nature

$$\tilde{\chi} \approx 0 \quad \text{and} \quad \tilde{\psi} \propto B_n^k P_n^k(\sin\varphi)\exp[i(k\lambda - \upsilon\tilde{t})]. \qquad (1.162)$$

Formula (1.162) is a well-known wave frequency derived by Haurwitz (1940) based on the nondivergent two-dimensional vorticity equation on a sphere extending the Rossby wave formula on the mid-latitude beta plane. Actually, Hough (1898) already derived even a higher-order approximation for the frequency of this kind of oscillations, but the meteorological significance of this wave mode had not been clarified until Rossby et al. (1939) discovered planetary wave motions through the analyses of upper-air data. A historical review of the second kind of oscillations given by Platzman (1968) is worth reading.

1.7.3 Eigenfunctions of the HSEs

A software package for calculating the Hough harmonic normal mode solutions is described by Swarztrauber and Kasahara (1985), which is referred to as SK85, and is available at National Center for Atmospheric Research, Boulder, CO. In this section we present the description of eigenfrequencies and eigenfunctions obtained from that software. The computing algorithm developed for the software package uses the vector spherical harmonics which is slightly different from that described in Section 1.7.1, but the algorithm can be written compactly and suitably for programming the codes. The vector spherical harmonics have been around in mathematical physics for sometime and its use was proposed to global meteorology and aeronomy by Moses (1974) and Jones (1970), but its merit in atmospheric science has never been apparent until Swarztrauber (1979, 1981, 1993) started to develop various formulas concerning the vector spherical harmonics for solving partial differential equations in spherical geometry.

1.7.3.1 Vector Spherical Harmonics

We will first describe briefly the concept of vector spherical harmonics. Since the horizontal velocity components u and v are vector components unlike geopotential height h which is a scalar, we need to represent horizontal velocity in terms of two scalar functions. To this end, it is natural to apply decomposition similar to the velocity potential χ and stream function ψ as in equations (1.114)-(1.115). Assuming that the variables are proportional to $\exp[i(k\lambda - \nu t)]$, the latitude-dependent variables $\hat{u}(\varphi)$ and $\hat{v}(\varphi)$ can be represented in terms of $\hat{\chi}(\varphi)$ and $\hat{\psi}(\varphi)$ as

$$\hat{u} = \frac{ik}{\cos\varphi}\hat{\chi} - \frac{\mathrm{d}}{\mathrm{d}\varphi}\hat{\psi}\,, \tag{1.163}$$

$$\hat{v} = \frac{\mathrm{d}}{\mathrm{d}\varphi}\hat{\chi} + \frac{ik}{\cos\varphi}\hat{\psi}\,. \tag{1.164}$$

Similarly, geopotential height is transformed to $\hat{h}(\varphi)$ which is already scalar.

The above consideration can suggest how to define the vector spherical harmonics to solve the HSEs (1.78)–(1.80). They are given in SK85 as

$$y_{n,1}^k = \begin{bmatrix} \frac{ik}{\cos\varphi}P_n^k \\ \frac{\mathrm{d}P_n^k}{\mathrm{d}\varphi} \\ 0 \end{bmatrix} \frac{\exp(ik\lambda)}{\sqrt{n(n+1)}}, \quad y_{n,2}^k = \begin{bmatrix} \frac{-\mathrm{d}P_n^k}{\mathrm{d}\varphi} \\ \frac{ik}{\cos\varphi}P_n^k \\ 0 \end{bmatrix} \frac{\exp(ik\lambda)}{\sqrt{n(n+1)}}, \quad y_{n,3}^k = \begin{bmatrix} 0 \\ 0 \\ P_n^k \end{bmatrix} \exp(ik\lambda). \tag{1.165}$$

By comparing (1.163), (1.164) and (1.165), one realizes that the vector $y_{n,1}^k$ corresponds to the case of setting $\hat{h} = 0$, $\hat{\psi} = 0$, $\hat{\chi} = P_n^k \exp(ik\lambda)\,[n(n+1)]^{-1/2}$ and $y_{n.2}^k$ corresponds to $\hat{h} = 0$, $\hat{\chi} = 0$, $\hat{\psi} = P_n^k \exp(ik\lambda)\,[n(n+1)]^{-1/2}$. Since the height $\hat{h}$ is a scalar, it is simply expanded in spherical harmonics with $\hat{\psi} = \hat{\chi} = 0$.

That is the third vector $y_{n,3}^k$. Therefore, we can see that the process of transforming the velocity components to the stream function and velocity potential described in the beginning of Section 1.7.1 is already taken care of by the definitions of vector spherical harmonics.

Note that the associated Legendre functions used in SK85 are slightly different from (1.127) and their use of P_n^k yields the normalization of vector spherical harmonics as

$$\int_0^{2\pi} \int_{-\pi/2}^{\pi/2} \left(y_{n,j}^k \right)^* y_{n,j}^k \cos\varphi \, d\lambda = 2\pi \quad \text{for} \quad j = 1, 2, 3 . \tag{1.166}$$

We should note that the algorithm of SK85 assumes the equivalent height D to be positive and the velocity components and height are scaled as given by (1.100). The solution of the HSEs (1.78)–(1.80) is given by the vector $\mathbf{W}$ according to (1.105). The Hough harmonic function, given by (1.111), expanded in spherical vector harmonics assumes the following form

$$\mathbf{H}_n^k(\lambda, \varphi) = \mathbf{\Theta}_n^k(\varphi) \exp(i\lambda) = \sum_n \left(iA_n^k \, y_{n,1}^k + B_n^k \, y_{n,2}^k - C_n^k \, y_{n,3}^k \right) . \tag{1.167}$$

It is important to note that the coefficients A_n^k, B_n^k, and C_n^k in (1.167) are similar, but they are not identical to those used in (1.126). By substituting (1.167) into (1.108), using the identities for $\mathbf{L}y_{n,1}^k$, $\mathbf{L}y_{n,2}^k$, and $\mathbf{L}y_{n,3}^k$, (1.31), (1.34), and (1.37), respectively, of SK85, and equating the coefficients of $y_{n,j}^k$ for j =1, 2, and 3, respectively, we obtain the three equations for A_n^k, B_n^k, and C_n^k, as given by (1.40)–(1.42) in SK85. The derivation of these equations requires considerable algebraic manipulations, however, the procedure is very similar to the one leading to (1.138)–(1.140), although the definitions of p_n^k, q_n^k, and r_n^k of SK85 are slightly different from those in (1.136) and (1.137). Therefore, we can derive two subsystems for symmetric and antisymmetric modes as the matrix problems of $\mathbf{AX} = \upsilon\mathbf{X}$ and $\mathbf{BY} = \upsilon\mathbf{Y}$ similar to (1.143) and (1.146), respectively. One merit of the matrices $\mathbf{A}$ and $\mathbf{B}$ of SK85 is that they are real symmetric pentadiagonal matrices, while matrices (1.142) and (1.145) are real and pentadiagonal, but not symmetric. Another merit of the SK85 scheme is that the Hough vector function $\mathbf{\Theta}_n^k(\varphi)$ can be obtained directly from (1.167), since the vector spherical harmonics are applied to the velocity components and height directly.

1.7.3.2 Eigenvalues and Eigenfunctions

Eigenvalues (wave frequencies) and eigenfunctions are calculated using the software package of SK85 which consists of four FORTRAN subroutines explained in the table form in SK85. All programs require the user to specify the total number of meridional modes in addition to specific input data needed for each program.

Table 1, SIGMA calculates the frequencies as functions of the square-root of Lamb's parameter $\sqrt{\varepsilon}$.

Table 2, ABCOEF calculates the expansion coefficients A_n^k, B_n^k, and C_n^k of vector spherical harmonics. See (1.167).

Table 3, UVH evaluates the meridional profiles of eigenfunctions $U_n^k(\varphi)$, $V_n^k(\varphi)$, and $Z_n^k(\varphi)$ as in (1.112).

Table 4, UVHDER tabulates the meridional profiles of stream function ψ, velocity potential χ, divergence $\nabla^2\chi$, vorticity $\nabla^2\psi$, gradients of U, V, Z as $\mathrm{d}(\cos\varphi\, U)/\mathrm{d}\varphi$, $\mathrm{d}(\cos\varphi\, V)/\mathrm{d}\varphi$, $\cos\varphi\, \mathrm{d}Z/\mathrm{d}\varphi$, respectively. (Note that, in the write-up of Table 4 in SK85, the term $\cos\varphi$ is missing in front of $\mathrm{d}Z/\mathrm{d}\varphi$.)

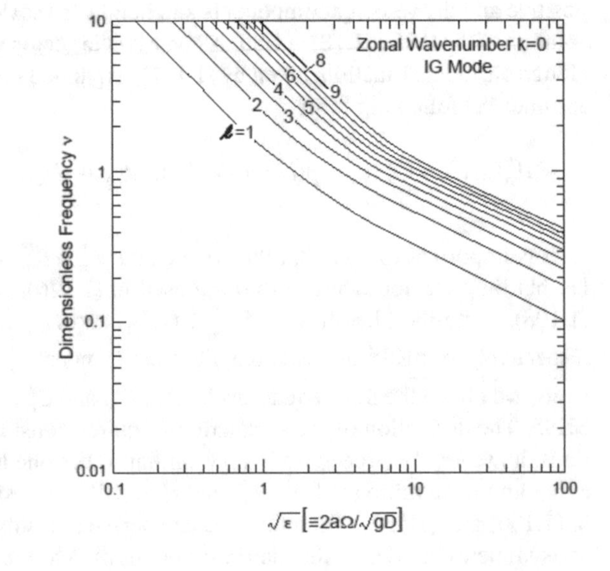

Fig. 1.6 Dimensionless frequency υ versus $\sqrt{\varepsilon}$ of inertio-gravity modes for zonal wavenumber $k = 0$ and meridional modes $l = 1, 2, \ldots, 9$. From Swarztrauber and Kasahara (1985), Fig. 1. Copyright ©1985 Society for Industrial and Applied Mathematics. Reprinted with permission. All rights reserved.

The dimensionless frequency υ is tabulated for zonal wavenumber $k = 0, 1, 2, 3$ in Tables 5 through 8 in SK85, respectively, with the exception of certain entries in Table 5 which we will discuss later. The first column in the tables indicate the meridional modal index l which is defined as $l = n - k$. Because the order of spherical harmonics n starts from k, the lowest meridional mode becomes $l = 0$ and the second mode is $l = 1$ and so on.

Figure 1.6 shows the dimensionless frequency υ for $k = 0$ as the function of $\sqrt{\varepsilon}$ for nine inertio-gravity (IG) modes, $l = 1, 2, \ldots 9$. The case of $k = 0$ is unique in that the frequencies of IG motions (first kind) appear as pairs of positive and negative values of the same magnitudes, and the frequencies of rotational (ROT) motions (second kind) are all zero. The meaning of eastward and westward propagation is lost in this case, but the meridional index for positive frequency is denoted by l_{EG} and for negative frequency by l_{WG}. Moreover, there is no mode corresponding to $l = 0$.

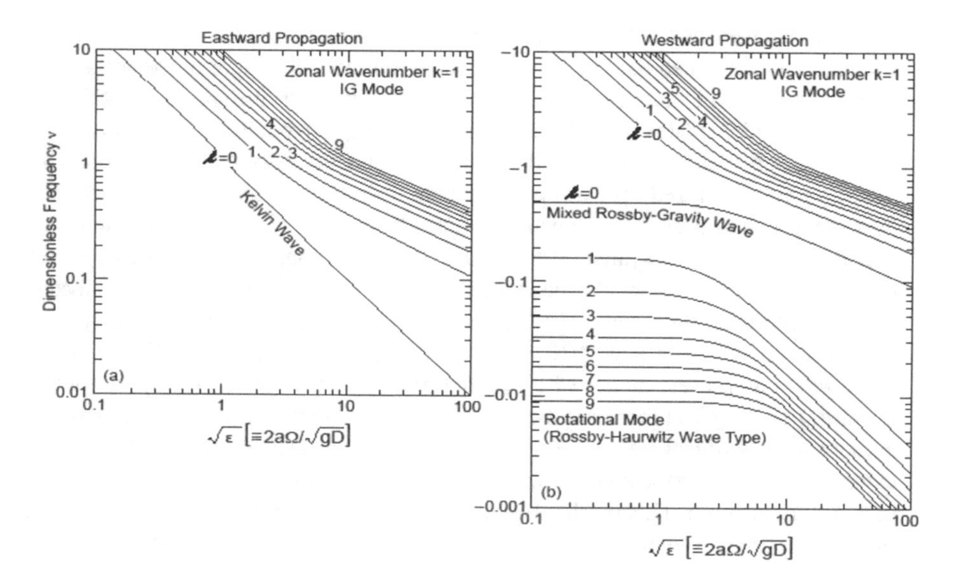

Fig. 1.7 Dimensionless frequency υ versus $\sqrt{\varepsilon}$ for zonal wavenumber $k = 1$ and meridional modes $l = 1, 2, \ldots, 9$. (a) Eastward-propagating modes, and (b) Westward-propagating modes. From Swarztrauber and Kasahara (1985), Fig. 2. Copyright ©1985 Society for Industrial and Applied Mathematics. Reprinted with permission. All rights reserved.

With respect to the eigenfunctions for $k = 0$, Fig. 1 of Kasahara (1978), which will be abbreviated as K78, shows an example of the meridional profiles of Hough vector function Θ_l^0 for $k = 0$ and $D = 10$ km for the Earth's atmosphere which corresponds to the value of $\sqrt{\varepsilon} = 2.9$. We should note that the sign for the value of V used in K78 is opposite of that calculated by the program of SK85. In Fig. 1, the mode corresponding to odd integer l is antisymmetric and denoted AY whereas the modes with even integer l are symmetric and denoted SY.

Figure 1.7 shows the value of υ for $k = 1$. The left panel (a) shows the eastward propagating modes of IG motions. The mode $l= 0$ is symmetric and behaves differently from the rest of IG modes. This is referred to as Kelvin mode whose property was first clarified by the normal mode analysis of Matsuno (1966) using the equatorial beta-plane. The right panel (b) shows the westward propagating modes of IG and rotational (ROT) waves. The ROT $l= 0$, which is antisymmetric, behaves like the rotational mode for small values of ε, but behaves more like the IG mode for large values of ε. For this reason, this particular mode is referred to as mixed Rossby-gravity (MRG) wave (Matsuno, 1966). Unlike the IG modes the odd (or even) ROT index l_R corresponds to SY (or AY). The reader is referred to similar diagrams in Longuet-Higgins (1968) for zonal wavenumber up to $k = 5$.

It may be worthwhile to comment on observational evidences on the existence of these global normal modes in the atmosphere. Immediately following theoretical works by Matsuno (1966), Lindzen (1966), Longuet-Higgins (1968) and others, many researchers began to investigate global wave motions in both observational analyses and simulated data by general circulation models in light of the properties of these normal modes. There were two notable research groups to identify the Kelvin and MRG modes in the tropics: Yanai and his collaborators at the University of Tokyo, and Wallace and his collaborators at the University of Washington. The Yanai's group analyzed special upper-air data over a dense observational network over the Marshall Islands in the central tropics from March to July 1958 (Yanai and Maruyama, 1966; Maruyama, 1967). Later, Yanai's group analyzed tropospheric data in the tropical western Pacific in April-July 1962 and Wallace's group analyzed in July-December 1963 and 1964 (Wallace, 1971). Both groups indeed confirmed the existence of Kelvin and MRG waves as described in detail by Wallace (1973). Both waves as gravest modes of all global normal modes which transport energy upward, implying that these are forced waves having a tropospheric energy source, perhaps, due to organized cumulus convection. Wallace (1973) noted that these are capable of modifying the zonal flow in the stratosphere through the vertical transport of zonal momentum. This mechanism of wave-mean flow interactions is believed to play an important role for producing the observed quasi-biennial oscillation of zonal flow in the tropics.

Likewise, there have been extensive studies to identify the ROT waves in the atmosphere as more general forms of the Rossby waves. We already mentioned earlier that the Rossby waves are similar to Haurwitz waves on the sphere which can be expressed in spherical harmonics. Perhaps, the earliest attempt to analyze the ROT modes in the observations was carried out by Eliasen and Machenhauer (1965) who used the spherical harmonics to analyze the stream function patterns from the 500 mb geopotential height fields and compared results with the solutions of the barotropic vorticity equation on the sphere with divergence effects. Clearly, the use of 3D NMFs is more appropriate for the identification of ROT waves as free global oscillations. Since this topic has its own long history as well as merits as a separate research subject, we only cite here one recent article by Madden (2007) who made an extensive review on the observation evidence on the existence of planetary-scale external ROT modes in the atmosphere using a 40-yr record of NCEP/NCAR Reanalysis data in comparison with the Hough solutions corresponding the 10 km equivalent height.

Table 1.1 summarizes the symmetry for the first four modes expressed by the meridional index l for a different value of zonal wavenumber k, indicating SY for symmetric type, and AY for antisymmetric type. The entry N denotes nonexistence, IG inertio-gravity, and ROT rotational modes. While the nonexistence of IG mode for $k = 0$ and $l = 0$ creates no problem, the nonexistence of ROT modes $k = 0$ and $l \geq 1$ is inconvenient for representation of the zonal flow by the normal mode expansion. We shall come back to a remedy of this problem later.

The Hough vector functions for $k = 1$, 2, and 3 for equivalent height $D = 10$ km are illustrated in K76. Again we caution that the sign for the value of V used in K78 is opposite of that calculated by the program of SK85. Longuet-Higgins (1968) shows extensive diagrams for the meridional profiles of Hough functions for various values of ε. For small values of ε the function profiles extend global, but for

Table 1.1 Parity of modes expressed by index l for different value of zonal wavenumber k. SY for symmetric mode and AY for antisymmetric mode. Entry N denotes nonexistence, IG inertio-gravity, and ROT rotational modes. SY: U and Z are symmetric, and V antisymmetric. AY: U and Z are antisymmetric, and V symmetric.

Mode	IG		ROT	
l	$k = 0$	$k > 0$	$k = 0$	$k > 0$
0	N	SY (Kelvin)	N	AY (Kelvin)
1	AY	AY	N	SY
2	SY	SY	N	AY
3	AY	AY	N	SY

large value ε, they look equatorially trapped though the shapes are the same. In fact, their circulation patterns are very much similar to those on the equatorial plane as illustrated by Matsuno (1966).

Rotational normal modes for k= 0

We now come back to discuss the rotational normal modes for $k = 0$ which we pointed out to be nonexistent. In the case of $k = 0$ and steady state, Laplace's tidal equations for the rotational motions reduce to the linear balance equation for zonal flows on the sphere. Since the balance equation is a generalized form of the geostrophic equation, we shall refer to the Hough harmonics of the rotational modes for $k = 0$ as geostrophic modes. K78 constructed a set of meridional functions corresponding to the geostrophic modes by a series of Legendre polynomials, and applied the Gram-Schmidt procedure to obtain an orthonormal set. These will be referred to as the K-modes. Tribbia (1979) adopted a similar procedure to construct geostrophic modes for his study of data initialization using the equatorial beta-plane shallow-water model. Since the geostrophic modes are created for the completeness of expansion functions, they are not unique. In fact, Silva Dias and Schubert (1979) constructed an orthogonal set of geostrophic modes for the equatorial beta-plane shallow-water model by taking the limit of the eigenfunctions of the rotational modes as the zonal wavenumber k approaches to zero, and applying L'Hospital's rule to derive the eigenfunctions.

The software program of SK85 provides the geostrophic modes following the idea proposed by Shigehisa (1983) who obtained the geostrophic modes of Laplace's tidal equations as the limit of rotational modes for $k \rightarrow 0$. This procedure is essentially the same as used by Silva Dias and Schubert (1979). However, unlike the equatorial beta-plane model, in which the zonal wavenumber is a real number, the zonal wavenumber for a sphere is an integer. Nevertheless, the limit of rotational modes is calculated by considering k as a continuous parameter and the ratio between k and the corresponding eigenfrequency υ to be finite. The latter condition ensures the phase speed υ/k to be continuous with respect to k. We shall refer the geostrophic modes in the SK85 program as the S-modes.

Tanaka and Kasahara (1992), which is referred to as TK92, conducted a performance comparison of the S- and K-modes using real atmospheric data. The zonal-mean components of atmospheric data from the FGGE IIIb reanalysis were projected onto the S- and K-modes separately, in addition to the IG modes. Both types of modes show similar meridional structures for a large equivalent height, say $D = 10$ km. But, for a small value of D, their structures are considerably different from each other. Even for a small value of D the K-modes have globally extended structures, whereas the S-modes have equatorially trapped structures. It was shown that the K-mode representation captures the majority of observed energy with much fewer meridional components than the S-mode representation. This is because the K-mode representation gives a faster convergence in the expansion series than the use of S-modes, especially for internal modes. However, it is important to note that the S-modes share the properties of the rotational modes of nonzero wavenumbers, while this is not case for the K-modes. Therefore, for illustration of the energy spectra with respect to, say, the phase speed $c = \upsilon/k$, the S-mode representation is useful. Therefore, each modal representation has own unique merits and we should make both programs available in the software package. In the following we present the method of calculating the K-modes.

Remembering that $\upsilon = 0$ and $k = 0$, Eq. (1.120) is reduced to

$$(\mu \nabla_0^2 + \mathcal{L})\tilde{\psi} = \nabla_0^2 \tilde{Z}, \tag{1.168}$$

which is referred to as the linear or geostrophic balance equation on a sphere (Merilees, 1966). The velocity potential $\tilde{\chi}$ vanishes due to the absence of meridional flow.

We construct a complete set of expansion functions for $\tilde{\psi}$ and $\tilde{Z}$ in terms of a finite series of Legendre polynomials $P_n(\mu)$ as

$$(\tilde{\psi},\ \tilde{Z}) = \sum_{n=0}^{2N+1} (\, B_n,\ C_n\,)\, P_n(\mu), \tag{1.169}$$

in parallel with (1.126). Useful formulas involving the Legendre functions can be obtained from (1.128-1.131) and (1.132) by substituting $k = 0$. From (1.168) and (1.169) we get

$$\sum_{n=0}^{2N+1} B_n\,(\mu \nabla_0^2 + \mathcal{L}) P_n = \sum_{n=0}^{2N+1} C_n\, \nabla_0^2\, P_n\,. \tag{1.170}$$

By using the formulas mentioned earlier, the above equation can be written as

$$\sum_{n=0}^{2N+1} B_n \left[\tfrac{n(n-1)(n+1)}{2n+1} P_{n-1} + \tfrac{n(n+2)(n+1)}{2n+1} P_{n+1} \right] = \sum_{n}^{2N+1} C_n\, n(n+1)\, P_n\,.$$

By collecting the coefficients of $P_n(\mu)$, we get

$$B_{n+1} \frac{(n+1)\,n\,(n+2)}{2n+3} + B_{n-1} \frac{(n-1)(n+1)\,n}{2n-1} = n\,(n+1) C_n\,.$$

By simplifying the above, we get

$$B_{n+1}\left(\frac{n+2}{2n+3}\right)+B_{n-1}\left(\frac{n-1}{2n-1}\right)=C_n\,. \tag{1.171}$$

Symmetric modes

Corresponding to (1.141), the coefficient vector for the symmetric case is expressed by

$$[B_1,\ C_0,\ B_3,\ C_2,\ \ldots,\ B_{2N+1},\ C_{2N}]^T\,. \tag{1.172}$$

We write each coefficient equation (1.171) for even integer n in explicit form from $n=0$, remembering that $B_{-1}=0$, as follows

$$\begin{aligned} n=0:&\quad B_1\left(\frac{2}{3}\right)+0=C_0,\\ n=2:&\quad B_3\left(\frac{4}{7}\right)+B_1\left(\frac{1}{3}\right)=C_2,\\ n=4:&\quad B_5\left(\frac{6}{11}\right)+B_3\left(\frac{3}{7}\right)=C_4,\end{aligned}$$

and so on.

For an even integer n in $[0,\ 2N-2]$, a simple set of N independent solutions of (1.171) can be obtained by choosing a single coefficient $B_{n+1}=1$ and all other B are zero. For example,

$n=0: B_1=1$ and all other B are zero. Then, we get $C_0=\frac{2}{3}$, $C_2=\frac{1}{3}$, and all other C are zero,

$n=2: B_3=1$ and all other B are zero. Then, we get $C_2=\frac{4}{7}$, $C_4=\frac{3}{7}$, and all other C are zero.

Therefore, we can say that N symmetric solutions can be written in a general form as

$$B_{n+1}=1\,,\quad C_n=\frac{n+2}{2n+3}\,,\quad C_{n+2}=\frac{n+1}{2n+3}\,, \tag{1.173}$$

for $n=0,\ 2,\ 4,\ \ldots,\ 2N-2$.

Antisymmetric modes

Corresponding to (1.144), the coefficient vector for the antisymmetric case is expressed by

$$[B_0,\ C_1,\ B_2,\ C_3,\ \ldots,\ B_{2N},\ C_{2N+1}]^T\,. \tag{1.174}$$

By going through the same derivation procedure as done for the symmetric case, we obtain the following coefficient equation,

$$B_{n+1}=1\,,\quad C_n=\frac{n+2}{2n+3}\,,\quad C_{n+2}=\frac{n+1}{2n+3}\,, \tag{1.175}$$

for $n=1,\ 3,\ 5,\ \ldots,\ 2N-1$. The form (1.175) looks identical to (1.173) except that the index n applies to an odd integer in $[1,\ 2N-1]$ for antisymmetric case.

Once the coefficients B_n and C_n are determined, the zonal stream function $\tilde{\psi}$ and zonal height $\tilde{Z}$ are obtained using (1.169), and the zonal velocity $\tilde{U}$ is obtained from (1.148) with $\tilde{\chi}= 0$ and $k= 0$, namely $\tilde{V}= 0$ from (1.149). The proportional factor for $\tilde{U}$ and $\tilde{Z}$ is still arbitrary and should be determined by the normalization

$$\int_{-1}^{1} \left(\tilde{U}^2 + \tilde{Z}^2\right) \mathrm{d}\mu = 1 . \tag{1.176}$$

At this stage, however, they are not orthogonal and need to be transformed into an orthogonal set. A vector orthogonalization routine based on the Gram-Schmidt method (e.g. Lanczos 1961) works well for this purpose.

The geostrophic modes are still not complete unless we add a vector corresponding to $\tilde{U} = 0$, $\tilde{V} = 0$, and $\tilde{Z} =$ constant to represent a constant height. With the same normalization condition (1.176), $\tilde{Z}$ should be chosen as $1/\sqrt{2}$. Using this vector as the lowest-order Hough vector ROT function ($l_R= 0$), we orthonomalize the vector functions to satisfy

$$\int_{-1}^{1} (\tilde{U}_l \tilde{U}_m + \tilde{Z}_l \tilde{Z}_m)\, \mathrm{d}\mu = \delta_{lm} \tag{1.177}$$

consistently with (1.113).

The meridional profiles of orthonormalized geostrophic modes ($k = 0$) for $l_R \geq 1$ are seen in Fig. 2 of K78.

1.8 Energy Considerations

One area of application of the 3D normal mode representation to global atmospheric modelling is to study the mechanism of general circulation through energy consideration. Energy exchanges take place between the kinetic, internal, and potential energies in response to forcing, heating, and dissipation. Global energetics analyses will be extensively discussed in later chapters. In this section, we simply demonstrate that the linear baroclinic system presented in subsection 1.3 is an energy conserving system and how the various forms of energy are represented by means of the normal modes.

Multiplying (1.30-1.33) by u', v', P', and $\partial P'/\partial \sigma$ respectively, and adding the resulting equations, we obtain

$$\frac{\partial}{\partial t}\frac{1}{2}\left[u'^2 + v'^2 + \frac{\sigma}{R\Gamma_0}\left(\frac{\partial P'}{\partial \sigma}\right)^2\right] = -\frac{\partial}{\partial \sigma}(P'\,W') - \nabla \cdot (P'\,V') . \tag{1.178}$$

Integration of the above equation on the sphere of radius a with longitude λ, latitude φ, and vertical coordinate σ and using the boundary conditions (1.27) and (1.29) yields

$$\frac{\partial}{\partial t}\int_0^{2\pi}\int_{-1}^{1}\int_0^1 \frac{1}{2}\left[u'^2 + v'^2 + \frac{\sigma}{R\Gamma_0}\left(\frac{\partial P'}{\partial \sigma}\right)^2 + \frac{P_s{'}^2}{R(T_0)_s}\right] a^2 \mathrm{d}\sigma\, \mathrm{d}\mu\, \mathrm{d}\lambda = 0, \tag{1.179}$$

where $\mu = \sin\varphi$ and $P_s{'}$ denotes the value of P' at $\sigma = 1$. This equation is the statement of the conservation of total energy as the sum of kinetic and potential energies being invariant in time. The first two terms in the brackets represent the kinetic energy. The 3rd and 4-th terms denote the potential energy, but it is more appropriate to call this quantity as available potential energy which represents the available portion of the sum of the potential and internal energies of the system associated with wave motions (Lorenz, 1955).

We express the variables u', v', and P' by the sum of m vertical modes as

$$(u', v', P') = \sum_m (\tilde{u}_m, \tilde{v}_m, g\tilde{h}_m)\, G_m(\sigma). \tag{1.180}$$

By substitution of (1.180) into (1.179), the time invariant of the total energy is expressed as

$$\int_0^{2\pi}\int_{-1}^{1}\int_0^1 \frac{1}{2}\left\{\left(\sum_m \tilde{u}_m G_m\right)^2 + \left(\sum_m \tilde{v}_m G_m\right)^2 + \frac{g^2\sigma}{R\Gamma_0}\left(\sum_m \tilde{h}_m \frac{dG_m}{d\sigma}\right)^2 + \frac{g^2}{R(T_0)_s}\left(\sum_m \tilde{h}_m G_m(1)\right)^2\right\} a^2\, \mathrm{d}\sigma\, \mathrm{d}\mu\, \mathrm{d}\lambda = \text{const.}. \tag{1.181}$$

The vertical structure function $G_m(\sigma)$ is the eigenfunction of the vertical structure equation

$$\frac{\mathrm{d}}{\mathrm{d}\sigma}\left(\frac{\sigma g}{R\Gamma_0}\frac{\mathrm{d}G_m}{\mathrm{d}\sigma}\right) + \frac{1}{D_m}G_m = 0 \tag{1.182}$$

according to (1.43). By multiplying (1.182) by G_n, integrating the resulting equation with respect to σ from 0 to 1, and using the boundary conditions (1.44) and (1.45), we obtain

$$\int_0^1 \frac{\sigma g}{R\Gamma_0}\frac{\mathrm{d}G_m}{\mathrm{d}\sigma}\frac{\mathrm{d}G_n}{\mathrm{d}\sigma} d\sigma + \frac{g}{R(T_0)_s}G_m(1)G_n(1) = \int_0^1 \frac{G_m G_n}{D_m} d\sigma. \tag{1.183}$$

Using the above identity and the orthogonality condition (1.56), Eq. (1.179) can be expressed as

$$\frac{\partial}{\partial t}\int_0^{2\pi}\int_{-1}^{1}\sum_m \frac{1}{2}\left(\tilde{u}_m^2 + \tilde{v}_m^2 + \frac{g}{D_m}\tilde{h}_m^2\right) a^2\, \mathrm{d}\mu\, \mathrm{d}\lambda = 0. \tag{1.184}$$

This equation demonstrates that the representation (1.180) in terms of the sum of m vertical modes permits the partition of total energy into the kinetic and available potential energies of each vertical mode separately.

1.9 Expansion of Three–Dimensional (3D) Global Data

Global atmospheric data produced by models and objective analyses can be represented by the sum of the normal mode functions (NMFs) obtained thus far. The advantage of using the 3D NMFs is that the horizontal velocity components and the geopotential height are simultaneously expressed by a single series of 3D vector functions consisting of products of the Hough harmonics and the vertical normal modes. The auxiliary fields such as the vorticity, divergence, and vertical velocity can be obtained optimally by using the entire input data.

Input data are the zonal and meridional components of velocity u^I, v^I, and geopotential height z^I at the 3D grid of longitude λ, latitude φ, and vertical level σ with the surface pressure p_s. The data in the vertical are placed at level j shown in Fig. 5 and are expressed as

$$u^I(\lambda, \varphi, \sigma_j) = u^I, \tag{1.185}$$

$$v^I(\lambda, \varphi, \sigma_j) = v^I, \tag{1.186}$$

$$P^I(\lambda, \varphi, \sigma_j) = gz^I + RT_0(\sigma_j) \ln p_s \,. \tag{1.187}$$

We then represent the input data vector by a series of the vertical structure function $G_m(j)$, as defined by (1.70).

$$\left(u^I, v^I, g^{-1}P^I\right)^T = \sum_{m=1}^{M} \mathbf{S}_m \mathbf{X}_m(\lambda, \varphi) G_m(j) \quad . \tag{1.188}$$

Here, the scaling matrix $\mathbf{S}_m$ is introduced and defined by

$$\mathbf{S}_m = \begin{bmatrix} \sqrt{gD_m} & 0 & 0 \\ 0 & \sqrt{gD_m} & 0 \\ 0 & 0 & D_m \end{bmatrix} . \tag{1.189}$$

The integer subscript m is used to identify the vertical mode which spans from the external mode $m = 1$ to the total number of vertical modes M for the rest of the internal modes. The introduction of scaling matrix $\mathbf{S}_m$ makes the horizontal coefficient vector $\mathbf{X}_m(\lambda, \varphi)$ dimensionless as seen from (1.100).

The horizontal coefficient vector $\mathbf{X}_m(\lambda, \varphi)$ can be calculated by the reverse transform of (1.188) through multiplication of (1.188) by $G_l(j)$ and summation of the result from $j = 1$ to J with the use of the orthogonality condition (1.70). The result becomes

$$\mathbf{X}_m(\lambda, \varphi) = \mathbf{S}_m^{-1} \sum_{j=1}^{J} \left(u^I, v^I, g^{-1}P^I\right)^T G_m(j) \,. \tag{1.190}$$

The horizontal coefficient vector $\mathbf{X}_m(\lambda, \varphi)$ for a given vertical mode m is going to be projected onto the Hough harmonics defined as

$$\mathbf{H}_r^k(\lambda, \varphi; m) = \mathbf{\Theta}_r^k(\varphi; m) \exp(ik\lambda) \quad . \tag{1.191}$$

This definition is slightly modified from that of (1.111). Instead of using subscript n for the meridional mode, we use the subscript r to indicate all combined meridional normal modes of three different species, consisting of the rotational (ROT), and eastward and westward propagating inertio-gravity (EG and WG) modes. The subscript r is a tag index ranging from 1 to integer R defined by

$$R = L_R + L_{WG} + L_{EG}, \tag{1.192}$$

where ROT mode: $l_R = 0, 1, 2, \ldots . L_R - 1$, the EG mode: $l_{EG} = 0, 1, 2, \ldots L_{EG} - 1$, and the WG mode: $l_{WG} = 0, 1, 2, \ldots L_{WG} - 1$.

The addition of integer m in the parentheses in (1.191) is to indicate a specific vertical mode m. Therefore, the Hough harmonics are characterized by the two indices for combined meridional mode r and vertical mode m, in addition to the zonal wavenumber k.

The horizontal coefficient vector $\mathbf{X}_m(\lambda, \varphi)$ for a given vertical mode m is now projected onto the Hough harmonics as

$$\mathbf{X}_m(\lambda, \varphi) = \sum_{r=1}^{R} \sum_{-K}^{K} X_r^k(m)\, \mathbf{H}_r^k(\lambda, \varphi; m), \tag{1.193}$$

where the zonal wavenumber k spans from a negative integer $-K$ to positive K, including zero for the zonal mode.

To obtain the scalar coefficient $X_r^k(m)$, we multiply (1.193) by $[\mathbf{H}_r^k]^*$, the complex conjugate of $\mathbf{H}_r^k$, integrate the resultant equation with respect to λ from 0 to 2π, and with respect to φ from $-\pi/2$ to $+\pi/2$, and then use the orthonormality condition (1.113). The result is

$$X_r^k(m) = \frac{1}{2\pi} \int_0^{2\pi} \int_{-1}^{1} \mathbf{X}_m(\lambda, \varphi) \cdot \left(\mathbf{H}_r^k\right)^* \mathrm{d}\mu\, \mathrm{d}\lambda, \tag{1.194}$$

where $\mu = \sin\varphi$. The double integral of (1.194) can be split into two transforms by using the definition of $\mathbf{H}_r^k$ given by (1.191):

$$X_r^k(m) = \int_{-1}^{1} \mathbf{Y}(\varphi; m) \cdot \left(\Theta_r^k(\varphi, m)\right)^* \mathrm{d}\mu \quad , \tag{1.195}$$

$$\mathbf{Y}(\varphi; m) = \frac{1}{2\pi} \int_0^{2\pi} \mathbf{X}_m(\lambda, \varphi) \exp(-ik\lambda)\, \mathrm{d}\lambda \quad . \tag{1.196}$$

Formula (1.195) is the Hough transform and the integration can be performed using the Gaussian quadrature. Formula (1.196) is the Fourier transform and a fast Fourier transform is available for calculation of the integral.

1.10 Hough Spectral 3D Primitive Equation Model

Current spectral methods for solving the 3D primitive equation (PE) model on the sphere that is described in Section 1.2, adopt the spherical harmonics as the basis functions to represent the dependent variables (Machenhauer, 1979). In this section we describe an alternative method, called the Hough spectral method, using the 3D normal mode vector functions as the basis functions instead of the spherical harmonics. According to Eckart (1960, Chap. 6), the most general limited solution of the time-depedent nonlinear prediction equation system such as the PE model can be constructed out by using the normal mode functions of the respective system. Here, limited solutions imply solutions corresponding to finite total energy which is the kind of the solutions we are seeking. Therefore, we can apply Eckart's general principle to construct the Hough spectral model. One notable advantage is that we only get one time-dependent equation to solve for the coefficients of the expansion functions even though the original system contains three time-dependent equations. Therefore, the evolution history of three dependent variables is represented by a single series of the expansion functions instead of three separate series. Another advantage is that we can select the type of normal mode functions as the basis functions both in the horizontal and vertical directions depending on the desire to simplify the physics of the PE model. For example, if we choose only the rotational modes for the basis functions, the resulting model becomes essentially a quasi-geostrophic model. Likewise, if we choose only two vertical modes, the continuous model becomes effectively a two-parameter model. Thus, flexibility in the selection of normal mode function terms can provide an insight into the physics of the PE model.

First we list the basic equations by extracting from Section 7.1.2 with the same notation. The equations of horizontal motion with velocity components u and v are given by (1.17) and (1.18) as

$$\frac{\partial u}{\partial t} - 2\Omega \sin\varphi\, v + \frac{1}{a\cos\varphi}\frac{\partial P}{\partial \lambda} = C_1 \quad , \tag{1.197}$$

$$\frac{\partial v}{\partial t} + 2\Omega \sin\varphi\, u + \frac{1}{a}\frac{\partial P}{\partial \varphi} = C_2 \quad , \tag{1.198}$$

where C_1 and C_2 are defined by (1.19) and (1.20). The thermodynamic equation for mass variable P is given by

$$\frac{\partial}{\partial t}\left[\frac{\partial}{\partial \sigma}\left(\frac{\sigma}{R\Gamma_0}\frac{\partial P}{\partial \sigma}\right)\right] - \nabla \cdot \mathbf{V} = \frac{\partial C_3}{\partial \sigma} + \mathbf{V} \cdot \nabla(\ln p_s) \quad , \tag{1.199}$$

where C_3 is defined by (1.25). Equation (1.199) is derived by differentiating (1.24) with respect to σ and then eliminating $\partial W/\partial \sigma$ using (1.22). In addition to those three equations for the velocity and mass variables, we need the prediction equation for the surface pressure p_s, which is given by (1.10), to complete the prediction system.

When the right-hand side of (1.197)-(1.199) vanishes, the system becomes identical in form with (1.30), (1.31) and (1.34), respectively. Therefore, we can represent the dependent variables, u, v, and P as the products of the horizontal and vertical structure functions:

$$\big(u,\ v,\ P\big) = \sum_{m} (u_m,\ v_m,\ g\, h_m)\, G_m(\sigma) \quad , \tag{1.200}$$

where the vertical structure function $G_m(\sigma)$ satisfies the vertical structure equation (1.43) which is reproduced as

$$\frac{d}{d\sigma}\left(\frac{\sigma g}{R\Gamma_0}\frac{dG_m}{d\sigma}\right) + \frac{G_m}{D_m} = 0 \quad . \tag{1.201}$$

Here, D_m is the equivalent height as the eigenvalue of (1.201) and is determined using the boundary conditions (1.44) and (1.45). The eigensolutions of (1.201) are discussed in detail in Section 1.4. The integer subscript m is used to identify the vertical mode which spans from the external mode $m = 1$ to the rest of internal modes $m > 1$ up to maximum M.

First, we substitute the expression (1.200) into (1.197) - (1.199), and replace the second derivative term in (1.199) using (1.201). Then, we multiply all three equations by the vertical structure function $G_n(\sigma)$ and integrate them from $\sigma = 0$ to $\sigma = 1$. Because of the orthogonality condition of the vertical structure functions,

$$\int_0^1 G_m(\sigma)\, G_n(\sigma)\, d\sigma = \delta_{m\,n} \quad , \tag{1.202}$$

where $\delta_{m\,n} = 1$ if $m = n$ and zero otherwise, the vertical dependence of the three resulting equations vanishes in the case of $m \neq n$. Hence we get the following horizontal structure equations

$$\frac{\partial u_m}{\partial t} - 2\Omega\sin\varphi\; v_m + \frac{g}{a\cos\varphi}\frac{\partial h_m}{\partial\lambda} = N_u \quad , \tag{1.203}$$

$$\frac{\partial v_m}{\partial t} + 2\Omega\sin\varphi\; u_m + \frac{g}{a}\frac{\partial h_m}{\partial\varphi} = N_v \quad , \quad \text{and} \tag{1.204}$$

$$\frac{\partial h_m}{\partial t} + D_m \nabla\cdot\mathbf{V}_m = D_m N_h \quad , \tag{1.205}$$

where

$$N_u = \int_0^1 C_1\, G_m(\sigma) \mathrm{d}\sigma \quad , \tag{1.206}$$

$$N_v = \int_0^1 C_2\, G_m(\sigma) \mathrm{d}\sigma \quad , \quad \text{and} \tag{1.207}$$

$$N_h = -\int_0^1 \left[\frac{\partial C_3}{\partial \sigma} + \mathbf{V} \cdot \nabla(\ln p_s) \right] G_m(\sigma) d\sigma \quad . \tag{1.208}$$

Now, we scale Eqs. (1.203)-(1.205) by introducing the following dimensionless variables:

$$\tilde{u}_m = \frac{u_m}{\sqrt{g D_m}}, \quad \tilde{v}_m = \frac{v_m}{\sqrt{g D_m}}, \quad \tilde{h}_m = \frac{h_m}{D_m}, \quad \tilde{t} = 2\Omega t \quad . \tag{1.209}$$

Then equations (1.203)–(1.205) can be expressed in the vector form (Kasahara, 1977, 1978)

$$\frac{\partial}{\partial \tilde{t}} \mathbf{W}_m + \mathbf{L}\mathbf{W}_m = \mathbf{F}_m (\lambda, \varphi, \tilde{t}) \quad , \tag{1.210}$$

where the dependent vector variable is defined as

$$\mathbf{W}_m = (\tilde{u}_m, \tilde{v}_m, \tilde{h}_m)^T \quad . \tag{1.211}$$

Also, $\mathbf{L}$ is the differential operator

$$\mathbf{L} = \begin{vmatrix} 0 & -\sin(\varphi) & \frac{\gamma_m}{\cos(\varphi)} \frac{\partial}{\partial \lambda} \\ \sin(\varphi) & 0 & \gamma_m \frac{\partial}{\partial \varphi} \\ \frac{\gamma_m}{\cos(\varphi)} \frac{\partial}{\partial \lambda} & \frac{\gamma_m}{\cos(\varphi)} \frac{\partial}{\partial \varphi} [\cos(\varphi)(.)] & 0 \end{vmatrix} , \tag{1.212}$$

in which

$$\gamma_m = \frac{\sqrt{g D_m}}{2a\Omega} \tag{1.213}$$

is a dimensionless parameter. The inhomogeneous vector term on the right-hand side of (1.210) is expressed by

$$\mathbf{F}_m (\lambda, \varphi, \tilde{t}) = (2\Omega)^{-1} \begin{bmatrix} (g D_m)^{-1/2} N_u \\ (g D_m)^{-1/2} N_v \\ N_h \end{bmatrix} . \tag{1.214}$$

When the inhomogeneous term (1.214) is absent, Eq. (1.210) is reduced to the Laplace's tidal equation (1.101) for vertical mode m.

We now seek the solution of (1.210) and express it as

$$\mathbf{W}_m(\lambda, \varphi, \tilde{t}) = \sum_{r=1}^{R} \sum_{k=-K}^{K} W_r^k(\tilde{t}; m) \mathbf{H}_r^k(\lambda, \varphi; m) \quad , \tag{1.215}$$

where $W_r^k(\tilde{t}; m)$ is the expansion coefficient which depends on time $\tilde{t}$ for vertical mode index m. Also, the Hough harmonics are given by

$$\mathbf{H}_r^k(\lambda, \varphi; m) = \Theta_r^k(\varphi; m) \exp(ik\lambda) \quad , \tag{1.216}$$

as defined by (1.111). Note that zonal wavenumber k runs from a negative integer $-K$ to a positive integer K, including $k = 0$ as the zonal mode. The summation for serial number r from 1 to R should include all meridional modes of the first and second kinds, namely westward and eastward moving IG and westward moving rotational waves.

The meridional structure function $\mathbf{\Theta}_r^k(\varphi;\, m)$ has three components

$$\mathbf{\Theta}_r^k(\varphi; m) = \begin{bmatrix} U_r^k(\varphi; m) \\ iV_r^k(\varphi; m) \\ Z_r^k(\varphi; m) \end{bmatrix} \tag{1.217}$$

as defined by (1.112). This function is also referred to as a Hough vector function and is associated with the dimensionless frequency $\nu_r^k(m)$, which is the eigenvalue of

$$\left(\mathbf{L} - i\nu_r^k(m)\mathbf{I}\right)\mathbf{H}_r^k(\lambda, \varphi, m) = 0 \tag{1.218}$$

as shown by (1.108), where $\mathbf{I}$ is the identity matrix.

Substituting (1.215) into (1.210), integrating the resultant equation over the entire globe after multiplied by the complex conjugate of Hough harmonics $\left(\mathbf{H}_r^k\right)^*$ and using the fact that $\mathbf{H}_r^k$ and ν_r^k satisfy (1.218) and the orthogonality condition of Hough harmonics (1.110), we obtain the spectral prediction equation

$$\frac{\mathrm{d}W_r^k(\tilde{t}; m)}{\mathrm{d}\tilde{t}} + i\nu_r^k(m)\, W_r^k(\tilde{t}; m) = A_r^k(\tilde{t}; m) \quad , \tag{1.219}$$

where

$$A_r^k(\tilde{t}; m) = \frac{1}{2\pi}\int_0^{2\pi}\int_{-1}^{1} \mathbf{F}_m \cdot \left(\mathbf{H}_r^k\right)^*(\lambda, \varphi, m)\, \mathrm{d}\mu \mathrm{d}\lambda \quad , \tag{1.220}$$

with $\mu = \sin\varphi$.

The form of spectral equation (1.219) corresponds to Eq. (3) of Eckart (1960, p. 86). This is the first order ordinary differential equation with respect to time $\tilde{t}$ that gives the evolution of $W_r^k(\tilde{t}; m)$ in time starting from its initial condition. Once $W_r^k(\tilde{t}; m)$ is determined, the 3D physical space solutions (u, v) and P as functions of $(\lambda, \varphi, \sigma)$ and $\tilde{t}$ can be obtained by using (1.215), summing up with respect to the vertical mode m from 1 to M and remembering that $\mathbf{H}_r^k(\lambda, \varphi; m)$ is expressed by (1.216) and the scaling used in (1.209):

$$(u, v, P)^T = \sum_{m=1}^{M} \mathbf{S}_m \sum_{r=1}^{R}\sum_{k=-K}^{K} W_r^k(\tilde{t}; m)\, \Theta_r^k(\varphi; m) \exp(ik\lambda)\, G_m(\sigma) \quad , \tag{1.221}$$

where $\mathbf{S}_m$ is the scaling matrix defined by (1.189).

In addition to calculating the solutions of the velocity and mass variables, we need to solve Eq. (1.10) to predict the surface pressure p_s. That completes the σ-coordinate PE spectral model using the 3D normal mode functions. The conservation of total energy in this system when the right-hand side of (1.219) is absent has been discussed in Section 1.7. See Eq. (1.184).

The globally averaged total energy TE is expressed by

$$TE = \frac{1}{2\pi} \int_0^{2\pi} \int_{-1}^{1} \sum_{m=1}^{M} TE_m \, \mathrm{d}\mu \, \mathrm{d}\lambda \quad , \tag{1.222}$$

where

$$TE_m = \frac{1}{2}(u_m^2 + v_m^2 + \frac{g}{D_m} h_m^2) = \frac{1}{2} g D_m (\tilde{u}_m^2 + \tilde{v}_m^2 + \tilde{h}_m^2) = \frac{1}{2} g D_m \mathbf{W}_m \cdot \mathbf{W}_m^* \tag{1.223}$$

with the aid of (1.209) and (1.211). By substituting (1.223) into (1.222) and performing the global integration, we obtain

$$TE = \frac{1}{2} \sum_{m=1}^{M} \left(g D_m \sum_{r=1}^{R} \sum_{k=-K}^{K} W_r^k \, W_r^{k*} \right) . \tag{1.224}$$

Actual calculations with this general formulation of 3D Hough spectral PE model in σ-coordinates, similar to operational forecasting models, have not yet been performed. However, the same principle has been applied as a research tool to investigate various dynamical problems using simpler settings of this general formulation. We will present some of the earlier dynamical studies using simplified procedures.

The feasibility of Hough spectral formulation has been tested to solve the nonlinear shallow water equations (barotropic primitive equations) over the sphere and the solutions were in good agreement with those of a fourth-order finite-difference method (Kasahara, 1977, 1978). One advantage in using the normal modes as the spectral expansion functions is that, because the basis functions are physical modes which consist of the rotational and inertio-gravity (IG) components, we can investigate the contribution of IG motions in the total motions. For example, in another studies Kasahara (1984) investigated the generation of IG motions in a stratified atmosphere due to tropical diabatic heating. For a parabolic profile of heating in the vertical, the internal modes with the equivalent height of a few hundred meters are favourably excited which is consistent with observations (e.g. Silva Dias, 1985). Of course, the atmospheric response to tropical heating is different depending on the temporal and spatial scales of heating. For example, for the case of stationary tropical heating, most of excited energy goes into the rotational modes, but a significant portion also goes to the Kelvin modes, while other IG modes play insignificant roles in general. For the case of transient tropical heating, the generation of IG modes, except for the Kelvin modes, depends strongly on the time scale of heating, while the rotational modes and the Kelvin modes are dependent only weakly on the heating rate. The unique behavior of the Kelvin modes may be explained by the resemblance of the heating pattern to the horizontal structure of Kelvin modes and the closeness of their frequencies to those of rotational modes. This type of numerical experiments may be useful for understanding the structure and properties of short-range forecast errors in a global model. Investigating the balance properties of the short-range forecast errors in the ECMWF 4D-Var ensemble, Žagar et al. (2013) found that nearly

50% of the global forecasting error variances at short-range are due to the IG motions that have equatorially trapped horizontal structures with the equivalent heights of a few tens to few hundred meters. In the eastward IG modes the variance growth is most noticeable in the large-scale equatorial Kelvin mode.

The study of a response of planetary waves to tropical heating just mentioned was further extended by Kasahara and da Silva Dias (1986) by including a basic zonal flow with meridional and vertical shear to investigate the dynamical mechanism of "teleconnection" (e.g. Horel and Wallace, 1981). The vertical shear of zonal flow gives rise to the coupling of the external mode with the internal vertical modes. As a result of this coupling, a significant response occurs in the external mode in the mid-latitudes due to the excitation of the "baroclinic" internal modes by stationary heating over the tropical ocean.

In another example, Tanaka (1985) formulated the 3D Hough spectral model with pressure coordinates to perform a diagnostic analysis of global atmospheric energetics and investigated energy spectrum distributions and energy transformations-exchanges among various normal modes partitioned into vertical and meridional indices and zonal wavenumbers. He used the First GARP (Global Atmospheric Research Program) analysis data for his study.

The examples that we just presented are concerned with the time-dependent problems from initial conditions and/or with diabatically imposed forcing. The Hough spectral modeling is also useful to investigate the structural stability of basic states such as mean zonal flows, climatological patterns and the baroclinicity. One of the simplest examples will be presented below to explain the basic principle of methodology. During the 60s and 70s, many observational studies were conducted to study traveling planetary waves in the atmosphere at long periods in excess of a few days. Among them, Madden (1978) examined the westward propagating zonal wavenumber one planetary waves having periods of 5 days and near 16 days and concluded that they are free oscillations of the external mode with the equivalent height of 10 km. However, while a 5-day period agrees with that derived from the Laplace tidal equations, there is a discrepancy in periods between 16 days and predicted 12.3 days. Since the predicted period is derived from the theory without the background flow, a study was conducted to calculate the normal modes of free oscillations including the effect of zonal flows and explained the "discrepancy" in the wave period. Following Kasahara (1980, referred to as K80), we present the methodology of such an approach using the Hough spectral formulation.

We seek the solutions of the global shallow water equations for the perturbation velocity components and height linearized around a steady zonal flow and its corresponding basic state height which are expressed in the following vector form

$$\frac{\partial}{\partial \tilde{t}}\mathbf{W} + \mathbf{L}\mathbf{W} = -i\mathbf{B}\mathbf{W}, \tag{1.225}$$

where $\mathbf{W}$ stands for the dimensionless velocity components and height as defined by (1.102), $\mathbf{L}$ denotes the tidal operator (1.103), and $\mathbf{B}$ is the basic state structural matrix which contains various terms representing the effects of steady basic zonal flow and its corresponding height. The explicit form of this 3×3 matrix is shown by (1.11) of K80, but it is not necessary to spell it out here for explaining the methodology.

We now employ the Hough spectral method to solve (1.225). Because (1.225) is a quasi-linear vector equation having constant coefficients in time and longitude, we represent its solution by a a product of harmonic function in time and longitude and a series of latitude-dependent Hough vector functions in meridional direction for the external mode in this example:

$$\mathbf{W}(\varphi; \tilde{t}, k) = \exp[i\,(k\lambda - \alpha\tilde{t})] \sum_{r=1}^{R} C_r^k\, \mathbf{\Theta}_r^k\,, \tag{1.226}$$

where k denotes the zonal wavenumber and α the frequency. The summation for serial number r from 1 to R includes all meridional modes of the first and second kinds.

Substituting (1.226) into (1.225) and using the fact that the eigenfunctions and eigenvalues corresponding to the Laplace tidal operator $\mathbf{L}$ are the Hough harmonics and frequencies $i\nu_r^k$, we obtain

$$-i\,\alpha \sum_{r=1}^{R} C_r^k\, \mathbf{\Theta}_r^k + i \sum_{r=1}^{R} \nu_r^k\, C_r^k\, \mathbf{\Theta}_r^k + i\mathbf{B} \sum_{r=1}^{R} C_r^k\, \mathbf{\Theta}_r^k = 0\,. \tag{1.227}$$

Taking the inner product of (1.227) with $\left(\mathbf{\Theta}_r^k\right)^*$, integrating the resulting equations from South to North Pole and using the orthogonality condition (1.113), we obtain

$$(\nu_r^k - \alpha)\, C_l^k + \sum_{r=1}^{R} b_{l\,r}^k\, C_r^k = 0, \tag{1.228}$$

where $l = 1, 2, ..., R$, and

$$b_{l\,r}^k = \int_{-1}^{1} \mathbf{B}\,\mathbf{\Theta}_r^k \cdot \mathbf{\Theta}_l^k\, d\mu \quad . \tag{1.229}$$

Since there are the same number R of the equations and the terms of expansion coefficient series, if we define the vector

$$\mathbf{X} = (C_1^k,\ C_2^k,\ \ldots,\ C_R^k)^T\,, \tag{1.230}$$

and the matrix

$$\mathbf{M} = \begin{bmatrix} (\nu_1^k + b_{1\,1}^k) & b_{1\,2}^k & - & - & b_{1\,R}^k \\ b_{2\,1}^k & (\nu_2^k + b_{2\,2}^k) & - & - & b_{2\,R}^k \\ - & - & - & - & - \\ b_{R\,1}^k & b_{R\,2}^k & - & - & (\nu_R^k + b_{R\,R}^k) \end{bmatrix}, \tag{1.231}$$

we can write (1.228) as a standard eigenvalue-eigenfunction problem

$$\mathbf{MX} = \alpha\mathbf{X} \quad , \tag{1.232}$$

where α denotes the eigenvalues of $\mathbf{M}$ and $\mathbf{X}$ represents the corresponding eigenvectors. Therefore, the solution (1.226) is expressed by

$$\mathbf{W}(\varphi;\ \tilde{t},\ k) = \exp[i\,(k\lambda - \alpha\tilde{t})] \sum_{r=1}^{R} \mathbf{X} \cdot \mathbf{\Theta}_r^k \quad . \tag{1.233}$$

The effect of the basic zonal flow on the free oscillations in a barotropic global atmosphere has been discussed in K80.

Before leaving this section we should note that, since the basic zonal flow in this calculation has a meridional shear, the wave frequency α may become complex, meaning that the basic zonal flow can become unstable to small perturbations. This is the case of barotropic instability. The same methodology can be extended to investigate classical problems of baroclinic instability of perturbations on the westerlies using primitive equation models on the sphere (e.g. Simmons and Hoskins 1976; Frederiksen 1978). In this case the solutions of the 3D primitive equation model are expanded in term of the 3D normal mode functions consisting of Hough harmonics in the horizontal together with the vertical structure functions. Again the task is to determine the series coefficients of the expansion functions for each vertical mode. Therefore, other than the involvment of vertical modes denoted by index m in the matrix calculation, the formalism of determining the expansion coefficients is similar to (1.228), namely

$$[\nu_r^k(m) \ - \ \alpha]C_l^k(m) + \sum_{n=1}^{M} \sum_{r=1}^{R} b_{l\,r}^k(m,n)\, C_r^k(n) \ = \ 0 \tag{1.234}$$

for meridional index $l = 1, 2, ..., R$ and vertical mode index $m = 1, 2, ..., M$.

Although the explicit form of the matrix elements is much more complicated due to the involment of vertical dimension, Eq. (1.234) can be cast in the standard form of eigenfunction-eigenvalue problem like (1.232) to obtain the series coefficients $C_r^k(m)$ and the frequency α. When the frequency α becomes complex for a certain range of zonal wavenumber k, the basic state is said to be unstable and the growth rate of perturbation can be obtained from the imaginary-part of the frequency α. The details of this calculation, applied to invstigate the stability problems of the 30° jet studied by Simmons, Hoskins and Frederiksen mentioned earlier, have been discussed by Kasahara and Tanaka (1989). A similar study of stability analysis using the method of 3D normal mode expansion was conducted also by Tanaka and Kung (1989).

1.11 Concluding Remarks

In this chapter we described 3D normal mode functions (NMFs) of the global baroclinic primitive-equation (PE) model. The 3D NMFs are the eigenmodes of adiabatic, inviscid, and linearized version of the PE model with respect to a resting basic state, and are expressed by the products of the horizontal structure functions, called the Hough harmonics, and the vertical structure functions (VSF). The 3D NMFs are orthogonal in the vertical as well as in the horizontal. The Hough harmonics, which are associated with the frequencies of the first kind are called the inertio-gravity (IG) waves, while the ones associated with the frequencies of the second kind are referred to as the rotational (ROT) Rossby-Haurwitz waves.

The VSFs are associated with the eigenvalues which are referred to as the equivalent heights (EHs). Values of the EH reflect the degree of atmospheric thermal stratification of the resting state. The number of EHs depends on the number of vertical levels in the model used to generate data and the calculated values of EH also vary depending on the value of thermal stratification, except for its largest value which is approximately 10 km and referred to as the external (or barotropic) mode. The remaining EH values represent the internal (or baroclinic) modes.

Main application of the 3D NMFs is to express the horizontal velocity and mass fields of global atmospheric data by a single series of custom-made orthogonal functions rather than three separate series of ordinary functions. The number of the spectral coefficients in the normal mode space thus obtained is very much smaller than a vast number of data in the original grid-point space. Moreover, because the entire data is involved in the calculation of NMF coefficients, the diagnostics of data, such as the calculation of vertical motion, diabatic heating, and energetics, becomes optimum and efficient. We already mentioned in the Introduction the application of 3D NMFs to the initialization of PE prediction model. Likewise, a new area of application to the analyses of short-range forecast errors for data assimilation begins to emerge.

Application of the 3D NMFs is not limited to data analysis. Since the eigen-solutions of the linear part of PE model are already obtained, the time-dependent solutions of the PE model can be advantageously calculated by setting up a spectral model using the 3D NMFs as its basis functions. Although such a spectral model may not be as efficient as operational forecasting models in terms of computational practice, the 3D NMF spectral model is ideal for various theoretical investigations related to atmospheric dynamics, such as the interactions between time-averaged flows and associated disturbances (baroclinic instability, teleconnections, life cycle of blocking, and etc.). One advantage in using the 3D NMFs as basis functions in the spectral model is that we have a freedom of selecting not only the number of basis functions, but also the species of wave modes, such as the first vs. second kind modes and the external vs. internal modes. Such a simplification is especially useful to investigate the variability of climate in a very long time scale focusing on a particular dynamical mechanism.

In the following chapters, various authors will present examples of applying the 3D NMF expansion methods and Hough spectral 3D primitive equation models to study the variability of weather and climate on global scales.

References

NASA (1962). U.S. standard atmosphere.

Baer, F. (1977). Adjustments of initial conditions required to suppress gravity oscillations in nonlinear flows. *Contrib. Atmos. Phys.*, 50:350–366.

Baer, F. and Tribbia, J. (1977). On complete filtering of gravity modes through nonlinear initialization. *Mon. Wea. Rev.*, 105:1536–1539.

Bolin, B. (1955). Numerical forecasting with the barotropic model. *Tellus*, 7:27–49.

Butler, S. T. and Small, K. A. (1963). The excitation of atmospheric oscillations. *Proc. Roy. Soc. of London, Ser. A*, 274:91–121.

Castanheira, J. M., DaCamara, C. C., and Rocha, A. (1999). Numerical solutions of the vertical structure equation and associated energetics. *Tellus*, 51A:337–348.

Chapman, S. and Lindzen, R. S. (1970). Atmospheric tides: Thermal and gravitational. *Gordon and Breach/Sci. Pub.*, 200 pp.

Cohn, S. E. and Dee, D. P. (1989). An analysis of the vertical structure equation for arbitrary thermal profiles. *Q. J. R. Meteorol. Soc.*, 115:143–171.

Courant, R. and Hilbert, D. (1953). Methods of mathematical physics, vol. 1.

Cressman, G. P. (1958). Barotropic divergence and very long atmospheric waves. *Mon. Wea. Rev*, 85:293–297.

Daley, R. (1979). The application of nonlinear normal mode initialization to an operational forecast model. *Atmos.-Ocean*, 17:97–124.

Daley, R. (1991). Atmospheric data analysis. *Cambridge University Press, Cambridge, UK*, 460 pp.

Dickinson, R. E. and Williamson, D. L. (1972). Free oscillations of a discrete stratified fluid with application to numerical weather prediction. *J. Atmos. Sci.*, 29:623–640.

Dikii, L. A. (1965). The terrestrial atmosphere as an oscillating system. *Izv. Acad. Sci. USSR, Atmos. Ocean. Phys. (translated by C.M. Wade)*, 1:469–489.

Eckart, C. (1960). Hydrodynamics of oceans and atmospheres.

Eliasen, E. and Machenhauer, B. (1965). A study of the fluctuation of atmospheric planetary flow patterns represented by spherical harmonics. *Tellus*, 17:220–238.

Rossby, G. G. and Collaborators (1939). Relation between variations in the intensity of the zonal circulation of the atmosphere and the displacements of the semi-permanent centers of action. *J. Marine Res.*, 2:38–55.

Flattery, T. W. (1967). Hough functions. Tech. Rep. 21, Dept. Geophys. Sci., University of Chicago, 175 pp.

Flattery, T. W. (1970). Spectral models for global analysis and forecasting. U.S. Naval Academy, Air Weather Service Tech. Rep., 242, 42-53.

Frederiksen, J. S. (1978). Growth rates and phase speeds of baroclinic waves in multi-level models on a sphere. *J. Atmos. Sci.*, 35:1816–1826.

Fulton, S. R. and Schubert, W. H. (1980). Geostrophic adjustment in a stratified atmosphere. Atmos. Sci. Paper No. 326, Dept. of Atmos. Sci., Colorado State Univ.

Gavrilin, B. L. (1965). On the description of vertical structure of synoptical processes, translated by C. M. Wade. *Izv. Acad. Sci. USSR, Atmos. Ocean. Phys.*, 1:8–17.

Haurwitz, B. (1940). The motion of atmospheric disturbances on the spherical earth. *J. mar. Res*, 3:254–267.

Haurwitz, B. (1965). The diurnal surface-pressure oscillation. *Arch. Meteor, Geophys. Bioklimatol.*, A14:361–379.

Hildebrand, F. B. (1958). Methods of applied mathematics. Prentice-Hall, Inc.

Hinkelmann, K. (1951). Der mechanismus des meteorologischen Lärmes. [The mechanism of meteorogical noise. Translated by P. Rasch et al., NCAR Technical Note, NCAR/TN-203+STR, 1983, National Center for Atmospheric Research, Boulder, CO]. *Tellus*, 3:285–296.

Holl, P. (1970). The completeness of the orthogonal system of the Hough functions. Translation by B. Haurwitz from Nachrichten der Akademie der Wissenschaften in Göttingen II. Mathematisch-Physikalische Klasse, Jahrgang 1970, No. 7, 159-168.

Horel, J. D. and Wallace, J. M. (1981). Planetary-scale atmospheric phenomena associated with the Southern Oscillation. *Mon. Wea. Rev.*, 109:813–829.

Hough, S. S. (1898). On the application of harmonic analysis to the dynamical theory of the tides - Part II. On the general integration of Laplace's dynamical equations. *Phil. Trans. Roy. Soc. London*, A191:139–185.

Jacobs, S. J. and Wiin-Nielsen, A. (1966). On the stability of a barotropic basic flow in a stratified atmosphere. *J. Atmos. Sci.*, 23:682–687.

Jones, M. N. (1970). Atmospheric oscillations I. *Planet. and Space Sci.*, 18:1393–1416.

Jordan, C. L. (1958). Mean soundings for the West Indies area. *J. Meteor.*, 15:91–97.

Kasahara, A. (1974). Various vertical coordinate systems used for numerical weather prediction. *Mon. Wea. Rev.*, 102:509–522. (Corrigendum, 1975, Mon. Wea. Rev., 103, 664.).

Kasahara, A. (1976). Normal modes of ultralong waves in the atmosphere. *Mon. Wea. Rev.*, 104:669–690.

Kasahara, A. (1977). Numerical integration of the global barotropic primitive equations with hough harmonic expansions. *J. Atmos. Sci.*, 34:687–701.

Kasahara, A. (1978). Further studies on a spectral model of the global barotropic primitive equations with hough harmonic expansions. *J. Atmos. Sci.*, 35:2043–2051.

Kasahara, A. (1980). Effect of zonal flows on the free oscillations of a barotropic atmosphere. *J. Atmos. Sci.*, 37:917–929. Corrigendum, J. Atmos. Sci., 38 (1981), 2284–2285.

Kasahara, A. (1984). The linear response of a stratified global atmosphere to a tropical thermal forcing. *J. Atmos. Sci.*, 41:2217–2237.

Kasahara, A. and da Silva Dias, P. L. (1986). Response of planetary waves to stationary tropical heating in a global atmosphere with meridional and vertical shear. *J. Atmos. Sci.*, 43:1893–1911.

Kasahara, A. and Puri, K. (1981). Spectral representation of three-dimensional global data by expansion in normal mode functions. *Mon. Wea. Rev.*, 109:37–51.

Kasahara, A. and Tanaka, H. (1989). Application of vertical normal mode expansion to problems of baroclinic instability. *J. Atmos. Sci.*, 46:489–510.

Kato, S. (1966). Diurnal atmospheric oscillation: 1. Eigenvalues and Hough functions. *J. of Geophys. Res.*, 71:3201–3209.

Lamb, H. (1932). Hydrodynamics. Dover Pub.

Lanczos, C. (1961). *Applied analysis.*

Leith, C. E. (1980). Nonlinear normal mode initialization and quasi-geostrophic theory. *J. Atmos. Sci.*, 37:958–968.

Lindzen, R. S. (1966). On the theory of the diurnal tide. *Mon. Wea. Rev.*, 94:295–301.

Longuet-Higgins, M. S. (1968). The eigenfunctions of Laplace's tidal equations over a sphere. *Phil. Trans. Roy. Soc. London, Series A. Mathematical and Physical Sciences*, 262:511–607.

Lorenz, E. N. (1955). Available potential energy and the maintenance of the general circulation. *Tellus*, 7:157–167.

Machenhauer, B. (1977). On the dynamics of gravity oscillations in a shallow water model, with applications to normal mode initialization. *Contrib. Atmos. Phys.*, 50:253–271.

Machenhauer, B. (1979). The spectral method. *Chapter 3 in Numerical Methods used in Atmospheric Models. GARP Publications Series No. 17, Ed. A. Kasahara, World Meteorological Organization*, pages 121–275.

Madden, R. (1978). Further evidence of traveling planetary waves. *J. Atmos. Sci.*, 35:1605–1618.

Madden, R. A. (2007). Large-scale, free Rossby waves in the atmosphere – an update. *Tellus*, 59A:571–590.

Margules, M. (1892). Air motions in a rotating spheroidal shell. Max Margules wrote three papers, "Luftbewegungen in einer rotierenden Sphäroidschale bei zonaler Druckverteilung", Sitz.-Ber. kaiserl. Akad. Wissensch. Wien, Math.-Nat. Cl., Abt. IIa, vol. 101 (1892), 597-626, "Luftbewegungen in einer rotierenden Sphäroidschale (II. Theil)", Ibid., vol. 102 (1893), 11-56, and "Luftbewegungen in einer rotierenden Sphäroidschale (III. Theil)", Ibid., vol. 102 (1893), 1369-1421. These three articles were translated by B. Haurwitz as an NCAR technical report "Air motions in a rotating spheroidal shell", NCAR/TN-156+STR, National Center for Atmospheric Research, Boulder, CO.

Maruyama, T. (1967). Large-scale disturbances in the equatorial lower stratosphere. *J. Meteor. Soc. Japan*, 45:391–408.

Matsuno, T. (1966). Quasi-geostrophic motions in the equatorial area. *J. Meteor. Soc. Japan*, 44:25–42.

McLandress, C. (2002). The seasonal variation of the propagating diurnal tide in the mesosphere and lower thermosphere. Part I: The role of gravity waves and planetary waves. *J. Atmos. Sci.*, 59:893–906.

Merilees, P. E. (1966). On the linear balance equation in terms of spherical harmonics. *Tellus*, 20:200–202.

Monin, A. S. and Obukhov, A. M. (1959). A note on general classification of motions in a baroclinic atmosphere. *Tellus*, 11:159–162.

Moses, H. E. (1974). The use of vector spherical harmonics in global meteorology and aeronomy. *J. Atmos. Sci.*, 31:1490–1499.

Platzman, G. W. (1968). The Rossby wave. *Q. J. R. Meteorol. Soc.*, 94:225–248.

Platzman, G. W. (1972). Two-dimensional free oscillations in natural basins. *J. Phys. Ocean.*, 2:117–138.

Platzman, G. W. (1996). The S-1 chronicle: A tribute to Bernhard Haurwitz. *Bull. Amer. Meteor. Soc.*, 77:1569–1577.

Robert, A., Henderson, J., and Turnbull, C. (1972). An implicit time integration scheme for baroclinic models of the atmosphere. *Mon. Wea. Rev.*, 100:329–335.

Sasaki, Y. K. and Chang, L. P. (1985). Numerical solution of the vertical structure equation in the normal mode method. *Mon. Wea. Rev.*, 113:782–793.

Shigehisa, Y. (1983). Normal modes of the shallow water equations for zonal wavenumber zero. *J. Meteor. Soc. Japan*, 61:479–493.

Shuman, F. G. and Hovermale, J. B. (1968). An operational six-layer primitive equation model. *J. Appl. Meteor.*, 7:525–547.

Siebert, M. (1961). Atmospheric tides. *Adv. Geophys.*, 7:105–187.

Silva Dias, P. L. (1985). A preliminary study of the observed vertical mode structure of the summer circulation over tropical South America. *Tellus* , 37A:185–195.

Silva Dias, P. L. and Schubert, W. H. (1979). The dynamics of equatorial mass-flow adjustment. Atmos. Sci. Paper No. 287, Dept. Atmos. Sci., Colorado State Univ.

Simmons, A. J. and Hoskins, B. J. (1976). Baroclinic instability on the sphere: Normal modes of the primitive and quasi-geostrophic equations. *J. Atmos. Sci.*, 33:1454–1477.

Simons, T. J. (1968). A three-dimensional spectral prediction equation. Atmos. Sci. Paper No. 127, Dept. Atmos. Sci., Colorado State Univ.

Staniforth, A., Beland, M., and Coté, J. (1985). An analysis of the vertical structure equation in sigma coordinates. *Atmos.-Ocean*, 23:323–358.

Staniforth, A. N. and Daley, R. W. (1977). A finite-element formulation for the vertical discretization of sigma-coordinate primitive equation models. *Mon. Wea. Rev.*, 105:1108–1118.

Swarztrauber, P. N. (1979). On the spectral approximation of discrete scalar and vector functions on the sphere. *SIAM. J. Numer. Anal.*, 16:934–949.

Swarztrauber, P. N. (1981). The approximation of vector functions and their derivatives on the sphere. *J. Numer. Anal.*, 18:191–210.

Swarztrauber, P. N. (1993). The vector harmonic transform method for solving partial differential equations in spherical geometry. *Mon. Wea. Rev.*, 121:3415–3437.

Swarztrauber, P. N. and Kasahara, A. (1985). The vector harmonic analysis of Laplace tidal equations. *SIAM J. Stat. Comput.*, 6:464–491.

Tanaka, H. (1985). Global energetics analysis by expansion into three-dimensional normal-mode functions during the FGGE winter. *J. Meteor. Soc. Japan*, 63:180–200.

Tanaka, H. and Kung, E. (1989). A study of low-frequency unstable planetary waves in realistic zonal and zonally varing basic states. *Tellus*, 41A:179–199.

Tanaka, H. L. and Kasahara, A. (1992). On thenormal modes of Laplace's tidal equations for zonal wavenumber zero. *Tellus*, 44A:18–32.

Taylor, G. I. (1936). The oscillations of the atmosphere. *Proc. Roy. Soc., London*, A156:318–326.

Temperton, C. and Williamson, D. L. (1981). Normal mode initialization for a multilevel grid point model. Part I: Linear aspects. *Mon. Wea. Rev.*, 109:729–743.

Terasaki, K. and Tanaka, H. (2007). An analysis of the 3-D atmospheric energy spectra and interactions using analytical vertical structure functions and two reanalyses. *J. Meteor. Soc. Japan*, 85:785–796.

Tribbia, J. (1979). Non-linear initialization on an equatorial beta plane. *Mon. Wea. Rev.*, 107:704–713.

Tribbia, J. (1982). On variational normal mode initialization. *Mon. Wea. Rev.*, 110:455–470.

Žagar, N., Isaksen, L., Tan, D., and Tribbia, J. (2013). Balance and flow-dependency of background-error variances in the ECMWF 4D-Var ensemble. *Q. J. R. Meteorol. Soc.*, 139:1229–1238.

Žagar, N., Tribbia, J., Anderson, J. L., and Raeder, K. (2009a). Uncertainties of estimates of inertio-gravity energy in the atmosphere. Part I: Intercomparison of four analysis datasets. *Mon. Wea. Rev.*, 137:3837–3857. Corrigendum, Mon. Wea. Rev., 138, 2476–2477.

Žagar, N., Tribbia, J., Anderson, J. L., and Raeder, K. (2009b). Uncertainties of estimates of inertio-gravity energy in the atmosphere. Part II: Large-scale equatorial waves. *Mon. Wea. Rev.*, 137:3858–3873. Corrigendum, Mon. Wea. Rev., 138, 2476–2477.

Wallace, J. M. (1971). Spectral studies of tropospheric wave disturbances in the tropical western Pacific. *Rev. Geophys. Space Phys.*, 9:557–612.

Wallace, J. M. (1973). General circulation of the tropical lower stratosphere. *Rev. Geophys.*, 11:191–222.

Washington, W. and Kasahara, A. (2010). The evolution and future research goals for general circulation models. [Chapter 3 in The Development of Atmospheric General Circulation Models, edited by L. J. Donner et al., Cambridge Univ. Press.].

White, A. A., Hoskins, B. J., Roulstone, I., and Staniforth, A. (2005). Consistent approximate models of the global atmosphere: shallow, deep, hydrostatic, quasi-hydrostatic and non-hydrostatic. *Q. J. R. Meteorol. Soc.*, 131:2081–2107.

Wiin-Nielsen, A. (1971a). On the motion of various vertical modes of transient, very long waves: Part I. Beta plane approximation. *Tellus*, 23:87–98.

Wiin-Nielsen, A. (1971b). On the motion of various vertical modes of transient, very long waves: Part II. The Spherical Case. *Tellus*, 23:207–217.

Wilkes, M. V. (1949). *Oscillations of the Earth's Atmosphere*. Cambridge Univ. Press.

Williamson, D. L. and Dickinson, R. E. (1976). Free oscillations of the NCAR global circulation model. *Mon. Wea. Rev.*, 104:1372–1391.

Williamson, D. L. and Temperton, C. (1981). Normal mode initialization for a multilevel grid point model. Part II: Nonlinear aspects. *Mon. Wea. Rev.*, 109:744–757.

Yanai, M. and Maruyama, T. (1966). Stratospheric wave disturbances propagating over the equatorial Pacific. *J. Meteor. Soc. Japan*, 44:291–294.

Chapter 2
Normal Mode Functions and Initialization

Joe Tribbia

2.1 Introduction

The problem of initialization in numerical weather prediction had its birth with the very first attempt by Richardson (1922) to use numerical computations to predict the weather. As is well-known, Richardson's forecast was a failure predicting surface pressure changes larger that 1000 hPa over a six hour interval. Richardson rightly attributed this to errors in the the wind divergence in the initial state of his forecast. In retrospect this error can be viewed as an improper initial state having inertial gravity waves compounded by the use of a large time increment extrapolating surface pressure tendencies. A time step of a few minutes would have resolved the oscillatory nature of the of the inertial gravity wave and resulted in a far better six hour weather forecast albeit increasing the number of computations necessary by an order of magnitude. Thus the first successful experimental numerical weather forecast relied on a simplified and less accurate set of equations as compared to those solved by Richardson, equations in which the problematic inertia-gravity oscillations are not represented, the equivalent barotropic vorticity equation used by Charney et al. (1950). By the late 1960's sufficient understanding had been developed so that the more accurate primitive equations could be used for numerical weather forecasting, however the techniques used for analyzing atmospheric observations to create initial conditions for primitive equation models still suffered from the fact that some sort adjustments to this state were necessary to control the amplitude of inertia gravity oscillations in the initial analysis; i.e. an initialization step. Quasi-geostrophic and linear balance initializations (Phillips, 1960) were proposed and used but were seen to eliminate or underestimate the important ageostrophic components of the flow that led to development and amplification of synoptic systems. A similar criticism was made to the use of Hough functions for analysis of observations (Flattery, 1967) since this method used only rotational or Rossby mode Hough functions to fit atmospheric observations.

The application of Normal Mode Functions to the problem of the initialization numerical models used for weather prediction began in the mid-1970's concomitant

N. Žagar and J. Tribbia (eds.), *Modal View of Atmospheric Variability*,
Mathematics of Planet Earth 8, https://doi.org/10.1007/978-3-030-60963-4_2

with advances made in relation to the original work of Kasahara (1976) describing the computation of Hough Normal Mode functions using spherical harmonics. This was a complement to an earlier study by Dickinson and Williamson (1972) who computed the equivalent discrete normal modes for a finite difference grid point primitive equation model in spherical coordinates with the motivation of model initialization in which the high frequency gravity oscillations were projected out of the initial state vector for the model. Williamson (1976) subsequently used this normal mode initialization in an experiment with a global finite difference shallow water model retaining only the the slow rotational Rossby modes in the initial state and discarding any projection onto the gravitational modes. This experiment demonstrated the benefit of using model normal modes in initialization although the method of projecting the gravitational oscillations out of the initial state vector exhibited small residual high frequency oscillations in the forecast height field. These studies set the stage for an advance that not only provided an effective, nearly perfect solution to the long standing problem of weather forecast model initialization but also elucidated the theoretical concept of atmospheric balance and introduced the concept of an atmospheric "slow manifold" embedded in the dynamics of the primitive equation system.

2.2 Nonlinear Normal Mode Initialization

2.2.1 Model Normal Modes

Before going into the mathematical details of Nonlinear Normal Mode Initialization, it is worthwhile to note that there are differences between normal mode functions derived from a numerical model and the analytic normal mode functions discussed previously in this volume and that these differences are advantageous for the specific application of model initialization. The main difference is that model normal modes are the algebraic eigen-vectors of the discretized model using the model numerics, finite difference, spectral, finite element, etc. So these mathematical structures are numerically compatible with evolution equations used for weather prediction which minimizes the numerically generated noise in the imposition of any initialization constraint. A second significant difference is that all the eigen-modes are completely resolved by the numerical model since they result from the algebraic manipulation of the the linearized dynamics of model equations themselves and, if the numerics is energy conserving the algebraic manipulations lead to hermitian symmetric matrices with orthogonal eigen-vectors using the total energy based inner product. A significant disadvantage of model normal modes for analyzing data and dynamical modeling is that a good fraction of the model normal mode functions are not physical, being either coarse inaccurate representations of true Hough Normal Mode Functions or numerical artifacts of the particular discretization scheme used by the model.

The construction of model normal modes follows the same essential features of the analytical developments used to construct the Hough Normal Mode Functions described earlier: separation of variables to obtain a vertical structure equation with vertically dependent eigen-vectors and eigen-values corresponding to a discrete set of equivalent depths which form vertical basis vectors spanning the vertical model dimension (Kasahara and Shigehisa, 1983; Temperton, 1984, hereafter KS83 and T84 respectively). Using the vertical structure function to eliminate the vertical dependence produces a set of linear shallow water equations for each value of the equivalent depth separation constant. In a grid point model the construction of the horizontally dependent normal modes then follows from Dickinson and Williamson (1972) while for a spherical harmonic based spectral transform model the construction is as in Kasahara (1976) with the spherical harmonic expansion of the stream function $\hat{\psi}$, velocity potential $\hat{\phi}$ and geopotential $\hat{Z}$, being restricted to the triangular or rhomboidal truncation used by the model.

2.2.2 Equations in Modal Variables

As demonstrated in KS83 and T84, the linearized model equations using the standard prognostic variables of the hydrostatic primitive equations, (u, v, T, p_s) can be algebraically manipulated to three prognostic equations for u, v, P , where P corresponds to the variable defined in chapter 1, $P \equiv \Phi + RT_o \ln(p_s)$, where $\Phi = gh$ is the geopotential, p_s is the surface pressure, the globally averaged temperature on model levels denoted by T_o, R is the gas constant and g is gravity. Defining $\mathbf{Z}$ as the state vector $(u, v, P)^T$, the linearized equations can be written as

$$\mathrm{d}\mathbf{Z}/\mathrm{d}t + \mathcal{L}\mathbf{Z} = \mathbf{0} \tag{2.1}$$

and the fully nonlinear primitive equations can be be recast as

$$\mathrm{d}\mathbf{Z}/\mathrm{d}t + \mathcal{L}\mathbf{Z} = \mathbf{N}(\mathbf{Z}). \tag{2.2}$$

The model state vector, $\mathbf{Z}$, can be expanded exactly in terms of the model normal mode Hough vectors which comprise a complete, orthonormal set since the operator, $\mathcal{L}$, in equation (1), above is a skew symmetric matrix with a sole degeneracy at zonal wavenumber $k = 0$. This degeneracy can easily be handled by taking the limit $k \rightarrow 0$ as shown in Shigehisa (1983). Using the model normal modes that state vector can be expressed as:

$$\begin{vmatrix} u\left(\lambda_i, \varphi_l, \sigma_j, t\right) \\ v\left(\lambda_i, \varphi_l, \sigma_j, t\right) \\ h\left(\lambda_i, \varphi_l, \sigma_j, t\right) \end{vmatrix} = \sum_{m=1}^{M} \sum_{n=1}^{R} \sum_{k=-K}^{K} \chi_n^k(m, t) \mathbf{H}_n^k\left(\lambda_i, \varphi_l; m\right) G_m(j) \quad . \tag{2.3}$$

Projecting the model governing equations onto the model normal modes and using the $\chi_n^k(m,t)$ coefficients as the state vector, the prognostic equation (2.2) can be transformed into:

$$d\boldsymbol{\chi}/dt + i\mathcal{D}\boldsymbol{\chi} = \mathbf{N}(\boldsymbol{\chi}), \tag{2.4}$$

in which $\boldsymbol{\chi}$ is the vector made up of all the $\chi_n^k(m,t)$ coefficients stacked vertically and $\mathcal{D}$ is a diagonal matrix with the frequencies of the corresponding normal modes as the diagonal entries.

For the purpose of initializing to minimize the 'high-frequency noise', normal mode initialization requires that the modal state vector, $\boldsymbol{\chi}$, be split into two partitions, a fast portion, $\boldsymbol{\chi}_f$, made up of gravity-inertia mode amplitudes, and a slow portion, denoted $\boldsymbol{\chi}_s$, made up of rotational (i.e. Rossby) modes. The governing equations are then partitioned into two vector equations:

$$d\boldsymbol{\chi}_f/dt + i\mathcal{D}_f\boldsymbol{\chi}_f = \mathbf{N}_f(\boldsymbol{\chi}_f, \boldsymbol{\chi}_s) \tag{2.5}$$

and

$$d\boldsymbol{\chi}_s/dt + i\mathcal{D}_s\boldsymbol{\chi}_s = \mathbf{N}_s(\boldsymbol{\chi}_f, \boldsymbol{\chi}_s), \tag{2.6}$$

where $\mathcal{D}_f$ and $\mathcal{D}_s$ are diagonal matrices with elements corresponding to the fast and slow normal mode frequencies respectively and, similarly, $\mathbf{N}_f(*,*)$ and $\mathbf{N}_s(*,*)$ are the nonlinear terms in the equations projected onto the fast and slow normal modes respectively. If the dimension of the state vector $\boldsymbol{\chi}$ is N_{total}, the fast state vector has dimension $2/3\, N_{total}$ and the slow mode state vector has dimension $1/3\, N_{total}$. With the fast normal mode tendencies isolated and being computed in equation (5), the goal of initialization is then to choose initial conditions for $\boldsymbol{\chi}_f$ that inhibit $\boldsymbol{\chi}_f(t)$ from exhibiting fast, oscillatory temporal evolution.

2.2.3 Nonlinear Normal Mode Initialization at First Order

In the following we closely follow the development in Baer and Tribbia (1977). To elucidate the concepts of Nonlinear Normal Mode Initialization (NNMI) it useful to non-dimensionalize the governing equations using traditional Rossby scaling; i.e. $(2\Omega)^{-1}$ for time, the radius of the Earth, a for the horizontal length scale, H_0, the external mode equivalent depth for the vertical scale and U for the velocity scale. With a velocity scale of 10 m/sec the Rossby number, $\epsilon = U/(2\Omega a)$, is very small and the scaled rotational mode frequencies are order ϵ as are the nonlinear terms. The fast gravity modes, on the other hand, have order one frequencies. The Rossby scaling results in equations (2.5) and (2.6) being transformed to

$$d\boldsymbol{\chi}_f/dt + i\mathcal{D}_f\boldsymbol{\chi}_f = \epsilon\mathbf{N}_f(\boldsymbol{\chi}_f, \boldsymbol{\chi}_s) \tag{2.7}$$

and

$$d\boldsymbol{\chi}_s/dt + i\epsilon\mathcal{D}_s\boldsymbol{\chi}_s = \epsilon\mathbf{N}_s(\boldsymbol{\chi}_f, \boldsymbol{\chi}_s). \tag{2.8}$$

From the above it can be clearly seen that fast oscillations can only enter the dynamics through fast oscillations in $\boldsymbol{\chi}_f$, for if $\boldsymbol{\chi}_f$ evolves slowly on the nonlinear timescale $T = \epsilon t$ then $\boldsymbol{\chi}_s$ must also evolve on the slow time T. Because ϵ is small, the slowly evolving initial state for $\boldsymbol{\chi}_f$ can computed as an asymptotic expansion in ϵ.

First, at the ϵ^0 order equation (2.7) above becomes

$$\mathrm{d}\boldsymbol{\chi}_f^0/\mathrm{d}t + i\mathcal{D}_f\boldsymbol{\chi}_f^0 = 0 \tag{2.9}$$

and it is easily seen that $\boldsymbol{\chi}_f^0(0) = 0$ is the necessary initialization to guarantee that no high frequency oscillations occur for zeroth order (i.e. order unity amplitude) fast normal modes. Moving next to the order ϵ equation

$$\mathrm{d}\boldsymbol{\chi}_f^1/\mathrm{d}t + i\mathcal{D}_f\boldsymbol{\chi}_f^1 = \mathbf{N}_f(\boldsymbol{\chi}_f^0, \boldsymbol{\chi}_s) = \mathbf{N}_f(0, \boldsymbol{\chi}_s)\,. \tag{2.10}$$

Now, $\boldsymbol{\chi}_s$ is evolving on the slow time scale, T, and can be treated as constant to within the asymptotic approximation so, defining $\mathbf{N}_f^0 \equiv \mathbf{N}_f(0, \boldsymbol{\chi}_s(0))$, the general solution to equation (2.10) is given by

$$\boldsymbol{\chi}_f^1(t) = \exp(-i\mathcal{D}_f t)(\boldsymbol{\chi}_f^1(0) - (i\mathcal{D}_f)^{-1}\mathbf{N}_f^0) + (i\mathcal{D}_f)^{-1}\mathbf{N}_f^0\,, \tag{2.11}$$

and so to eliminate high frequency oscillations at order ϵ

$$\boldsymbol{\chi}_f^1(0) = (i\mathcal{D}_f)^{-1}\mathbf{N}_f^0\,, \tag{2.12}$$

is the necessary initial condition for the fast normal modes.

At this level of initialization there are two ways of proceeding to higher order, one described in Baer (1977) and Baer and Tribbia (1977), the second, developed independently by Machenhauer (1977). However, it is useful at the order ϵ to note some important conceptual and dynamical aspects of NNMI before discussing higher order nonlinear balance conditions. The dynamical geometric interpretation of NNMI developed after the initial development of NNMI in the aforementioned articles by Baer, Tribbia and Machenhauer (loc.cit.) was elegantly derived in the articles by Leith (1980) and Lorenz (1980). The use of Rossby scaling and the focus on the slow (advective) time scale in the derivation above is reminiscent of the scale analysis used in the derivation of the quasigeostrophic equations. Realizing this, Leith (loc.cit.) showed that for the primitive equations on the $f-$ or $\beta-$plane, order ϵ NNMI corresponded exactly to quasigeostrophy. Leith also introduces the modal phase space geometric concept describing slow dynamics in a dynamical system that permits fast and slow dynamics as dynamical behavior constrained to the slow manifold of the phase space.

Figure 2.1 from Leith's article is reproduced here, in which the slow manifold is depicted as a nonlinear surface in the fast and slow mode phase space, denoted by the G and R axes, respectively, in Leith's figure. Quasigeostrophy (order ϵ balance) appears as the first order approximation to the slow manifold where the fast modes are specified as determined by a quadratic (nonlinear) function of the slow mode

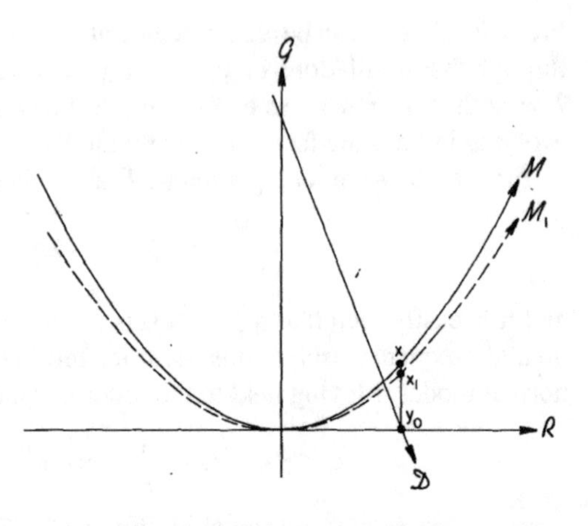

Fig. 2.1 Schematic diagram showing linear and nonlinear manifolds in dynamical phase space: $\mathcal{R}$, rotationl; $\mathcal{G}$, gravitational; $\mathcal{D}$, data; all linear; and $\mathcal{M}$, slow nonlinear with first approximation $\mathcal{D}_1$. From Leith (1980), Fig. 1. ©American Meteorological Society. Used with permission.

amplitudes. This also leads directly to the idealized geometric goal of initialization as being to ensure that the model initial state be a point on the model's slow manifold.

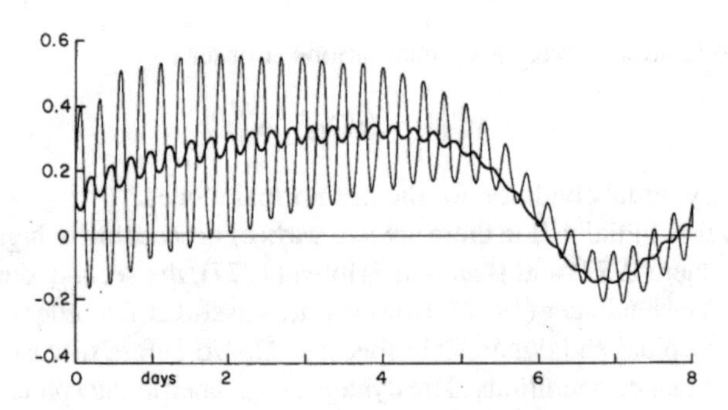

Fig. 2.2 Variations of y_1 (heavy curve) and z_1 (thin curve) during the first eight days of first numerical solution of PE model. From Lorenz (1980), Fig. 2. ©American Meteorological Society. Used with permission.

In the Lorenz (1980), the concept of the slow manifold was further elaborated upon and demonstrated in numerical experiments using a nine-component truncation of the shallow water equations on the f−plane. In these experiments Lorenz demonstrated this forced dissipative dynamical system, for the parameters chosen, had an attracting set that was well-approximated by the quasigeostrophic version of the equations. As shown in Figures (2.2) and (2.3) from the article (reproduced here), initial transient gravity-inertia oscillations decay in time and the dynamical model's chaotic climate attractor is devoid of fast oscillatory temporal behavior. Lorenz also demonstrated that the slow manifold had embedded within it a slow, chaotic attractor set and this set could be approached with great and greater precision using higher and higher order approximations to the nonlinear balance conditions. Thus, Lorenz demonstrated that

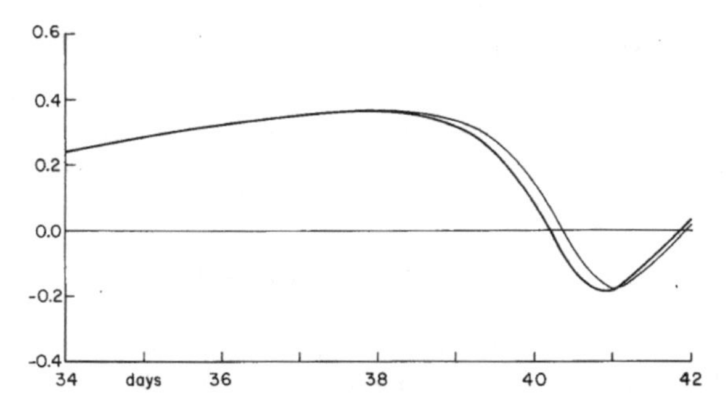

Fig. 2.3 Variations of y_1 (heavy curve) and z_1 (thin curve) from $t = 34$ days to $t = 42$ days in first numerical simulation of PE model. The difference between the curves is unresolvable from 34 to 38 days. From Lorenz (1980), Fig. 3. ©American Meteorological Society. Used with permission.

within this model, perfect initialization could be achieved just as in the geometric ideas put forward by Leith.

2.2.4 Higher Order Balance Conditions

As mentioned in the previous section, at the time nonlinear normal mode initialization was first published two slight variants were proposed and used. The development of initialization to first order was very close to quasigeostrophic balance and had been already utilized in the normal mode framework by Dickinson and Williamson (1972). Bennert Machenhauer (Machenhauer, 1977) noted that if the reasoning that the fast modes could be made slowly varying in time by balancing the linear and nonlinear terms in the fast mode equations, the first order balance expressed in equation (2.12) would be disrupted by the non-vanishing amplitude of the fast modes in the nonlinear terms. Machenhauer proceeded to iteratively insert the fast mode amplitude into the nonlinear terms until the sequence converged; i.e. iterate

$$\boldsymbol{\chi}_f^{n+1}(0) = (i\mathcal{D}_f)^{-1}\mathbf{N}_f(\boldsymbol{\chi}_f^n(0), \boldsymbol{\chi}_s(0)) \tag{2.13}$$

to convergence. In Machenhauer's experiments with a nonlinear shallow water model convergence of the iteration occurred in three iterations and the subsequent forward integration of the model was devoid of high frequency oscillations.

At the same time Baer and Tribbia (Baer and Tribbia, 1977) were pursuing initialization using normal modes with an asymptotic expansion approach including a formal expansion in powers of ϵ and the method of multiple time (Nayfeh 1975). In their method the fast mode initial condition is expressed as

$$\boldsymbol{\chi}_f(0) = \sum_{k=0}^{K} \epsilon^k \boldsymbol{\chi}_{f_k}(0) \tag{2.14}$$

To begin the asymptotic expansion solution, BT77 first define two times $t^* \equiv t$, the fast time and $T \equiv \epsilon t$, the slow time and treats each term in the asymptotic series as a function of both the fast and slow times; i.e. $\chi_{f_k}(t) = \chi_{f_k}(t^*, T)$, for every k. The fast mode prognostic equation (2.6) is then decomposed and sequentially solved term by term equating like powers of epsilon as follows: At order ϵ^0 we have:

$$\partial \chi_{f_0}/\partial t^* + i\mathcal{D}_f \chi_{f_0} = 0, \tag{2.15}$$

and if the objective of initialization is to suppress oscillations on the fast time scale, the obvious choice for fast mode initial condition is $\chi_{f_0}(0) = 0$.

At order ϵ^1 we have:

$$\partial \chi_{f_1}/\partial t^* + i\mathcal{D}_f \chi_{f_1} = \mathbf{N}_f(0, \chi_s(T)), \tag{2.16}$$

the solution of which was discussed above in the analysis of Leith's work. So, a solution devoid of t^* oscillations at order ϵ can only be obtained if

$$\chi_{f_1}(0) = (i\mathcal{D}_f)^{-1}\mathbf{N}_f(0, \chi_s(0)), \tag{2.17}$$

which corresponds to both the first iteration of the Machenhauer developed and (restricted to a midlatitude beta-plane) quasigeostrophic linear balance initialization.

At order ϵ^2 we have:

$$\partial \chi_{f_2}/\partial t^* + i\mathcal{D}_f \chi_{f_2} = \left(\mathbf{N}_f\left(\chi_{f_1}(T), \chi_s(T)\right) - \mathbf{N}_f\left(0, \chi_s(T)\right)\right) - \partial \chi_{f_1}/\partial T, \tag{2.18}$$

The potentially fast part of equation (18) is on the left-hand side and the slow forcing is on the right-hand side. And, as was the case at order ϵ, setting the partial derivative with respect to t^* to zero gives the balance condition required to eliminate oscillations at order ϵ^2; i.e.

$$\chi_{f_2}(0) = (i\mathcal{D}_f)^{-1}[\mathbf{N}_f(\chi_{f_1}(0), \chi_s(0)) - \mathbf{N}_f(0, \chi_s(0)) - \partial \chi_{f_1}(0)/\partial T] \tag{2.19}$$

is the order ϵ^2 contribution to the initial state of the fast modes. Two aspects of the BT77 scheme should be noted here. First, although the first term on the RHS is the same as the second iteration of the Machenhauer scheme, the subtraction of the second term changes this to give the second order increment, since BT77 is computing incremental changes at each order. The final term involving the slow time derivative of the first order fast mode amplitude requires some explanation as to its meaning. The asymptotic slowly varying time dependent solution of the first order fast mode amplitude is

$$\chi_{f_1}(T) = \mathbf{F}(T) \equiv (i\mathcal{D}_f)^{-1}\mathbf{N}_f(0, \chi_s(T)), \tag{2.20}$$

and so the final term on the RHS is $\partial \mathbf{F}/\partial T$ at $T = 0$.

In Tribbia (1979), it was shown that because of the the weak separation between the frequencies of planetary scale Kelvin and Mixed Rossby-gravity modes, second order Baer-Tribbia initialization was needed to reduce oscillatory behavior in these

Normal Mode amplitudes. Because of the computational difficulty of deriving and computing $\partial \mathbf{F}/\partial T$, Tribbia (1984) developed a simple scheme showing this term could be approximated using finite difference methods making second order Baer-Tribbia initialization computationally only slightly more demanding than second order Machenhauer initialization. Although higher order refinements can easily be derived, in practice second order initialization using Baer-Tribbia or Machenhauer schemes worked extremely well and the refinements of higher order only became used in answering theoretical concerns associated with the strict mathematical existence of the slow manifold (Lorenz, 1986; Lorenz and Krishnamurthy, 1987; Warn, 1997) and the exploration non-elliptic regions in the divergence of iterations for initializations in which only height information is specified and required to maintain its analyzed initial value, so-called height constrained initialization.

2.3 The Efficacy of NNMI

2.3.1 Operational Success and Issues

The theory of NNMI was developed in 1977 and by the next year two operational centers, the Canadian Atmospheric Environment Service and ECMWF, began testing and experimenting with NNMI for use in their operational analysis forecast system. This was an important next step since the modeling results that supported the theory had been in simplified shallow water systems which only suggested that NNMI would be useful in a practical operational setting. Using a shallow water version of the Canadian spectral model, Daley (1978) demonstrated the potential of NNMI showing that after two iterations of NNMI grid point values of field variables were free of high frequency oscillations as seen in Figure 2.4. Daley also noted that NNMI, as originally developed, would keep fixed only the portion of the initial state vector that projected onto the slow rotational normal modes, undermining some of the effort going into the objective analysis of the synoptic observations in the analysis cycle. Daley devised a variational method to produce initialized states that were as near as possible in a specified metric to the analyzed state. At the European Center, Temperton and Williamson (1981) used the ECMWF grid-point model and demonstrated similar success in eliminating high frequency grid point oscillations as seen in Figure 2.5. After these successful experiments showing that the use of model normal mode functions and offsetting the switch on forcing of the nonlinear terms were the keys to the successful solution to the problem of model initialization, ECMWF and Canada's AES shortly thereafter adopted NNMI into their operational forecast stream. One of the most notable benefits of the insertion into forecast analysis cycle was that since the short term forecasts were smoothly varying in time it became easier to assimilate new data in the analysis part of the cycle. The mismatch between data and first guess short term forecast did not include the high frequency noise of fast mode oscillations and therefore gave a more coherent forecast-observation difference from which the data assimilation method estimates analysis increments. This was particularly useful for the satellite infrared measurements that were frequent in time but had information only in the mass field, a field easily contaminated by fast mode noise.

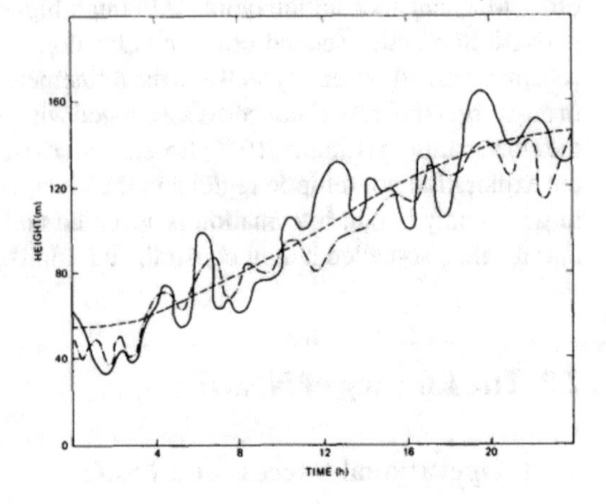

Fig. 2.4 Time evolution of point (46°N, 180°W) in 10 km model for 3 initial conditions. unadjusted (full line), non-linear adjustment (dashed line) and linear adjustment (dotted line). From Daley (1978), Fig. 4.

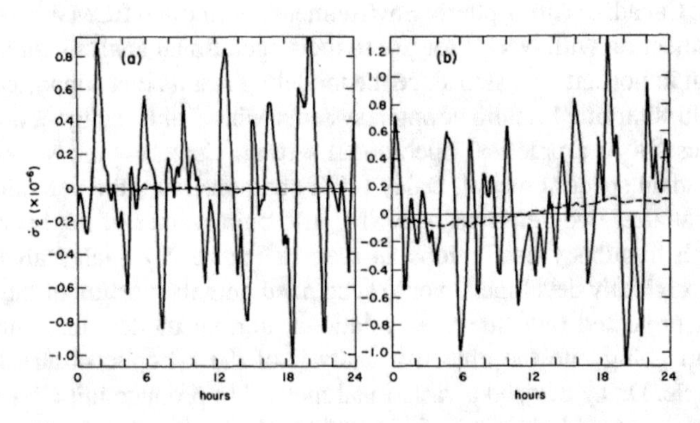

Fig. 2.5 $\dot{\sigma}$ versus time at (a) (40°N, 90°W) and (b) (30°N, 90°E) with no initialization (solid line and with two nonlinear iterations with the first five vertical modes (dashed line). From Temperton and Williamson (1981), Fig. 5. ©American Meteorological Society. Used with permission.

Despite these successful and beneficial aspects of NNMI, there arose subtle issues that involved aspects that have been ignored so far in the above presentation. It may be noted that we have been very careful to derive NNMI as a means of initialize fast degrees of freedom and leaving untouched the slow degrees of freedom. It has been implicitly implied that fast modes are identified all the inertia-gravity modes along with the Kelvin modes Mixed Rossby-Gravity modes and the slow modes being only the Rotational normal modes. This is an adequate demarcation in middle and high latitudes where the Coriolis frequency f ensures good separation between the slowest inertia-gravity modes and Rossby modes. This is patently untrue however in tropical latitudes. The need for higher order initialization to counter the shrinking gap in frequency was already identified in Tribbia's (1979) paper, however the problem is not just a shrinking gap but entire loss of any separability as normal modes become trapped closer and closer to the equator as the equivalent depth, h_m, associated with each vertical structure function G_m diminishes with increasing index m. Equatorially trapped fast modes remain fast compared to Rossby modes of the same equivalent depth but may be slower than some rotational normal modes with larger equivalent depths, especially for small longitudinal wavenumbers. Because of this only the deepest few modes with the largest equivalent depths were eventually subject to NNMI.

Yet another problem arose in the equatorial region which was due to quasi-stationary convective forcing. Because NNMI in its initial implementations only considered the adiabatic or dynamical tendency terms in the primitive equations. Initialization did not account for the fact that many fast waves in the tropics are quasi-stationary and have a significant fraction of their amplitude forced by the diabatic heating terms in the full model tendencies. Applying NNMI without diabatic forcing led to the collapse of the Hadley and Walker cells in the tropics and applying NNMI with diabatic terms led to grid-point storms as the convection intensified with each iteration. Puri and Burke (1982) solved this dilemma pragmatically by diagnosing beforehand which equatorial waves on which the diabatic terms were most projecting and not initializing those strongly forced fast normal modes.

As analysis forecast systems developed new problems arose for the use of NNMI due unforeseen consequences of more sophistication and complexity. One system improvement was the decision to move to a more frequent analysis updates, from a 12 hour to 6 hour interval between analyses. Clearly this was a useful development leading to more accurate analyses, but this caused problems because the atmospheric diurnal tide was now being resolved in the analysis. NNMI applied to 6-hourly analyses eliminated the diurnal tide which increased the mismatch between the first-guess forecast and the observations deteriorating both. Wergen (1989) rectified this problem by first removing the canonical diurnal tide signal at the proper phase for each analysis time from the analysis, then performing NNMI on the tide-filtered analysis. Lastly, the canonical tide signal was re-inserted into the initialized tide-filtered analysis and used as the models initial conditions for the next forecast analysis cycle.

2.3.2 Theoretical Concerns Regarding the Slow Manifold

Issues not only arose with the implementation of effective NNMI in the operational arena but very early on concerns were raised regarding one of the theoretical underpinnings of NNMI, the strict mathematical existence of the slow manifold. Doubts about the existence of a strict slow manifold were first raised by Warn (1997) which, as the preamble by Shepherd to this article notes, was originally submitted in 1983.

Warn argued that although the dynamics of the rotational modes that determines the slow nonlinear forcing of the fast modes might be slow, this dynamics should be chaotic for realistic forcing since it should exhibit similar dynamic behavior to that of the atmosphere. Chaotic temporal behavior would require that the nonlinear forcing of the fast modes would have a continuous temporal spectrum and therefore force each fast mode at its resonant frequency, albeit with small amplitude since the frequency spectrum the forcing would exhibit power law decay at high frequency. Since the published numerical experiments at the time of Lorenz (1980) did not exhibit any fast behavior and Warn did not back up his reasoning with definitive numerical results, publication of Warn's article was delayed for 14 years until evidence amassed supporting his arguments.

One of the first firm mathematical results suggesting that strict existence would be challenging to prove came in the the article by Vautard and Legras (1986), in which they proved that a necessary and sufficient condition for the convergence of any Taylor series expansion in time of the fast modes' slow temporal evolution was that the slow forcing be an entire function of time, meaning the forcing must be restricted to functions that are analytic everywhere in the complex t-plane, except $|t| \to \infty$. Because Tribbia (1984) had shown that higher order terms in the Baer-Tribbia scheme were asymptotically equivalent to balancing equivalent order terms in the Taylor series of the forcing function as a function of time, Vautard and Legras effectively proved that the slow manifold could not be fully constructed as a power series or iteration sequence if the slow nonlinear forcing exhibited any interesting temporal behavior, much less a chaotic temporal behavior.

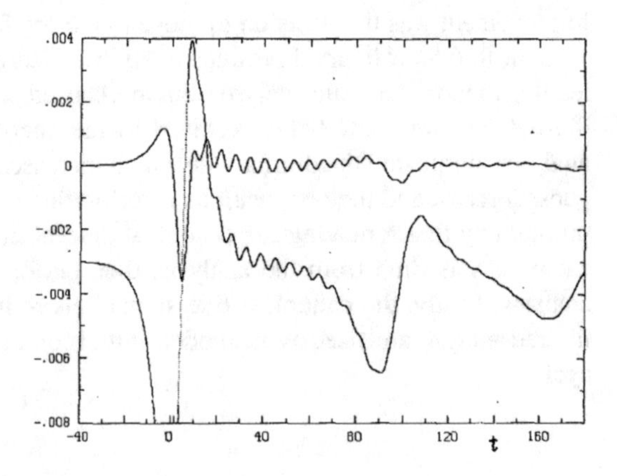

Fig. 2.6 The variations of Z (ipper curve) and X (lower curve) with t for the conditions of Fig. 1, for an extended range of t. The scale on the left is for Z; the curve for X has been displaced downward by 0.003 units to reduce the overlap. From Lorenz and Krishnamurthy (1987), Fig. 2. ©American Meteorological Society. Used with permission.

Lorenz (1986) showed in a 5 mode truncation of his original 9-mode shallow water that a slowest invariant manifold did exist in this system and could be defined independent of a series which seemed to end the controversy in favor of the existence of a slow manifold. However in Lorenz and Krishnamurthy (1987) a reconsideration of the numerics and experiments using double-double precision showed that a small amplitude high frequency oscillation was persistently present in the attractor of the 5-component system implying the non-existence of a slow manifold. Vanneste

(2008a) and Vanneste (2008b) used exponential asymptotics to show that the 5-component system possessed exponentially small amplitude fast wave emissions with amplitude proportional to $exp(1/\epsilon)$, where ϵ could be equated with the Rossby number in 5-mode model. This theoretically derived estimate appeared to be the last necessary step in proving once and for all that while an invariant slowest invariant manifold exists it is only nearly slow because there will always be a small residual fast oscillation. However, in 2017 Chekroun et al. (2017) argued that while the 5-mode truncation studied by Krishnamurthy, Lorenz and Vanneste possessed only a nearly slow manifold, the original 9-mode truncation in Lorenz (1980) possessed a truly slow invariant manifold. Thus the theoretical question remains open although for practical purposes, such as forecast initialization, exponentially small fast modes are not concerning.

2.4 The End of NNMI

Aside from high order initialization necessary to examine what is achievable with respect to asymptotic approaches to a slow or almost slow manifold, explicit NNMI has almost entirely ceased to be utilized and no longer appears in the operational forecast analysis cycle of global forecast systems. There are several reasons for this all having in common the fact that other simpler and less computationally demanding methods for initialization, inspired by NNMI, became pragmatically better suited to the new challenges that have arisen in response to advances in global and regional Numerical Weather Prediction. As supercomputing moved to massively parallel architecture the storage and movement of large datasets and computational memory became the main bottleneck that could slow the operational workflow. Because of this, methods of reducing the burden of storing and using three dimensional NMFs for initialization were necessary and as the horizontal and vertical resolution this bottleneck became more serious. One effective solution to this was called Implicit Normal Mode Initialization (Temperton, 1989) in which only approximate normal modes were computed for initialization. Temperton noted that moving parts of the linear operator $\mathcal{L}$ in a spectral model to the RHS and treating them iteratively as the nonlinear terms are treated in NNMI could result in highly simplified modified LHS matrix operators that could be easily inverted. This coupled with the fact that only the gravest few vertical modes were being initialized gave a satisfactory computationally efficient alternative to the original NNMI iteration schemes. The Temperton implicit norm mode method was used by Murakami and Matsumura (2007) to initialize the high resolution JMA model with grid spacing of 20km and 40 vertical layers.

Another alternative to NNMI that completely avoids computing normal modes is the digital filtering initialization method developed by Lynch and Huang (1992) and modified by Lynch (1997) to use a Dolph-Chebyshev window filter to filter high frequency oscillations. In this method the forecast model is integrated for a short period of time, 3hrs usually is sufficient, and the output state variable is stored for each time step and the filter is applied. Then, as in NNMI, the filtered initial condition is reinserted and the process is iterated, normally twice more, to ensure a slow evolution of the state variable absent of fast mode oscillations. The digital filter only uses the iterative insertion of information aspect of NNMI and has several advantages: 1) the procedure is much easier to implement, 2) it can easily be used in any numerical forecast model even limited area models with unstructured horizontal grids and 3) it can easily be used as part of a 4D-var assimilation method so that the modifications of the state vector due to initialization can be properly weighted in the analysis cycle.

The final reason for the disappearance of NNMI from the operational stream is that initialization is becoming less necessary. In addition to initialization constraints being built into the cost function of variational assimilation methods in ensemble Kalman filter (EnKF) assimilation and hybrid methods where the covariance is constructed out sampled model trajectories analysis increments are constrained to be on the model's slow manifold. This is because the major axes of the forecast covariance matrix is along the amplifying directions for forecast errors, largest Lyapunov vectors which are all on the slow manifold. This placement onto the model's slow manifold is slightly interfered with by the localization of the covariance in space which is needed to overcome the sampling error caused by the small number of ensemble members used in ensemble schemes, leading to erroneous distant correlations. The small residual fast mode oscillations can be initialized either using a temporal smoothness constraint for the hybrid assimilation methods or digital filtering for the EnKF.

2.5 Conclusion

Despite the fact that NNMI is no longer widely used in the operational stream of NWP, many of its component aspects, eliminating high frequency oscillations by discarding the portion of the initial state responsible and the nonlinear iteration in particular, form the basis of initial state vector preparation to this day. The breakthrough in the solution to the problem of initialization that came with NNMI was the result of the clarity that normal mode methods brought to the initialization problem. This clarity coupled with the geometric picture of the slow manifold the rapid adoption of normal mode methods which inspired further studies using NMFs in the analysis and modeling as reported elsewhere in this volume. However, it is important to note the important contribution of the earlier studies of Dickinson and Williamson and Kasahara, which inspired researchers to use Hough Normal Mode Functions and to develop a new 'Normal Mode' approach to problems.

References

Baer, F. (1977). Adjustments of initial conditions required to suppress gravity oscillations in nonlinear flows. *Contrib. Atmos. Phys.*, 50:350–366.

Baer, F. and Tribbia, J. (1977). On complete filtering of gravity modes through nonlinear initialization. *Mon. Wea. Rev.*, 105:1536–1539.

Charney, J. G., Fjörtoft, R., and von Neumann, J. (1950). Numerical integration of the barotropic vorticity equation. *Tellus*, 4:237–254.

Chekroun, M., Liu, H., and McWilliams, J. (2017). The emergence of fast oscillations in a reduced primitive equation model and its implications for closure theories. *Computers and Fluids*, 151:3–22.

Daley, R. (1978). Variational non-linear normal mode initialization. *Tellus*, 30:201–218.

Dickinson, R. E. and Williamson, D. L. (1972). Free oscillations of a discrete stratified fluid with application to numerical weather prediction. *J. Atmos. Sci.*, 29:623–640.

Flattery, T. W. (1967). Hough functions. Tech. Rep. 21, Dept. Geophys. Sci., University of Chicago, 175 pp.

Kasahara, A. (1976). Normal modes of ultralong waves in the atmosphere. *Mon. Wea. Rev.*, 104:669–690.

Kasahara, A. and Shigehisa, Y. (1983). Orthogonal vertical normal modes of a vertically staggered discretized atmospheric model. *Mon. Wea. Rev.*, 111:1724–1735.

Leith, C. E. (1980). Nonlinear normal mode initialization and quasi-geostrophic theory. *J. Atmos. Sci.*, 37:958–968.

Lorenz, E. N. (1980). Attractor sets and quasi-geostrophic equilibrium. *J. Atmos. Sci.*, 37:1685–1699.

Lorenz, E. N. (1986). On the existence of a slow manifold. *J. Atmos. Sci.*, 43:1547–1557.

Lorenz, E. N. and Krishnamurthy, V. (1987). On the nonexistence of a slow manifold. *J. Atmos. Sci.*, 43:2940–2950.

Lynch, P. (1997). The Dolph-Chebyshev window: a simple optimal filter. *Mon. Wea. Rev.*, 125:655–660.

Lynch, P. and Huang, X. Y. (1992). Initialization of the HIRLAM model using a digital filter. *Mon. Wea. Rev.*, 120:1019–1034.

Machenhauer, B. (1977). On the dynamics of gravity oscillations in a shallow water model, with applications to normal mode initialization. *Contrib. Atmos. Phys.*, 50:253–271.

Murakami, H. and Matsumura, T. (2007). Development of an effective non-linear normal mode initialization method for a high-resolution global model. *J. Met. Soc. Japan*, 85:187–208.

Phillips, N. (1960). On the problem of initial data for the primitive equations. *Tellus*, 12:121–126.

Puri, K. and Burke, W. (1982). A scheme to retain the Hadley circulation during nonlinear normal mode initialization. *Mon. Wea. Rev.*, 110:327–335.

Richardson, L. F. (1922). *Weather Prediction by Numerical Process*. Cambridge University Press.

Shigehisa, Y. (1983). Normal modes of the shallow water equations for zonal wavenumber zero. *J. Meteor. Soc. Japan*, 61:479–493.

Temperton, C. (1984). Orthogonal vertical normal modes for a multilevel model. *Mon. Wea. Rev.*, 112:503–509.

Temperton, C. (1989). Implicit normal mode initialization for spectral models. *Mon. Wea. Rev.*, 117:436–451.

Temperton, C. and Williamson, D. L. (1981). Normal mode initialization for a multilevel grid-point model. Part I: Linear aspects. *Mon. Wea. Rev.*, 109:729–743.

Tribbia, J. (1979). Non-linear initialization on an equatorial beta plane. *Mon. Wea. Rev.*, 107:704–713.

Tribbia, J. (1984). A simple scheme for high-order nonlinear normal mode initialization. *Mon. Wea. Rev.*, 112:278–284.

Vanneste, J. (2008a). Asymptotics of a slow manifold. *SIAM J. Appl. Dyn. Syst.*, 7:1163–1190.

Vanneste, J. (2008b). Exponential smallness of inertia-gravity wave generation at small Rossby number. *J. Atmos. Sci.*, 65:1622–1637.

Vautard, R. and Legras, B. (1986). Invariant manifolds, quasi-geostrophy and initialization. *J. Atmos. Sci.*, 43:565–584.

Warn, T. (1997). Nonlinear balance and quasi-geostrophic sets. *Atmosphere-Ocean*, 35:135–145.

Wergen, W. (1989). Normal mode initialization and atmospheric tides. *Q. J. R. Meteorol. Soc.*, 115:535–545.

Williamson, D. L. (1976). Normal mode initialization procedure applied to forecasts with the global shallow water equations. *Mon. Wea. Rev.*, 104:195–206.

Chapter 3
Normal Mode Functions and Atmospheric Data Assimilation

Nedjeljka Žagar

Abstract The normal-mode function framework is applied for the formulation of the background-error covariance matrix for data assimilation, for the quantification of the information content of observations in data assimilation at many scales, for the systematical evaluation of the role of equatorial waves on mass-wind coupling near the equator, and for the discussion of the growth of forecast uncertainties associated with the Rossby and inertio-gravity modes.

3.1 Introduction

The history of Numerical Weather Prediction (NWP) is intimately related to the application of the Normal Mode Functions (NMFs) in data assimilation and initialization of NWP models. Many details of the historical developments are contained in the book by R. Daley (1991).

There are several reasons for repeated efforts to make the use of the Hough functions in data assimilation. First of all, the Hough functions are physically meaningful, as they are the eigensolutions of the horizontal part of the linearized primitive equations (Chapter 1 by A. Kasahara). Being both waves and the orthogonal normal basis functions, the Hough functions can be the basis functions for a prognostic model as derived in Chapter 1, section 10 (Kasahara, 1977, 1978; Callaghan et al., 1999). Second, they can be used for the initialization to filter unwanted frequencies as discussed in the chapter by J. Tribbia (Baer and Tribbia, 1977; Machenhauer, 1977; Tribbia, 1984), as well as to check the numerical solutions of the discretized shallow-water equations (Dee and Silva, 1986). Finally, the Hough functions represent a meteorologically meaningful approach to multivariate analysis, as shown by their early application for objective analysis at NCEP (Flattery, 1970). Known as the Flattery analysis scheme, this early method fitted the observations of temperature and winds at horizontal levels to the prescribed Hough functions of the leading vertical mode. The vertical structure functions were empirical orthogonal functions (EOFs). A later

N. Žagar and J. Tribbia (eds.), *Modal View of Atmospheric Variability*,
Mathematics of Planet Earth 8, https://doi.org/10.1007/978-3-030-60963-4_3

study by Halberstam and Tung (1984) combined the Hough function with a first guess in data void areas to remove the ill-conditioning of the analysis problem.

In this chapter, we apply the Hough functions for multivariate data assimilation with focus on the tropics, and for the diagnostic of modern data assimilation systems. The chapter is organised in six sections. The first section introduces the data assimilation problem and its solution using the variational approach. Section 2 discusses the application of the Hough functions for the representation of the background-error covariances. In Section 3, the NMF representation is used to formulate several diagnostics of the global assimilation and prediction systems. The application of the normal modes of the linearized shallow-water equations on the equatorial β−plane in tropical data assimilation is presented in Section 4 and conclusions are given in Section 5.

3.2 Atmospheric Data Assimilation Modelling

3.2.1 Basics of Data Assimilation

Data assimilation combines observations and model forecasts (prior or first-guess information) to prepare an improved estimate of the initial state (called analysis) for the next forecast. A properly solved analysis solution is characterized by a smaller uncertainty than input observations and prior information.

To illustrate the main properties of data assimilation, let us consider a single model level or a 2D (horizontal) model with its prognostic variables discretized on a regular model grid with totally N points. The short-range forecast (background or first guess or prior field) on this grid makes a vector denoted $\mathbf{x}_b$, with elements $x_b(i), i = 1, ..., N$. We want to correct $\mathbf{x}_b$ by the latest observations, denoted $y(j)$, $j = 1, ..., M$. The observation vector is denoted $\mathbf{y}$. In general, locations of observations do not coincide with the model grid and we need an operator to transform the first-guess values to their observed equivalents at the locations of observations. For simplicity, we can assume this operator to be linear and denote it $\mathbf{H}$. Then the difference between the background and observations at the observation location is $\mathbf{y} - \mathbf{H}\mathbf{x}_b$. Normally, the observation operator is not linear, and it can include complex models to recompute model equivalents such as radiances. The nonlinear observation operator is denoted H.

The observations and the background field contain errors. If the vectors with the (unknown) true values of the background field and observations are denoted $\mathbf{x}_t$ and $\mathbf{y}_t$, respectively, then the errors for the i−th element of $\mathbf{x}_b$ and j−th element of $\mathbf{y}$ are $\epsilon_b(i) = x_b(i) - x_t(i)$ and $\epsilon_o(j) = y(j) - y_t(j)$, respectively. The vectors with the background and observation errors are $\boldsymbol{\epsilon}_b = \mathbf{x}_b - \mathbf{x}_t$ and $\boldsymbol{\epsilon}_o = \mathbf{y} - \mathbf{y}_t$, respectively.

Assuming that the background and observation errors are unbiased, i.e. $\overline{\boldsymbol{\epsilon}_b} = 0$ and $\overline{\boldsymbol{\epsilon}_o} = 0$, the covariance between errors in the first-guess field at points i and j,

$$\epsilon_b(i,j) = \left\langle (\epsilon_b(i) - \overline{\epsilon}_b(i))\,(\epsilon_b(j) - \overline{\epsilon}_b(j)) \right\rangle,$$

becomes simply $\langle \epsilon_b(i)\, \epsilon_b(j) \rangle$. The $\langle\,\rangle$ and $\overline{(\,)}$ operators denote averaging over a number of the error samples. The matrix of the covariances, known as the "background-error covariance matrix" is denoted $\mathbf{B}$, and it has $N \times N$ elements:

$$\mathbf{B} = \langle \boldsymbol{\epsilon}_b \boldsymbol{\epsilon}_b^T \rangle\,. \tag{3.1}$$

The information about the covariances among the errors in the first guess is a very important component for data assimilation as it defines how the observed information is spread in the model space.

Similarly, we define the observation-error covariance matrix with $M \times M$ elements as

$$\mathbf{R} = \langle \boldsymbol{\epsilon}_o \boldsymbol{\epsilon}_o^T \rangle\,. \tag{3.2}$$

The minimum-variance, linear unbiased estimate is derived assuming that the errors in the background and observations are normally distributed, that the background errors and observation errors are un-correlated and that error covariances are known (Daley, 1991). The solution for the analysis vector $\mathbf{x}_a$ with the minimal error variance is the following:

$$\mathbf{x_a} = \mathbf{x_b} + \mathbf{BH^T}\left(\mathbf{HBH^T} + \mathbf{R}\right)^{-1}(\mathbf{y} - \mathbf{Hx_b})\,. \tag{3.3}$$

The matrix

$$\mathbf{K} = \mathbf{BH^T}\left(\mathbf{HBH^T} + \mathbf{R}\right)^{-1} = \left(\mathbf{B}^{-1} + \mathbf{H^T R^{-1} H}\right)^{-1} \mathbf{H^T R^{-1}}$$

is called the gain matrix. For a single observation, the column vector of $\mathbf{K}$ will spread the impact of the observation in model space. We can define analysis increment $\delta\mathbf{x}$, $\delta\mathbf{x} = \mathbf{x}_a - \mathbf{x}_b$, where $\left(\mathbf{B}^{-1} + \mathbf{H^T R^{-1} H}\right)\delta\mathbf{x} = \mathbf{H^T R^{-1}}\,(\mathbf{y} - \mathbf{Hx}_b)$.

The analysis error vector $\boldsymbol{\epsilon}_a$ is made of elements $\epsilon_a(i) = x_a(i) - x_t(i)$, and the analysis covariance matrix $\mathbf{A} = \langle \boldsymbol{\epsilon}_a \boldsymbol{\epsilon}_a^T \rangle$ associated with the solution (3.3) is given by

$$\mathbf{A}^{-1} = \mathbf{B}^{-1} + \mathbf{H^T R^{-1} H}\,. \tag{3.4}$$

For details of the derivation of equations (3.3)-(3.4) the reader is referred to Daley (1991).

Since $\mathbf{x}_t$ is not available, the matrix $\mathbf{K}$ is impossible to specify and some proxies of the error covariances are needed. But even with the approximate covariances specified, the problem (3.3) can hardly ever be solved directly as the large dimensions prohibit the computation of inverses of matrices $\mathbf{B}$ and $\mathbf{R}$. The covariance matrices have to be simplified with the goal to make them as close as possible to the diagonal forms.

Towards this goal, we notice that the solution (3.3) can be obtained also by minimizing the cost function J which measures the distance of the analysis to observations and to the first guess as follows:

$$J(\mathbf{x}) = J_b + J_o = \frac{1}{2}(\mathbf{x}_b - \mathbf{x})^T \mathbf{B}^{-1}(\mathbf{x}_b - \mathbf{x}) + \frac{1}{2}(\mathbf{y} - H\mathbf{x})^T \mathbf{R}^{-1}(\mathbf{y} - H\mathbf{x}) \ . \quad (3.5)$$

The function J is defined as a sum of the two terms: the term J_b which measures the distance to a background model state, whereas the distance to the observations is measured by J_o. The solution, $\mathbf{x}_a$, equivalent to (3.3), is obtained by minimizing J in model space which is denoted E:

$$\min_{x \epsilon E} J = J_b + J_o \ . \quad (3.6)$$

If other information is available, the cost function may have additional terms, i.e. J_c. The control term J_c measures the distance to the defined balanced state.

A more general formulation of the assimilation problem may be derived using Bayes theorem. This theorem describes the conditional probability of the event A given B, $P(A/B)$, in relation to probabilities of A and B, $P(A)$ and $P(B)$, respectively, as

$$P(A/B) = \frac{P(B/A)P(A)}{P(B)} \ .$$

In the case of data assimilation the background and observations probabilities are given and we search for the most probable analysis solution as:

$$\mathbf{x}_a = \arg\max_x (P(\mathbf{x}/\mathbf{y} \text{ and } \mathbf{x}_b)) \ .$$

In the case of Gaussian probabilities, the variational assimilation cost function is found as follows:

$$J = -\log(P(\mathbf{x}/\mathbf{y} \text{ and } \mathbf{x}_b)) + const. \ ,$$

and since log is a monotonic function,

$$\mathbf{x}_a = \arg\max_x (J(\mathbf{x})) \ .$$

In the above definition of J, all observations are assumed valid at a single time, the procedure known as the three-dimensional variational data assimilation (3D-Var). With observations available over a time interval, (3.5) is modified to take into account observations distributed among K different times in the frame of the four-dimensional variational data assimilation (4D-Var):

$$J(\mathbf{x}) = J_b + J_o = \frac{1}{2}(\mathbf{x}_b - \mathbf{x})^T \mathbf{B}^{-1}(\mathbf{x}_b - \mathbf{x}) + \frac{1}{2}\sum_{n=1}^{K}(\mathbf{y} - H\mathbf{x})_n^T \mathbf{R}^{-1}(\mathbf{y} - H\mathbf{x})_n \ . \quad (3.7)$$

The observation operator H in Eq. (3.7) includes also a forecast model that propagates the background variables from the start of the assimilation time window to times with observations.

The application of (3.7) and its minimization requires several steps. First we need an initial estimate of $\mathbf{x}$, the computation of J and its gradient with respect to $\mathbf{x}$. Then this information is passed to the minimization scheme that computes a more accurate estimate of $\mathbf{x}$. The gradient of J is given by

$$J' = \nabla_{\mathbf{x}} J(x) = \nabla_{\mathbf{x}} J_b + \nabla_{\mathbf{x}} J_o = \mathbf{B}^{-1}(\mathbf{x} - \mathbf{x}_b) + \mathbf{H}\mathbf{R}^{-1}(H\mathbf{x} - \mathbf{y}) \ . \tag{3.8}$$

where $\mathbf{H}$ is now the tangent linear operator to H in the vicinity of $\mathbf{x}$, and in the case of 4D-Var $\mathbf{H}$ is including also the linearized forecast model.

In the range of validity of the tangent linear operator, the Hessian, J'', is given by

$$\nabla^2_{\mathbf{x}} J = \mathbf{B}^{-1} + \mathbf{H}\mathbf{R}^{-1}\mathbf{H}. \tag{3.9}$$

This is the same as (3.4), the analysis error covariance matrix. The minimizing solution is searched by setting the expression (3.8) to zero. In practice (for example, at ECMWF and for the HIRLAM data assimilation), the minimization has been performed after the pre-conditioning is done with $\mathbf{B}^{1/2}$, meaning that the problem (3.7) is reformulated in such a way that a new control variable transforms $\mathbf{B}$ to a diagonal matrix. We shall apply the Hough functions for this purpose.

3.2.2 The Background Error Covariance Matrix

A reliable estimate of the properties of forecast errors in the first guess poses a real challenge since the whole truth will never be available. Thus the background errors, as the short-range forecast errors are commonly referred to, need to be represented by surrogate quantities with statistical and dynamical properties assumed similar to those of the unknown forecast errors (e.g. Parrish and Derber, 1992). Derived dependencies between the forecast errors are built into the $\mathbf{B}$ matrix; the purpose of these relationships is to spread observed information from a point to nearby grid-points and levels. Moreover, the observed information is also distributed to other variables. In this way observations of the temperature field carry information about the wind field, and vice versa. This is the fundamental reason why the balance relationships between the mass- and the wind-field variables are of such great importance for data assimilation, especially in regions where observations are sparse and in a global observing system (GOS) dominated by mass field information.

The matrix $\mathbf{B}$ is real and symmetric. Any symmetric matrix can be represented as $\mathbf{B} = \mathbf{P}\mathbf{D}\mathbf{P}^{-1}$. Here, $\mathbf{P}$ is a matrix composed of the eigenvectors of $\mathbf{B}$ and $\mathbf{D}$ is the diagonal matrix constructed from the corresponding eigenvalues. Also, $\mathbf{P}$ is a unitary matrix meaning that $\mathbf{P}^* = \mathbf{P}^{-1}$, where $*$ denotes complex conjugate. Using these properties,

$$J_b = \frac{1}{2}(\mathbf{x} - \mathbf{x}_b)^T \mathbf{B}^{-1}(\mathbf{x} - \mathbf{x}_b) \tag{3.10}$$

can be re-written as

$$\begin{aligned} J_b &= \tfrac{1}{2}(\mathbf{x}-\mathbf{x}_b)^T\,\mathbf{P}\mathbf{D}^{-1/2}\mathbf{D}^{-1/2}\mathbf{P}^{-1}(\mathbf{x}-\mathbf{x}_b) \\ &= \tfrac{1}{2}\left(\mathbf{D}^{-1/2}\mathbf{P}^{-1}(\mathbf{x}-\mathbf{x}_b)\right)^T\left(\mathbf{D}^{-1/2}\mathbf{P}^{-1}(\mathbf{x}-\mathbf{x}_b)\right). \end{aligned} \tag{3.11}$$

Introducing a new vector variable χ, such that

$$\chi = \mathbf{D}^{-1/2}\mathbf{P}^{-1}(\mathbf{x}-\mathbf{x}_b), \tag{3.12}$$

the background term of the cost function and its gradient, get their simplest form:

$$J_b = \frac{1}{2}\chi^T\chi \quad \text{and} \quad \nabla_\chi J_b = \chi. \tag{3.13}$$

Within the NMF framework, we aim at approximating $\mathbf{D}^{-1/2}\mathbf{P}^{-1}$ with a sequence of operators that will project analysis increments during the minimization of J onto the eigenmodes of the linearized primitive equations. With the assumption of spatially homogeneous background error variances and horizontal and vertical independence, the **B** can be made diagonal and the minimization accomplished fast. This assumes that the forecast errors can, for first order of approximation, be represented within a subspace E_b of model space E. This subspace can be defined by the geostrophic balance (Phillips, 1986) as done in the statistical interpolation algorithms in 1980s. In the 3D-Var NMF implementation of Parrish and Derber (1992), the subspace E_b is defined by the linear balance equation on the sphere. The goal (3.6) is now changed to

$$\min_{x\epsilon E_b} J = J_b + J_o. \tag{3.14}$$

This means that J_b is reformulated in the subspace E_b as

$$J_b = \frac{1}{2}\left[\mathbf{S}_b(\mathbf{x}-\mathbf{x}_b)\right]^T\mathbf{B}_b^{-1}\left[\mathbf{S}_b(\mathbf{x}-\mathbf{x}_b)\right], \tag{3.15}$$

where $\mathbf{S}_b$ represents the projection of $(\mathbf{x}-\mathbf{x}_b)$ onto the E_b subspace of selected modes. A more general notation of (3.15) is

$$J_b = \frac{1}{2}(\mathbf{x}-\mathbf{x}_b)^T\,\mathbf{S}_b^T\tilde{\mathbf{B}}_b^{-1}\mathbf{S}_b(\mathbf{x}-\mathbf{x}_b), \tag{3.16}$$

where $\tilde{\mathbf{B}}_b$ is identical to $\mathbf{B}_b$ on E_b but can be different outside this subspace (Heckley et al., 1993). In the language of the normal-mode decomposition in Rossby and inertio-gravity (IG) subspaces, the latter denoted E_{IG}, $\tilde{\mathbf{B}}_b$ can take any value on E_{IG}. The minimization (3.14) then becomes

$$\min_{x\epsilon E_b} J \Leftrightarrow \mathbf{S}_b\left[\min_{x\epsilon E} J(\mathbf{S}_b(\mathbf{x}))\right]. \tag{3.17}$$

Assuming that the initial point for the minimization is on the subspace (for example, subspace in which balance relationships are valid), than $\min_{x\epsilon E_b} J$ and

$\min_{x \in E} J(\mathbf{S}_b(\mathbf{x}))$ are algorithmically equivalent. Notice that for (3.16) to apply, the operator $\mathbf{S}_{IG}(\mathbf{x} - \mathbf{x}_b)$ is defined to be zero at the start of the minimization. The comparison of (3.15) with (3.10) tells us that using a balanced subspace of E for data assimilation means that the inverse of the effective background-error covariance matrix is given by $\mathbf{S}_b^T \tilde{\mathbf{B}}_b^{-1} \mathbf{S}_b$, with the $\tilde{\mathbf{B}}_b^{-1}$ matrix to be specified. This was the idea behind the NMF-based preconditioning of (3.5) in early 3D-Var at ECMWF (Heckley et al., 1993).

3.3 Background-Error Covariance Modelling Using the Hough Modes

The representation of 3D data in terms of the Hough functions was introduced in Chapter 1 by Akira Kasahara. The Hough functions are denoted $\mathbf{H}_n^k(\lambda, \varphi; m)$[1] and are defined by the zonal wavenumber k, meridional mode index n and vertical mode index m associated with equivalent depth D_m. In the derivation of the Hough functions by Swarztrauber and Kasahara (1985), which is implemented in MODES software (Žagar et al., 2015), the frequencies of the normal modes are obtained as the eigenvalues for a problem defined by the zonal wavenumber and equivalent depth. For each combination of k and m, the frequency $\nu_n^k(m)$ depends also on n which defines two kinds of oscillations, the first kind and the second kind of Hough solutions. The first kind describes the inertio-gravity (IG) waves whereas the solutions of the second kind describe low-frequency, westward propagating, balanced waves of Rossby-Haurwitz type. The IG solutions are divided into westward and eastward propagating IG modes, denoted EIG and WIG respectively.

3.3.1 Summary of the Normal Mode Decomposition

The decomposition of discrete 3D data in NMFs consists of two steps. In the first step, the input data vector $\mathbf{X}(\lambda, \varphi, \sigma_j) = (u, v, h)^{\mathrm{T}}$ at time step t on j-th σ level is decomposed using a series of M orthogonal vertical structure functions $G_m(j)$ as

$$\mathbf{X}(\lambda, \varphi, \sigma_j) = \sum_{m=1}^{M} G_m(j) \mathbf{S}_m \mathbf{X}_m(\lambda, \varphi) \quad , \tag{3.18}$$

where $\mathbf{S}_m$ is a 3×3 diagonal matrix with elements $\sqrt{gD_m}$, $\sqrt{gD_m}$, and D_m. The latitude and longitude are denoted φ and λ, respectively. The result of (3.18) is the vector $\mathbf{X}_m(\lambda, \varphi) = (\tilde{u}_m, \tilde{v}_m, \tilde{h}_m)^{\mathrm{T}}$ that describes non-dimensional oscillations of the horizontal wind and height fields in a shallow-water system with depth D_m, i.e.

[1] Notice that the linearized observation operator, which in the previous section was denoted $\mathbf{H}$, is no longer used.

$$\tilde{u}_m = \frac{u_m}{\sqrt{gD_m}}, \quad \tilde{v}_m = \frac{v_m}{\sqrt{gD_m}} \quad \text{and} \quad \tilde{h}_m = \frac{h_m}{D_m}. \tag{3.19}$$

The vertical mode index m varies from the external (barotropic) mode, $m = 1$, to the total number of vertical modes M, with $M \leq J$.

In the second step, the dimensionless horizontal motions for each m are represented by a series of the Hough harmonics $\mathbf{H}_n^k$ using the complex Hough expansion coefficients $\chi_n^k(m)$ as

$$\mathbf{X}_m(\lambda, \varphi) = \sum_{n=1}^{R} \sum_{k=-K}^{K} \chi_n^k(m)\, \mathbf{H}_n^k(\lambda, \varphi; m). \tag{3.20}$$

In (3.20), the maximal number of zonal waves in Eq. (3.18) is denoted by K, starting from $k = 0$ for the zonal mean state. The maximal number of meridional modes, R, combines equal numbers of balanced modes, N_R, the eastward-propagating inertio-gravity modes, N_E and the westward-propagating inertio-gravity modes, N_W, i.e. $R = N_R + N_E + N_W$. The dimension of the modal (or spectral) space is thus $N = M \times R \times (K+1)$. The three indices k, n, and m constitute the 3-component modal index $\nu = (k, n, m)$ of the Hough expansion coefficients, χ_ν. The same expansion coefficient $\chi_n^k(m)$ represents u, v and h because the expansion (3.20) is complete as the truncation limits (M, R, K) go to infinity. In practice, the zonal and meridional truncation criteria are defined by the data resolution and the vertical truncation M is limited by the number of vertical levels.

The practical application involves the inverses of equations (3.18) and (3.20), i.e. first computing $\mathbf{X}_m$ as

$$\mathbf{X}_m(\lambda, \varphi) = \mathbf{S}_m^{-1} \sum_{j=1}^{J} G_m(j)\, \mathbf{X}_j \quad , \tag{3.21}$$

followed by the computation of the Hough expansion coefficients χ_ν as

$$\chi_\nu = \chi_n^k(m) = \frac{1}{2\pi} \int_0^{2\pi} \int_{-1}^{1} \mathbf{X}_m \cdot \left[\mathbf{H}_n^k\right]^* d\mu\, d\lambda. \tag{3.22}$$

The zonal expansion is performed by using the fast Fourier transform while the Gaussian quadrature approximates the integration in the meridional direction. Numerical aspects of the horizontal expansion are described in Swarztrauber and Kasahara (1985). Equations (3.18) and (3.21) are the vertical transform pair whereas equations (3.20) and (3.22) are the horizontal transform pair. Further details are in Chapter 1 and references therein. The result of the expansion (3.21)-(3.22) is the time series of the non-dimensional coefficients $\chi_\nu(t)$. The dimensional form of the derived statistical quantities is obtained by the multiplication implied by $\mathbf{S}_m$ for each m.

3.3.2 Approximation of the Background-Error Covariance Matrix by the Hough Functions

Let denote the discrete Hough matrix operator for the vertical mode m by $\mathcal{H}_m$. It is defined as

$$\mathcal{H}_m = \mathbf{H}_n^k(\lambda, \varphi; m) = \mathbf{\Theta}_n^k(\varphi, m)\, e^{ik\lambda}, \tag{3.23}$$

where the definition of the Hough vector functions in the meridional direction $\mathbf{\Theta}_n^k(\varphi, m)$ were given in Chapter 1 (see also Swarztrauber and Kasahara, 1985). A single column of $\mathcal{H}_m$ contains one of the Hough functions for the $m-$th vertical mode evaluated on the model grid.

Using the data assimilation terminology, $\mathbf{x}_m$ denotes the vector with model variables on the model grid, after the vertical decomposition. The Hough transformation between physical space and spectral space is written as

$$\mathbf{x}_m = \mathcal{H}_m \boldsymbol{\chi}_m, \tag{3.24}$$

where $\boldsymbol{\chi}_m$ is a vector containing the amplitudes of the Hough functions for the given state vector $\mathbf{x}_m$.

The background-error covariance matrices in modal space and physical space are defined as (with the subscript m dropped)

$$\tilde{\mathbf{B}} = \left\langle (\chi - \chi_t)(\chi - \chi_t)^T \right\rangle \quad \text{and} \quad \mathbf{B} = \left\langle (\mathbf{x} - \mathbf{x}_t)(\mathbf{x} - \mathbf{x}_t)^T \right\rangle, \tag{3.25}$$

respectively, where $\mathbf{x}_t$ and χ_t denote true values in model and modal space, respectively. Using (3.24), the two matrices are related as

$$\mathbf{B} = \mathcal{H}\tilde{\mathbf{B}}\mathcal{H}^T. \tag{3.26}$$

If elements of $\tilde{\mathbf{B}}$ are organised so that all Rossby modes are included followed by the IG modes, the covariance matrix is made of four parts that constitute covariances between Rossby (or slow) modes (denoted $cov(R-R)$), between IG (or fast) modes ($cov(IG-IG)$) and covariances between the Rossby and IG modes ($cov(R-IG)$):

$$\tilde{\mathbf{B}} = \begin{bmatrix} cov(R-R) & cov(R-IG) \\ cov(IG-R) & cov(IG-IG) \end{bmatrix}. \tag{3.27}$$

Note that the Rossby mode subspace R in (3.27) is different from $\mathbf{R}$, the observation error covariance matrix.

In his application of the Hough functions for the representation of the $\mathbf{B}$ matrix in optimal interpolation using a global shallow-water model, Parrish (1988) applied the background variance equipartition among slow modes. In this case, equation (3.27) becomes

$$\tilde{\mathbf{B}} = \begin{bmatrix} \gamma\mathbf{I} & 0 \\ 0 & 0 \end{bmatrix}, \tag{3.28}$$

with γ a scalar representing the same level of the forecast-error variance assigned to all slow modes. Using $\delta\mathbf{x} = \mathbf{x}_a - \mathbf{x}_b$ for the analysis increment, and $\delta\mathbf{y} = \mathbf{y} - \mathbf{H}\mathbf{x}_b$ for the observation innovation, the analysis equation (3.3) is written as

$$\delta\mathbf{x} = \mathbf{B}\mathbf{H}^T\left(\mathbf{H}\mathbf{B}\mathbf{H}^T + \mathbf{R}\right)^{-1}\delta\mathbf{y}\,. \tag{3.29}$$

By transforming the analysis increment from modal space to grid-point space using the Hough operator $\mathcal{H}$, $\delta\mathbf{x} = \mathcal{H}\chi$, the analysis equation becomes

$$\left(\mathbf{I} + \tilde{\mathbf{B}}\mathcal{H}^T\mathbf{H}^T\mathbf{R}^{-1}\mathbf{H}\mathcal{H}\right)\chi = \tilde{\mathbf{B}}\mathcal{H}^T\mathbf{H}^T\mathbf{R}^{-1}\delta\mathbf{y}\,. \tag{3.30}$$

With $\tilde{\mathbf{B}}$ assumed a diagonal matrix, the sequence of operators in (3.30) becomes an inexpensive procedure with other advantages including global solution, balanced analysis increments, and the removal of complications involved with estimating the background-error statistics (Parrish, 1988).

Parrish (1988) also noticed the importance of the two special tropical modes, the Kelvin and mixed Rossby-gravity (MRG) mode on mass-wind correlations near the equator. As his paper was published in a conference proceedings hard to obtain, a few of his figures are shown in Fig. 3.1. The presented correlations include the impact of the MRG and Kelvin modes along with the Rossby modes. In the left panel of Fig. 3.1 the Rossby modes and the MRG mode were considered in the computation of correlations for a point at the equator. The correlations appear anisotropic, a significant difference from the mid-latitude case. The other two panels in this figure illustrate that the anisotropy at the equator is reduced when the MRG are excluded (Fig. 3.1, middle panel) and when the Kelvin waves are included (Fig. 3.1, right panel).

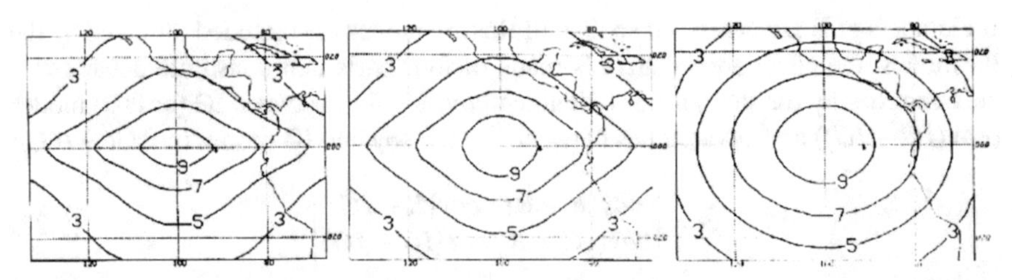

Fig. 3.1 Left: Correlations of height field at any point with the height field at 0°N, 100°W. Contour interval is 0.2. Middle: As in left panel, but mixed Rossby-gravity waves have been excluded from slow modes, Right: Kelvin waves with zonal wavenumbers 1-3 included among the slow modes. From Parrish (1988), Figures 3, 11 and 12. ©Dave Parrish. Used with permission.

Dave Parrish was the first to introduce the NMF representation in the background-error covariance modelling and the first to demonstrate the potentially important role of the large-scale Kelvin and MRG modes for the mass-wind coupling in tropical data assimilation. When contacted for his figure permission in summer 2020, he recalled his work on the Hough functions: "This work was inspired by a conversation with

Norman Phillips, then the chief scientist at what was then the National Meteorological Center Development Division. He was looking at a simple 1D shallow water model, trying to understand what background error should look like. I immediately got what he was talking about and began work on a global model version of this idea, using normal modes. I spent the next two years or so trying to explain this new idea about background error formulation. Always got blank faces from everyone until one meeting where a French guy (Talagrand perhaps) jumped up and shouted "I understand now!" or something close to that. That was an interesting time. Then I had a meeting with John Derber, then a post-doc at Princeton. I gave a seminar, with the usual blank stares. But later in John's office, he immediately understood. Not long after, he came to work at NMC. He always said he stole my work, and I had to keep pointing out to him how important his contribution was. That was an exciting time. " (D. Parrish, 2020, personal communication, 10 September).

The early version of ECMWF 3D-Var data assimilation system was formulated using the Hough harmonics. The description is given in Heckley et al. (1993). At ECMWF, the background term of the cost function was specified as

$$J_b = \frac{1}{2}\boldsymbol{\chi}^T \mathbf{D}^{-1/2} \boldsymbol{\chi} \qquad \text{with} \qquad \boldsymbol{\chi} = \mathbf{L}^{-1}(\mathbf{x} - \mathbf{x}_b), \tag{3.31}$$

and the transformation operator $\mathbf{L}$ was defined as

$$\mathbf{L} = \mathbf{W}^{-1}\mathbf{S}\mathbf{N}\mathbf{S}^{-1}\mathbf{W}\mathbf{h}\mathbf{P}^{-1} \quad . \tag{3.32}$$

The involved operators perform the following operations: $\mathbf{W}$ converted wind components in spectral space to vorticity and divergence, $\mathbf{S}$ performed transformation to spectral space, $\mathbf{N}$ was the normalization with the background errors in grid-point space, $\mathbf{h}$ was the horizontal background-error covariance matrix for spectral components, and $\mathbf{P}$ was the projection onto the eigenvectors of the vertical background error correlation matrices. In (3.31), $\mathbf{D}$ stands for a diagonal matrix of the eigenvalues of the vertical background error correlations.

The problem of balance and relative contribution of the Rossby and IG modes was treated by splitting J_b into three part: the Rossby part, the inertio-gravity part and a part that was analyzed multivariately. The weights assigned to the Rossby and IG parts were defined by parameter w:

$$J_b - \frac{1}{4(w-1)}\chi_R^T \mathbf{D}_R^{-1} \chi_R + \frac{1}{4w}\chi_G^T \mathbf{D}_G^{-1} \chi_G + \frac{1}{2}\chi_U^T \mathbf{D}_U^{-1} \chi_U \quad , \tag{3.33}$$

with $w = \frac{1}{2}$ representing the percentage of error variance associated by the IG wave part of the flow. The applied value of 0.1 represented the assumption that about 10% of the wind error variance was in the divergent part of the flow. The control of unwanted noise could be performed by formulating the cost function in terms of fields after the non-linear normal-mode initialization or by introducing a cost function J_c which measured the tendency of the gravity wave mode G (Heckley et al., 1993).

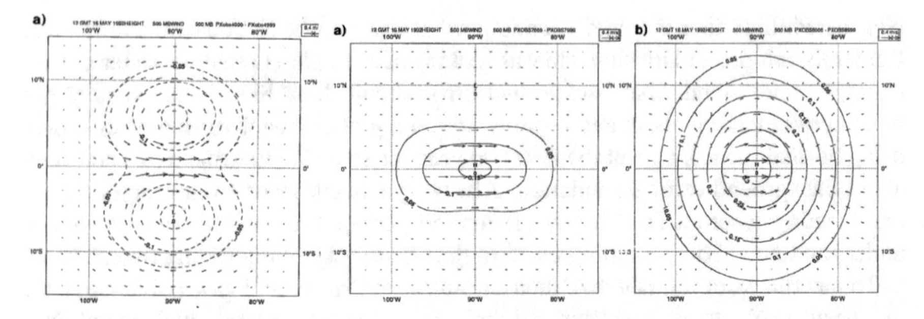

Fig. 3.2 Response of the 3D-Var analysis at 500 hPa to a positive zonal wind observation at 0° for (left) w=0.1, (middle) w = 0.5 and (right) w = 0.9. Contours represent height field and the arrows indicate vector wind. From Heckley et al. (1993), their Fig. 2a, Fig 4a, 4b, respectively. ©ECMWF. Used with permission.

Single observation experiments using the 3D-Var formulation (3.33) were presented in Heckley et al. (1993) for various w and a few of their results are shown in Fig. 3.2. Analysis increments associated with a single westerly wind observation at the equator shows the maximal impact on the associated height field increment in the subtropics. The solution in Fig. 3.2 (left panel) appears predominantly balanced as ϵ is small, with about 10% of the forecast error variance associated with the IG modes. A very similar solution is obtained by imposing the linear balance equator (not shown). When the strength of the Rossby balance is reduced, the analysis increments appear largely unbalanced and the relationship between the geopotential height and winds suggests the dominant role of the Kelvin wave coupling (Fig. 3.2, right panel).

The appropriate weighting between the Rossby and IG modes in the background cost function (3.33) was a largely unknown parameter at the time. With other unknowns including the choice of vertical modes to be included in the background-error term, the NNMI initialization in the tropics and the normal-mode decomposition with increasingly greater number of vertical levels and higher model top, the development of the above described background error term did not continue. The statistically estimated balance for the background-error covariance matrix was implemented based on the global linear balance equation. The balanced and unbalanced components of the control vector following (Parrish and Derber, 1992) were part of the incremental approach to 3D-Var and 4D-Var (Courtier et al., 1998; Rabier et al., 1998; Derber and Bouttier, 1999). Balanced initial conditions were ensured by the digital-filter initialization (Lynch and Huang, 1992), either as a strong or as a weak constraint in 4D-Var (Gustafsson, 1992).

3.4 Diagnostics of Data Assimilation Systems Using Hough Modes

Recent applications of NMFs in global data assimilation and prediction systems focused on the role of IG modes in the covariance matrices and the growth of forecast errors. The derived quantities include the information content of observations, the spectra of analysis uncertainties and forecast-error variance distribution in time, the reliability of the analysis and forecast ensembles and the bias in assimilation systems. Each of the quantities is derived in this section, followed by the results of their applications to the state of the art data assimilation and prediction systems. We discus the structure of analysis increments in relation to balance, the background-error variance reduction for the Rossby and IG modes in data assimilation and the growth of errors in prediction.

3.4.1 Representation of Covariances Using the Normal-Mode Decomposition of the Ensembles

In the ensemble data assimilation, background error covariances are derived at each analysis step from an ensemble of short-term forecasts. For an ensemble of size N_e, the $\mathbf{B}$ matrix can be estimated

$$\mathbf{B} \approx \frac{1}{N_e - 1} \sum_{i=1}^{N_e} \Delta \mathbf{x}_{b,i} \cdot \Delta \mathbf{x}_{b,i}^T , \tag{3.34}$$

where the elements of $\Delta \mathbf{x}_b$ are defined as the departures of individual ensemble members $\mathbf{x}_{b,i}$ from the ensemble mean $\overline{\mathbf{x}}_b$, $\Delta \mathbf{x}_{b,i} = \mathbf{x}_{b,i} - \overline{\mathbf{x}}_b$, at time step t.

Following the discussion in 3.3.2, we search for a sequence of operators, a matrix denoted $\mathbf{L}$, that will transform the background and analysis covariance matrices to diagonal forms. In modal space, this is achieved using the orthogonal matrix whose columns are the Hough eigenvectors.

Using the definitions of the NMF expansion from the previous sections, the complete transformation between the physical and modal space variables, $\Delta \mathbf{x}_b$ and χ, can be written with the help of a transformation operator $\mathbf{L}$ as

$$\chi = \mathbf{L}^{-1} \Delta \mathbf{x}_b , \tag{3.35}$$

where

$$\mathbf{L}^{-1} = \mathbf{D}\, \mathbf{T}_\varphi\, \mathbf{F}_\lambda \mathbf{G}_m . \tag{3.36}$$

Here, $\mathbf{G}_m$ is the projection on the vertical structure functions, $\mathbf{F}_\lambda$ is the forward Fourier transform in the zonal direction, $\mathbf{T}_\varphi$ denotes the projection on the meridionally dependent part of the Hough harmonics whereas $\mathbf{D}$ stands for the normalization by the modal variance density and it is an $N \times N$ matrix with diagonal elements γ_ν, each associated with an error variance in a single mode ν.

The pseudo-inverse of the transformation (3.36) can be used to move from the space of the control vector χ to the model physical space:

$$\Delta \mathbf{x}_b = \mathbf{L}\boldsymbol{\chi} = \mathbf{G}_m^{-1}\,\mathbf{F}_\lambda^{-1}\,\mathbf{T}_\varphi^{-1}\,\mathbf{D}^{-1}\tilde{\mathbf{T}}^{-1}\boldsymbol{\chi}\,. \tag{3.37}$$

We call it "pseudo-inverse" because Eq. (3.36) should in principle also include an operator $\tilde{\mathbf{T}}^{-1}$ representing the three truncations, i.e. the Fourier truncation, the meridional truncation for the Hough vectors and the vertical truncation. Since such an operator is not invertible, its pseudo-inverse consists of filling in zeros for the truncated components.

The orthogonality of the 3D structure functions applied in (3.36) can be formulated as

$$\langle \chi_\nu [\chi_{\nu'}]^* \rangle = \begin{cases} 0 & \text{if} \quad \nu \neq \nu' \\ \gamma_\nu^2 & \text{otherwise} \end{cases}\,. \tag{3.38}$$

The modal variance of an eigenmode χ_ν is thus given by γ_ν^2. The orthogonality is derived using the energy norm (Žagar et al., 2016).

3.4.2 Ensemble Variance in Modal Space

Following the vertical projection at a single time instant t, the variance of the ensemble of analyses or forecasts at the point $(\lambda_i, \varphi_j, m)$ can be defined as follows:

$$S^2(\lambda_i, \varphi_j, m) = \frac{1}{P-1}\sum_{p=1}^{P}\left(\Delta u_p^2(\lambda_i, \varphi_j, m) + \Delta v_p^2(\lambda_i, \varphi_j, m) + \frac{g}{D_m}\Delta h_p^2(\lambda_i, \varphi_j, m)\right)\,. \tag{3.39}$$

Here, Δu_p, Δv_p and Δh_p denote departures of the ensemble member p from the ensemble mean for wind components and geopotential height at location $(\lambda_i, \varphi_j, m)$. The size of the ensemble is denoted by P. The ensemble spread in a single point of physical space at time t is thus $S(\lambda_i, \varphi_j, m)$. Then the globally integrated variance in model space is the following sum:

$$\sum_i \sum_j \sum_m S^2(\lambda_i, \varphi_j, m)\,, \tag{3.40}$$

where the zonal index is denoted i and the meridional index is j. The summations along the latitudes on the regular Gaussian grid in physical space assumes the use of the Gaussian weights $w(\varphi_j)$.

Since the normal-mode projection is a complete projection, the global ensemble variance computed by (3.40) is equivalent to the integrated modal variance

$$\sum_k \sum_n \sum_m \left[\Sigma_n^k(m)\right]^2, \tag{3.41}$$

where the specific modal variance Σ^2 for the mode $\nu = (k, n, m)$ is defined as

$$\left[\Sigma_n^k(m)\right]^2 = \frac{1}{P-1} \sum_{p=1}^{P} g D_m \left(\Delta\chi_n^k(m;p) \left[\Delta\chi_n^k(m;p)\right]^* \right). \tag{3.42}$$

The coefficients $\Delta\chi_n^k(m; p)$ are obtained by projecting the differences between the ensemble members $p = 1, ..., P$ from the ensemble mean $\overline{\chi_n^k(m)}$ onto a predefined set of normal modes. The ensemble mean at time t is obtained by averaging over P ensemble members:

$$\overline{\chi_n^k(m)} = \frac{1}{P} \sum_{p=1}^{P} \chi_n^k(m;p). \tag{3.43}$$

The ensemble spread $\Sigma_n^k(m)$ is defined for each balanced and inertio-gravity mode and the total variance (4.116) in modal space is the summation over the zonal, meridional and vertical modes. Both modal space and physical space variance are in units $m^2 s^{-2}$ i.e. J/kg. The equivalence between (3.40) and (4.116) is easily derived using the orthogonality properties of the NMFs defined in Chapter 1, as shown for example in Žagar et al. (2015).

The growth of the global forecast uncertainties in an NWP model, which has often been represented in terms of the ensemble spread, can be represented by the variable $\Sigma_n^k(m)$ along any of the three orthogonal directions.

3.4.3 Modal Representation of Ensemble Reliability

Reliability is an essential property of an ensemble-based, probabilistic system (Wilks, 2011). A probabilistic system is reliable if, on average, events that are predicted with a probability P are also observed with a probability P, and this can be measured by assessing whether the ensemble variance is equal to the average mean-squared error of the ensemble mean. A reliable ensemble is characterized by an unbiased ensemble mean and provides samples of the future state of the atmosphere representative for the whole probability distribution of forecasts (Candille and Talagrand, 2005). The quality of operational ensemble systems has been routinely assessed by comparing the ensemble standard deviation (ensemble spread) with the average error of the ensemble-mean forecast verified against operational analyses (e.g. Hagedorn et al., 2012; Buizza, 2014).

Following the same ideas, we can represent the ensemble variance in modal space as defined in section (3.4.2) and inspect to what extent it agrees with the mean squared error of the ensemble mean. The comparison can be performed along any of the three modal dimensions as well as integrated along these directions. The difference between the ensemble mean at forecast time t, $\overline{\chi_n^k(m,t)}$, and the control analysis valid at the same time, $\chi_n^k(m,0)$, is $\|\chi_n^k(m,0) - \overline{\chi_n^k(m,t)}\|$, represents the (k,n,m)–th modal component of the error of the ensemble mean. The mean squared error of the ensemble mean at forecast step t, denoted $\left[\Delta_n^k(m,t)\right]^2$, is defined using the energy norm just like the ensemble variance (3.42):

$$\left[\Delta_n^k(m,t)\right]^2 = \left\langle gD_m \left(\chi_n^k(m,0) - \overline{\chi_n^k(m,t)}\right)\left(\chi_n^k(m,0) - \overline{\chi_n^k(m,t)}\right)^*\right\rangle \quad . \tag{3.44}$$

The averaging over a sample is denoted by $\langle\ \rangle$. For reliability to hold, the ensemble variance $[\Sigma_n^k(m)]^2$ should approximate the mean squared error of the ensemble mean $\Delta_n^k(m)$, i.e.

$$\left[\Delta_n^k(m)\right]^2 \approx \left[\Sigma_n^k(m)\right]^2 \quad . \tag{3.45}$$

In a reliable ensemble, the true state of the atmosphere is on average included in the range spanned by the ensemble members, while in an under-dispersive ensemble, i.e. in an ensemble with a spread that is on average smaller than the error of the ensemble mean, the true state can lay outside. Similarly, an under-dispersive ensemble will be characterized by a lack of spread in one of directions of modal space of the Rossby and IG modes. Depending on the scales and motion types, the under-dispersiveness can be associated with the tropical or midlatitude circulations.

3.4.4 Information Content of Observations in Modal Space

Several measures are used to estimate the value of various observing systems in operational data assimilation for NWP. They compare the uncertainty of the background (prior) state as described by the probability density $\mathcal{P}^B(\mathbf{x})$ with the probability density $\mathcal{P}^A(\mathbf{x})$ representing the uncertainty of the posterior state (analysis) obtained after all the observation information has been assimilated. The included assumptions are that the background and the observation errors are Gaussian and that the error propagation through the model is linear so that the analysis solution is also Gaussian. As earlier, the analysis and background estimation errors are characterized by the analysis and background error covariance matrices, $\mathbf{A}$ and $\mathbf{B}$, respectively.

One of the measures of the information content of observations in data assimilation is the degrees of freedom for signal (DFS) metric which is defined as (Fisher, 2003; Singh et al., 2013)

$$\mathcal{DFS} = n - \mathrm{trace}\left(\mathbf{B}^{-1}\mathbf{A}\right) , \tag{3.46}$$

where n denotes the dimension of the state vector $\mathbf{x}$. Another measure is the entropy, $\mathcal{H}$, which measures the average uncertainty with which one knows the state $\mathbf{x}$ of the system:

$$\mathcal{H}(\mathcal{P}) = \int_{R^n} \mathcal{P}(\mathbf{x}) \ln(\mathcal{P}(\mathbf{x}))\mathrm{d}x \, . \tag{3.47}$$

Its reduction due to the data assimilation, described by the difference in entropy between the prior and posterior probability densities can be shown to be (Rogers, 2000; Singh et al., 2013)

$$\mathcal{S} = \mathcal{H}(\mathcal{P}^B) - \mathcal{H}(\mathcal{P}^A) = \frac{1}{2}\ln\det(\mathbf{B}) - \frac{1}{2}\ln\det(\mathbf{A}) = \frac{1}{2}\ln\det(\mathbf{B}\,\mathbf{A}^{-1})\,. \quad (3.48)$$

Being dependent on the ratio between the analysis and forecast error covariance matrices, equations (3.46)-(3.48) measure the error reduction due to new information from the observations at time step t.

Fisher (2003) showed how to compute the entropy reduction and degrees of freedom for signal (DFS) for a large variational assimilation system. He defined degrees of freedom for signal by using a transformation $\mathbf{L}$ that reduces the background error covariance matrix to the identity matrix. Since the components of the new state vector are statistically independent with unit variance, they correspond to independent degrees of freedom.

In modal representation, using the definitions (3.36)-(3.37), the entropy reduction (3.48) for the control variable χ is written as

$$\mathcal{S}_\chi = \frac{1}{2}\ln\det(\mathbf{L}\,\mathbf{B}\,\mathbf{L}^T) - \frac{1}{2}\ln\det(\mathbf{L}\,\mathbf{A}\,\mathbf{L}^T) \approx \frac{1}{2}\ln\prod_{\nu=1}^{N}\frac{[\gamma_\nu^B]^2}{[\gamma_\nu^A]^2} \approx \frac{1}{2}\sum_{\nu=1}^{N}\ln\frac{[\gamma_\nu^B]^2}{[\gamma_\nu^A]^2}\,. \quad (3.49)$$

Similarly, the expression (3.46) for degrees of freedom for signal (DFS) for χ becomes

$$\mathcal{DFS}_\chi = N - \mathrm{trace}\left(\left(\mathbf{L}\mathbf{B}\mathbf{L}^T\right)^{-1}\left(\mathbf{L}\,\mathbf{A}\,\mathbf{L}^T\right)\right) = N - \sum_{\nu=1}^{N}\frac{[\gamma_\nu^A]^2}{[\gamma_\nu^B]^2}\,. \quad (3.50)$$

Each equation includes a summation of the variance ratio associated with individual modes of the background and analysis states. Žagar et al. (2016) defined a modal variance reduction parameter $\mathcal{I}_\nu$ as a ratio between the posterior and prior variance in mode ν as

$$\mathcal{I}_\nu = \mathcal{I}_n^k(m) = \frac{[\gamma_\nu^A]^2}{[\gamma_\nu^B]^2}\,. \quad (3.51)$$

For every ν, the quantity $\mathcal{I}_\nu$ measures the reduction of the prior variance in a spatial scale characterized by the zonal wavenumber k, meridional mode n and the vertical mode m due to new information from observations. The quantity $(1 - \mathcal{I}_\nu) \times 100\%$ represents the percentage of the prior variance reduced by the assimilation at the analysis step t. Equations (3.49-3.50) are now expressed in terms of $\mathcal{I}_\nu$ as

$$\mathcal{S}_\chi = \frac{1}{2}\sum_{\nu=1}^{N}\ln\mathcal{I}_\nu^{-1} \quad \text{and} \quad \mathcal{DFS}_\chi = N - \sum_{\nu=1}^{N}\mathcal{I}_\nu\,. \quad (3.52)$$

3.4.5 Systematic Analysis Increments in Modal Space

Systematic analysis increments are one way to evaluate the data assimilation system bias. The representation of the bias using NMFs was introduced by Žagar et al. (2010). In modal space, we denote a difference field between the analysis and first-guess fields in the modal space by $\Delta\chi_\nu$. The time-averaged analysis increment (bias) in mode ν is

$$\overline{\Delta\chi_\nu} = \frac{1}{N}\sum_{t=1}^{N}\left[\chi_\nu^a(t) - \chi_\nu^b(t)\right] \quad . \tag{3.53}$$

The superscripts 'a' and 'b' denote analysis and background fields, respectively. Equation (3.53) formally defines analysis system bias in terms of the $\overline{\Delta\chi_\nu}$ variable at various model levels, or summed across a single modal index. The averaged increment field transformed from modal space back to model space should ideally correspond to the averaged analysis increments in the model space; the degree of agreement depends on the accuracy of the projection.

Similarly, we can define the average difference between the analysis field and the background field variances in modal space:

$$\overline{\Delta V_\nu} = gD_{m,\nu}\left\langle \chi_\nu^a\left(\chi_\nu^a\right)^* - \chi_\nu^b\left(\chi_\nu^b\right)^*\right\rangle \quad , \tag{3.54}$$

where $\langle\,\rangle$ represents the time averaging. The $\overline{\Delta V_\nu}$ represents the average tendency of the assimilation system to place or subtract the energy to/from a particular mode.

Systematic analysis increments are indicative of the analysis system bias if they have amplitudes comparable to those of typical analysis increments (Dee, 2005). This can be inspected by comparing Eq. (3.53) with the following equation which defines the energy spectrum of analysis increments:

$$\Delta I_\nu(t) = gD_{m,\nu}\left\langle\left(\chi_\nu^a - \chi_\nu^b\right)\left(\chi_\nu^a - \chi_\nu^b\right)^*\right\rangle \quad . \tag{3.55}$$

The distribution of time-averaged energy, $\overline{\Delta I_\nu}$, shows parts of the modal space where the assimilation step on average makes most significant changes in the background.

3.4.6 Modal Diagnostics of Data Assimilation and Ensemble Prediction Systems

Before going to modal space, several properties of ensemble prediction systems are shown in model space using a perfect-model observing-system simulation experiment (OSSE) (Žagar et al., 2016). Figure 3.3 shows the spread of 12-hour forecast ensemble in the zonal wind at 3 levels in the upper troposphere. It illuminates differences between the midlatitudes and the tropics relevant for the discussion of the outputs of normal-mode decomposition. In the midlatitudes, uncertainty in the

12-hour forecasts is associated with developing baroclinic systems on the midlatitude westerly flow that is seen in the eastward propagation of the ensemble spread throughout the troposphere (Fig. 3.3c). On the contrary, the tropical spread contains both westward and eastward propagation directions as well as some stationary features. The latter are best seen at the top of the intense convection over the Indian ocean close to 100 hPa (Fig. 3.3a,b). We can also notice in Fig. 3.3 that the tropical spread has on average twice the amplitude of the midlatitude spread and that wide areas of tropical oceans are characterized by large forecast uncertainties.

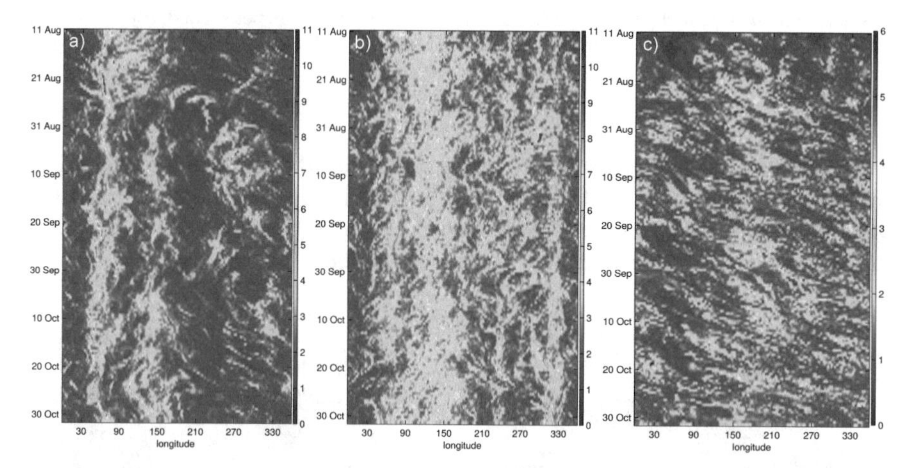

Fig. 3.3 Time evolution of the zonal wind spread in the upper troposphere in 12-hr forecast along the latitude circle in the (a,b) tropics and (c) midlatitudes. Presented are model levels (a) 10 (around 118 hPa), (b) 17 (around 369 hPa), and (c) 14 (around 227 hPa). The colourbar is in meters per second. From Žagar et al. (2016), Fig. 4. ©American Meteorological Society. Used with permission.

These properties, although simulated in a perfect-model framework of the observing-system sensitivity experiment (Žagar et al., 2016) hold also in the operational data assimilation system at ECMWF (Žagar et al., 2013). The zonally averaged, zonal wind spread in Fig. 3.4 shows that the maximum of the spread in the tropical upper troposphere is a dominant feature of both a perfect-model experiment and the state-of-the-art global NWP system which includes the model error.

The average difference between the spread of the analysis ensemble and 12-hour forecast ensemble, normalized by the spread of the forecast ensemble, is shown in Fig. 3.5. It shows that the largest information content of observations as represented by the reduction of the prior ensemble spread is in the midlatitude upper troposphere. The altitude of the maximal impact of observations is different for the zonal temperature and wind variables in agreement with the thermal-wind coupling between the mass and wind field errors. As the perfect-model experiment, which produced this result, applied a homogeneous observation network, globally constant observation errors and the perfect model, at least one-third smaller average fractional reduction of the spread in the tropics than in the midlatitudes is a consequence of data assimilation

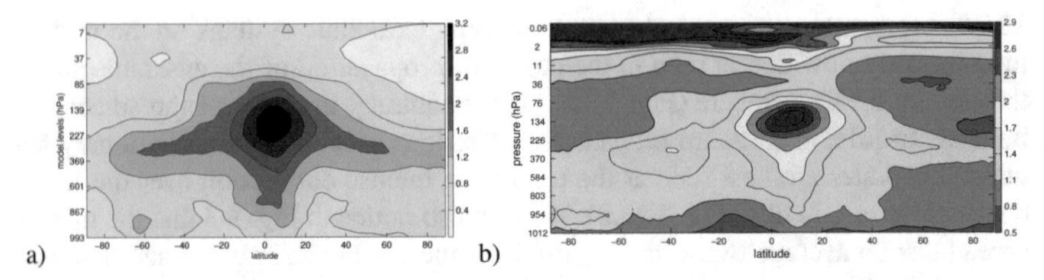

Fig. 3.4 Zonally averaged spread in zonal wind of (a) the 12-hour forecast ensemble with the perfect-model experiment, and (b) the 3-hr forecasts ensemble with the ECMWF data-assimilation ensemble based on model version cy32r3. The colorbars are in meters per second. (a) From Žagar et al. (2016), Fig. 5a. ©American Meteorological Society. (b) From Žagar et al. (2013), Fig. 2. Used with permission.

modelling and underlying dynamics. At the same time, the errors in the prior are on average twice as large in the tropics (Fig. 3.3). This is due to the large forecast error growth in the tropics associated with deep convection (e.g. Reynolds et al., 1994) and significant uncertainties in the posterior ensemble.

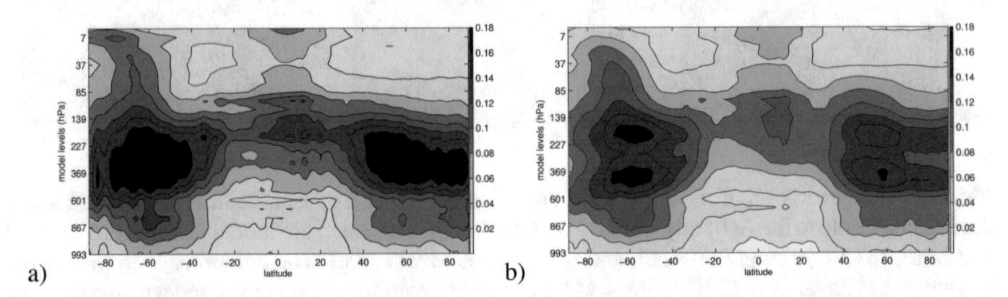

Fig. 3.5 Difference between the spread of the 12-hour forecast ensemble and analysis ensemble divided by the forecast ensemble spread for (a) zonal wind and (b) temperature. The computation is performed in every model grid point over the September-October period, then averaged in time and along the latitude circles. From Žagar et al. (2016), Fig. 6. ©American Meteorological Society. Used with permission.

3.4.6.1 Background-Error Variance Spectra

The 2D modal distribution of the vertically integrated ensemble variance Σ^2, at 12-hour forecast range is presented in Fig. 3.6 for a perfect-model OSSE of Žagar et al. (2016). It shows that the maximum of the global balanced variance is in the meridional modes $n = 1 - 3$ and several zonal wavenumbers centred around $k = 6$. This is the region of wavenumbers of the strongest wave activity in the global atmosphere associated with the baroclinic instability. The meridional mode $n = 0$ dominates the variance for the both the EIG and WIG modes and it corresponds to the Kelvin mode. However, the zonal scales of the maximal variance are $k = 1$

and $k = 4$ for the EIG and WIG modes, respectively. Different zonal scales point to different dynamical properties of the two mode types. The maximal Kelvin wave variance in $k = 1$ was also obtained in previous studies by Žagar et al. (2011) and Žagar et al. (2013) and it is related to the maximal activity of the tropical Kelvin wave in this scale.

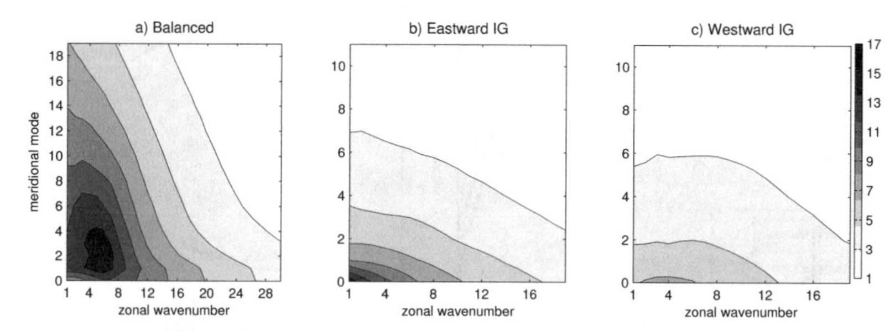

Fig. 3.6 2D distribution of the 12-hour forecast ensemble variance as a function of the zonal wavenumber and meridional mode in (a) balanced (Rossby), (b) EIG and (c) WIG modes. Values (in m^2s^{-2}) are multiplied by a factor 100. From Žagar et al. (2016), Fig. 7. ©American Meteorological Society. Used with permission.

The WIG mode variance is distributed across zonal wavenumbers 2 to 6. The relative shift in the maximal variance of forecast errors associated with the WIG modes in comparison to the EIG modes is associated with dynamical properties of forecast errors. In the tropics, the flow-dependent spread associated with the IG modes shows properties of both eastward and westward propagation, just like the convective system to which it is associated with. It can be more dominated by a certain direction depending on the background state; for example, during the active phase of the MJO oscillation, the tropical ensemble spread in the ECMWF system was dominated by the EIG modes in all scales (Žagar et al., 2013). In the midlatitudes, a part of the spread associated with the westward-propagating developing baroclinic disturbances will project to the WIG modes. There is a global dominance of the balanced spread and a prevalence of the WIG spread over the EIG spread in the midlatitudes.

If the variance in Fig. 3.6 is integrated meridionally, the resulting spectra present the distribution of the variance as a function of the zonal wavenumber (Fig. 3.7). Figure 3.7a shows that the largest forecast variance is at the largest scales. The balanced variance dominates over the IG variance except at the smallest scales. Furthermore, the EIG variance dominates over WIG variance on large scales up to around zonal wavenumber 8 (scale around 1700 km at the equator), in agreement with the idea of the large-scale tropical forecast errors being associated with the Kelvin mode. If the variance in each zonal wavenumber is split between the balanced and IG parts, the dominance of the balanced variance on planetary scales is visible more clearly (Fig. 3.7b). The ratio between the balanced and IG variance is at a nearly constant level of 70% : 30% for Rossby and IG modes for the zonal wavenumbers one to six (scales larger than 2000 km in the tropics). At the wavenumber around 70, the 12-hour forecast variance associated with IG modes exceeds the variance due to the Rossby modes.

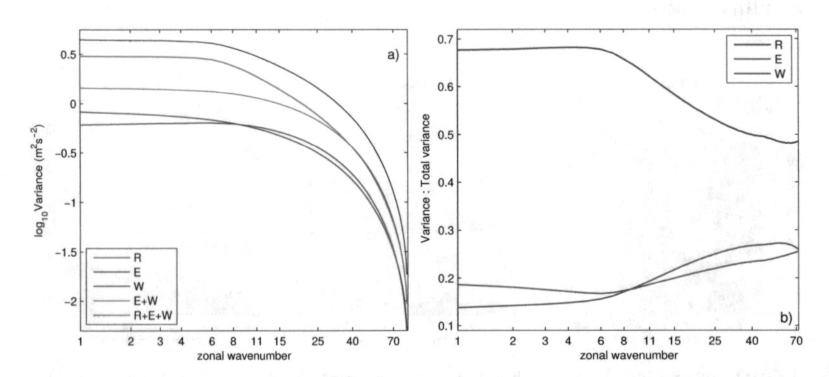

Fig. 3.7 Scale distribution of the prior ensemble variance as a function of the zonal wavenumber. (a) Variance in each zonal wavenumber integrated meridionally and vertically. (b) Division of variance among balanced and inertio-gravity modes in each zonal wavenumber. The three motion types are defined in legend as R for balanced, E for EIG and W for WIG modes. From Žagar et al. (2016), Fig. 9. ©American Meteorological Society. Used with permission.

Figure 3.7 is from an ensemble data assimilation system and perfect-model OSSE; but, the results from the ensemble of 4D-Var analyses of ECMWF, a real system with a model error, is not significantly different (Žagar et al., 2013; Cardinali et al., 2014). Based on ECMWF 4D-Var system, Fig. 3.8 shows that the variance growth between 3-hour and 12-hour forecast length is larger in balanced variance than in the IG variance and larger in EIG than in WIG modes. Among all IG modes, the greatest variance increase is in the Kelvin mode (meridional mode 0 in panel b). The total IG variance at 3-hour range dominates over the balanced variance at scales below the zonal wavenumber 6. At the 12-hour forecast range, the IG variance dominance is present only at $k = 1$. This is the consequence of the larger variance growth in the balanced modes at all but especially at the largest zonal scales. The overall similarity between the distribution of the short-range forecast-error variance in a perfect-model OSSE and the operational system of ECMWF suggests that the model error may not be the dominating factor responsible for the large-scale uncertainties in tropical analyses and short-range forecasts.

3.4.6.2 Diagnosing Model Biases

The modal framework was used to diagnose systematic deficiencies of the ECMWF analyses already in early 1980 by Cats and Wergen (1982). They applied free solutions of the linearized, multi-level ECMWF grid-point model. The results revealed

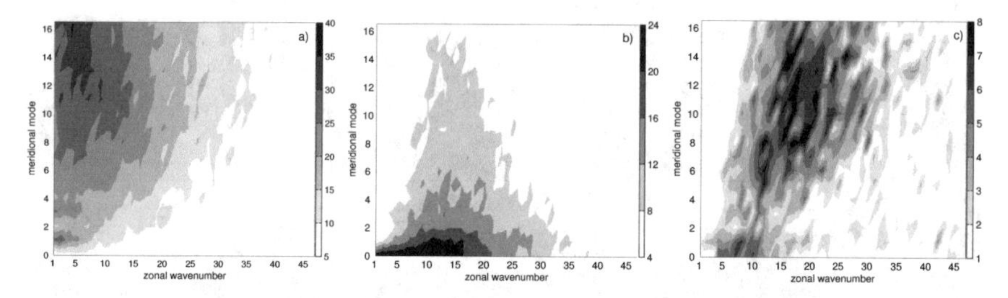

Fig. 3.8 Forecast-error variance growth in (a) balanced (b) EIG modes and (c) WIG modes as a function of the zonal wavenumber and meridional mode. The growth is expressed as a difference between the 12-hour and 3-hour variances normalized by the 3-hour variances and multiplied by 100. From Žagar et al. (2013), Fig. 5.

difficulties in the analysis of the largest balanced ($n = 1$ Rossby), Kelvin and mixed Rossby-gravity modes in the tropics at the time.

Almost four decades later, and using a much improved prediction system, the same method provides fine details of the assimilation and model errors in the tropics. For example, Žagar et al. (2010) presented systematic analysis increments that showed that the most significant tropospheric biases in the operational data assimilation systems of ECMWF and NCEP are in the tropical inertio-gravity modes. In the extra-tropics, the biases in the systems projected mostly onto the balanced motions and the winter hemisphere. A significant portion of the tropical biases projected onto the two most energetic IG modes, the Kelvin and the MRG mode.

A more recent example of systematic analysis increments in the ECMWF operational forecasts is shown in Fig. 3.9. The zonal wind analysis increments in autumn 2016 are largest at the 150 hPa level in the tropics in both balanced and unbalanced modes. This is the region with the largest initial-state uncertainties and the largest initial growth of the spread in ensemble forecasts. Lower in the troposphere, the increments are distributed more evenly and appear smoother, especially for the balanced component in the mid-latitudes. Smaller increments at the 150 hPa level in the extra-tropics are above the tropopause, where variability is significantly smaller.

In the tropics, a larger part of the analysis increments is associated with unbalanced modes than with balanced modes. Moreover, a significant part of tropospheric analysis increments projects onto unbalanced modes also in the extratropics. Overall, Fig. 3.9 confirms that tropics remains the most difficult part of data assimilation problem. The short-term forecast errors grow fastest in the tropics and a lack of direct wind observations makes it difficult to constrain circulation in the analyses (Žagar, 2017).

Furthermore, modal space diagnostics revealed how the covariance inflation affects the information content of observations in the ensemble Kalman filter assimilation (Žagar et al., 2011). The difference between the WIG and EIG variance, with the former being dominant, was shown to be associated with the inflation applied to the posterior ensemble. The short-term forecast-error variance maintained the scale distribution imposed by the inflated analyses which was produced using the

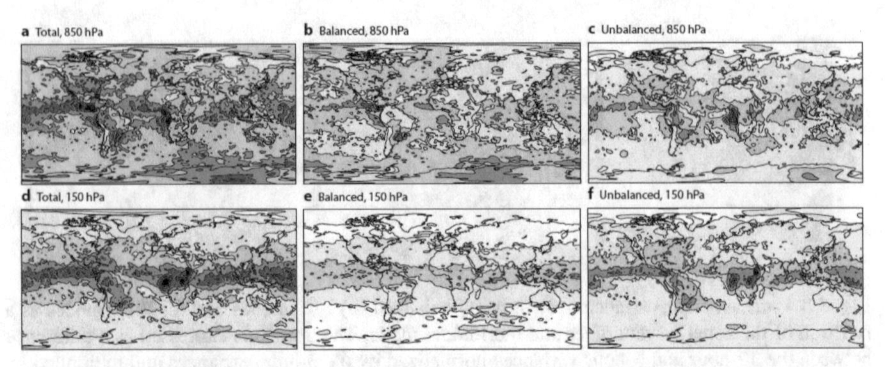

Fig. 3.9 Mean analysis increments of zonal wind from September to November 2016 showing (a) total increments at model level 114 (approx. 850 hPa), (b) balanced increments at level 114, (c) unbalanced increments at level 114, (d) total increments at level 68 (approx. 150 hPa), (e) balanced increments at level 68, and (f) unbalanced increments at level 68. Increments are computed as mean absolute differences between 12-hour forecasts and analyses valid at 18 UTC. From Žagar et al. 2016b, Fig. 6. ©ECMWF. Used with permission.

radio-sonde network and dominanted by zonal wavenumber $k = 2$. For this reason, the minimum forecast-error variance reduction occurred at $k = 2$ which was absent in the perfect-model experiment without covariance inflation.

3.4.6.3 Modal Representation of the Information Content of Observations

The information content of observations defined by Eq. (3.52) can be evaluated as the total variance reduction, $1 - \mathcal{I}_\nu$, along the zonal (k), meridional (n) and vertical (m) directions. For example, for the eastward-propagating inertio-gravity modes (EIG) the variance ratio $\mathcal{I}_\nu$ is defined as:

$$\mathcal{I}_k^{EIG} = \frac{\sum_{n=1}^{N_E}\sum_{m=1}^{M}[\gamma_\nu^A]^2}{\sum_{n=1}^{N_E}\sum_{m=1}^{M}[\gamma_\nu^B]^2} \quad \mathcal{I}_n^{EIG} = \frac{\sum_{k=0}^{K}\sum_{m=1}^{M}[\gamma_\nu^A]^2}{\sum_{k=0}^{K}\sum_{m=1}^{M}[\gamma_\nu^B]^2} \quad \mathcal{I}_m^{EIG} = \frac{\sum_{k=0}^{K}\sum_{n=1}^{N_E}[\gamma_\nu^A]^2}{\sum_{k=0}^{K}\sum_{n=1}^{N_E}[\gamma_\nu^B]^2} . \tag{3.56}$$

Equivalent expressions provide the variance reduction for the balanced and WIG modes. Similarly, the efficiency of data assimilation along a single dimension can be evaluated for each group of modes as follows:

$$\mathcal{I}_{kn}^{EIG} = \frac{\sum_{m=1}^{M}[\gamma_\nu^A]^2}{\sum_{m=1}^{M}[\gamma_\nu^B]^2} \quad \mathcal{I}_{km}^{EIG} = \frac{\sum_{n=1}^{N_E}[\gamma_\nu^A]^2}{\sum_{n=1}^{N_E}[\gamma_\nu^B]^2} \quad \mathcal{I}_{nm}^{EIG} = \frac{\sum_{k=0}^{K}[\gamma_\nu^A]^2}{\sum_{k=0}^{K}[\gamma_\nu^B]^2} . \tag{3.57}$$

Žagar et al. (2016) showed that the qualitative picture of the scale dependency of the variance reduction resulting from (3.56-3.57) is equivalent to that obtained by evaluating (3.52). For a given amount of available observations and their accuracy, the variance reduction defined by (3.56)-(3.57) can be discussed in terms of its scale dependency and dynamics.

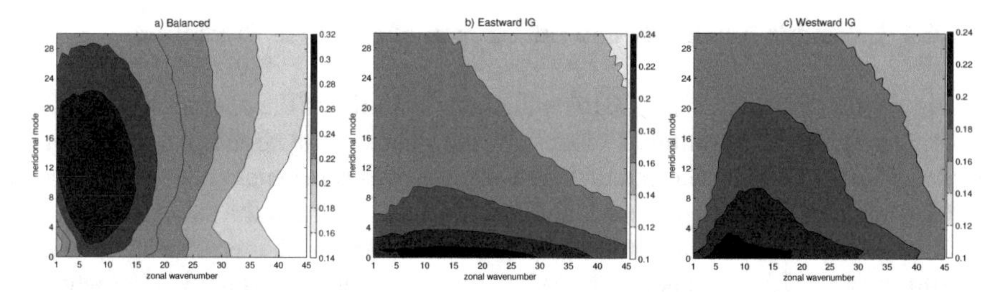

Fig. 3.10 Portion of the reduced the 12-hour forecast ensemble variance by data assimilation for (a) balanced (Rossby), (EIG), and (c) WIG modes. The reduction is shown as a function of the zonal wavenumber and meridional mode. From Žagar et al. (2016), Fig. 12 a,b,c. ©American Meteorological Society. Used with permission.

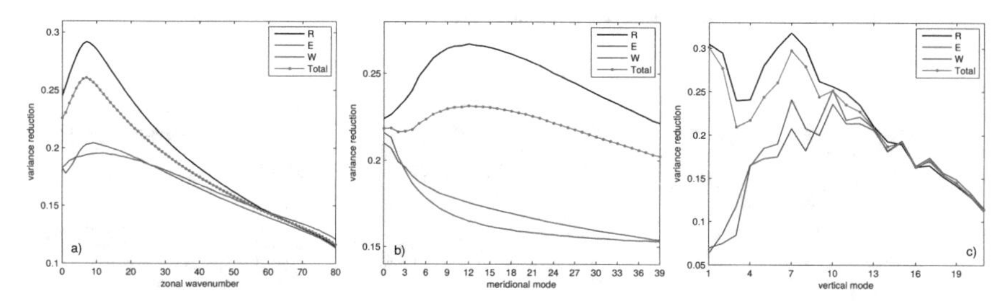

Fig. 3.11 As in Fig. 3.10, but for the variance reduction along the direction of (a) zonal wavenumber, (b) meridional mode and (c) vertical mode for the Rossby or balanced (R), EIG (E), WIG (W) and for all modes (Total=R+W+E). From Žagar et al. (2016), Fig. 13. ©American Meteorological Society. Used with permission.

Figures 3.10-3.11 show that the information content of observations is a strongly scale-dependent quantity with large differences between the horizontal and vertical scales as well as between the balanced and IG components. On average, the data assimilation is more efficient at reducing the prior variance in balanced modes than in the IG modes. This is especially well seen in Fig. 3.11 which shows the variance reduction along each direction of modal space.

The information content of observations as a function of the zonal wavenumber is maximal at the synoptic scales for the Rossby modes. The peak reduction is in the zonal wavenumber 7, with the variance reduction up to about 30% in the perfect-model experiment analyzed in Fig. 3.11. However, Žagar et al. (2016) found that the shape of the variance reduction curves is affected by the radius of the covariance localization. They showed that a localisation radius that may be suitable for the baroclinic scales in midlatitudes is too small to provide a reduction of the forecast variance at the planetary scales in the tropics. This is be particularly harmful for the IG modes.

3.4.6.4 Ensemble Reliability and the Growth of Errors in Prediction

Using Eq. (3.45), Žagar et al. (2015) evaluated reliability of the operational ECMWF ensemble forecasts in different modes for the planetary, synoptic and subsynoptic scales (Fig. 4.12). They showed that the ensemble was somewhat under-dispersive in the Rossby modes at the planetary and synoptic scales and in the subsynoptic-scale IG modes. The information about the scale distribution of the missing spread is shown in Fig. 4.12. In planetary scales, the missing spread was associated with the symmetric balanced modes with small n, especially the Kelvin modes ($n = 0$ EIG). The WIG modes in Fig. 4.12 have a similar distribution of reliability in the synoptic and subsynoptic scales as the balanced modes in agreement with previous figures and discussion. The mixed Rossby-gravity modes ($n = 0$ Rossby) was also lacking spread in the synoptic and subsynoptic scales.

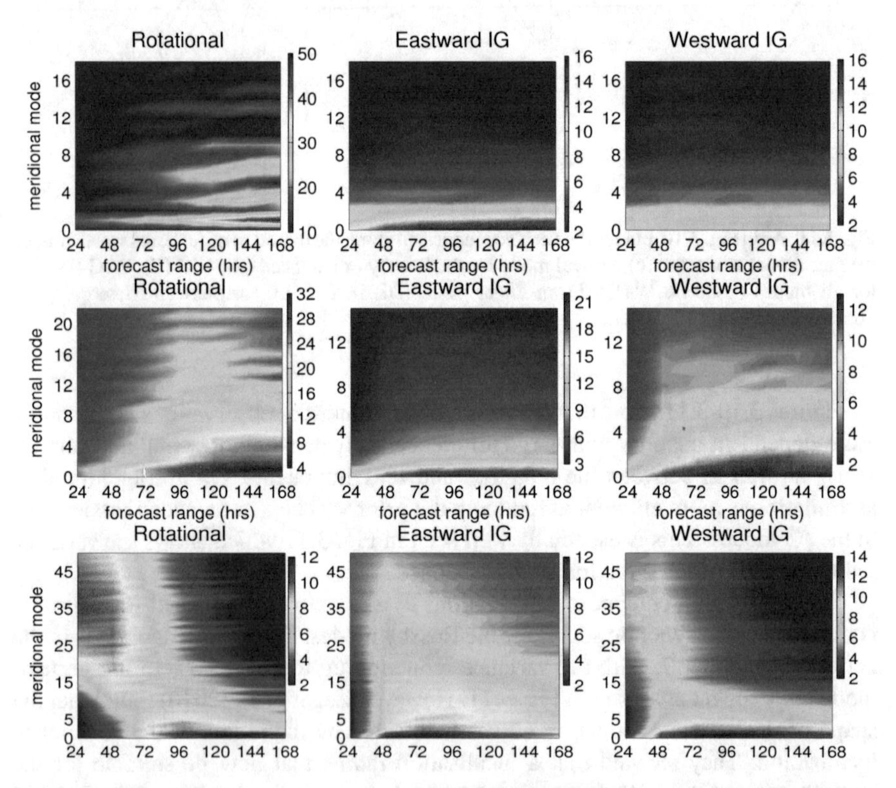

Fig. 3.12 Difference between root-mean-square difference and spread in m/s in (top row) planetary, (middle row) synoptic and (bottom row) subsynoptic scales for (left) balanced modes, (middle) EIG modes and (right) WIG modes as a function of the forecast range and meridional mode index. From Žagar et al. (2015), Fig. 14. ©American Meteorological Society. Used with permission.

However, an overall reliability of the ensemble variance in Fig. 4.12 justifies its use as a proxy for the growth of forecast errors. This is illustrated in Fig. 3.13 for the barotropic mode $m = 1$ that can be considered a modal space equivalent of the global mid-troposphere features including winds and geopotential height coupled through the linear balance equation.

For the initial state uncertainties (Fig. 3.13a), an important difference to the statistics based on the mid-latitude datasets is that a large part of the initial spread is contributed by uncertainties in the tropics that project to the lowest meridional modes. The largest uncertainties in the balanced modes are at the synoptic horizontal scales. Initially in small n (tropics), after 24 hours, the maximum of the spread is shifted to larger n (Fig. 3.13b), in agreement with a larger growth of the spread in the midlatitudes than in the tropics. In 7-day forecasts, the spread is largest in $n = 5$ that corresponds to the large forecast uncertainties in the storm-track regions of the mid-latitudes (Fig. 3.13c). There is also a shift from zonal wavenumbers $k = 6 - 8$ in initial states (Fig. 3.13a) to $k = 3 - 5$ in 7-day forecasts (Fig. 3.13c).

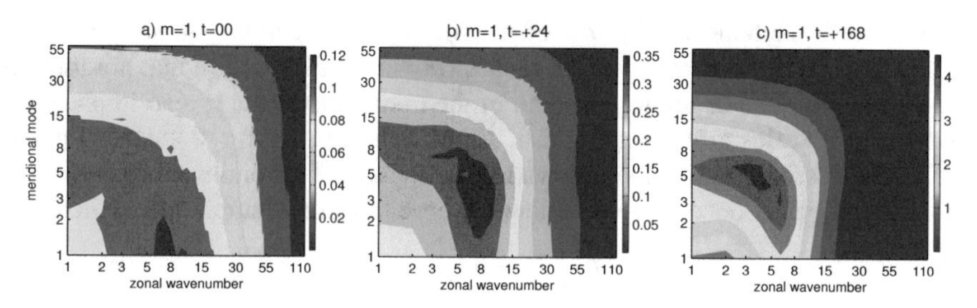

Fig. 3.13 Ensemble spread as a function of the zonal wavenumber and meridional mode for the balanced vertical mode $m = 1$ at (a) initial time, (b) 24-hour forecast, and (c) 7-day forecast. Colourbar is in meters per second. From Žagar (2017), Fig. 3.

The growth of uncertainties as a function of the zonal wavenumber is illustrated in Fig. 3.14. It shows that the spread in small scales saturates quickly while the spread in planetary and synoptic scales steadily increases. A quick saturation of the spread in small scales in the barotropic mode ($m = 1$, Fig. 3.14a) is associated with the distribution of initial perturbations in the leading vertical mode that contains most of variance in large zonal scales. In larger meridional modes, which contain larger initial perturbations in the subsynoptic scales such as mode $n = 10$ (Fig. 3.14c), the spread in the smallest scales is quickly growing in forecasts, especially in deep modes.

Figure 3.14a,b is an effective illustration of the scale-dependent amplitude of forecast uncertainties. It shows that the larger the zonal scale, the greater the amplitude of the ensemble spread. At initial time, this applies to all zonal wavenumbers and vertical modes. Later in the forecasts, the growth of spread in synoptic scales in the upper troposphere, which largely projects on the barotropic vertical mode, dominates over the growth in planetary scales (Fig. 3.14a).

Figures 3.15 presents the evolution of spread in several meridional and vertical Rossby modes. It shows that the spread grows in all scales from the start of the forecasts. The barotropic mode in Fig. 3.15,b is similar to the results obtained by

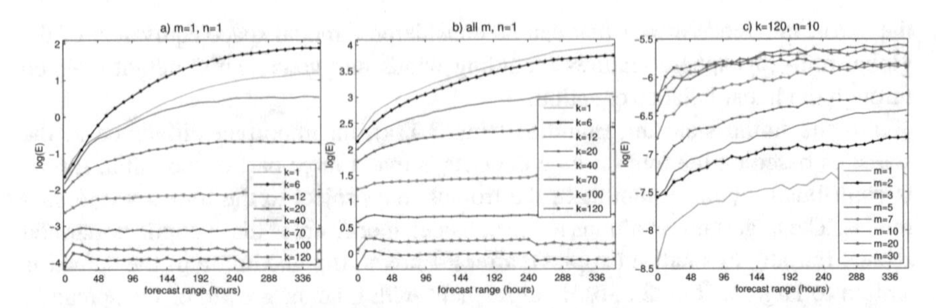

Fig. 3.14 The growth of estimated forecast errors in selected meridional modes and zonal scales as a function of the zonal wavenumber and vertical mode (as specified in the legend). (a) meridional mode $n = 1$ and vertical mode $m = 1$, (b) meridional mode $n = 1$ and all vertical modes integrated and (c) meridional mode $n = 10$ and zonal wavenumber $k = 120$. From Žagar (2017), Fig. 10.

other analysis methods usually applied to extratropical forecasts (e.g. Satterfield and Szunyogh, 2011). In higher vertical modes ($m = 4$ in Fig. 3.15c), that are initially more representative of the tropics than of the midlatitudes, the spread in planetary and synoptic scales grows equally fast. An exception is the $n = 0$ balanced mode, the mixed Rossby-gravity mode shown in Fig. 3.15a, that is dominated by the spread in zonal wavenumbers $k = 4 - 6$, known to have the largest amplitude of the MRG wave signal.

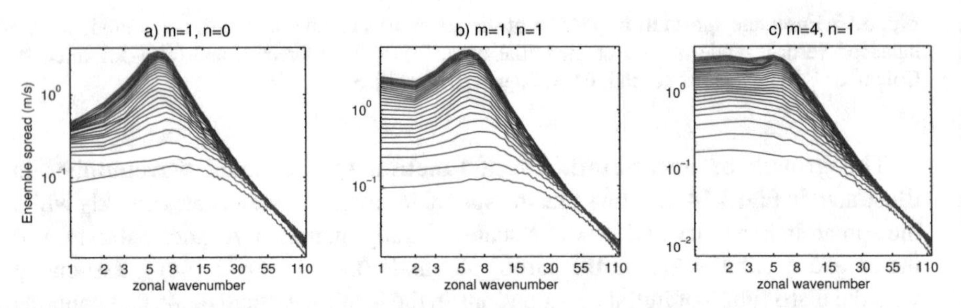

Fig. 3.15 The growth of ensemble spread in selected balanced modes as a function of the zonal wavenumber. The bottom curve in each panel shows spread at the initial time whereas every next curve applies to 12-hour longer forecasts. (a) meridional mode $n = 0$ and vertical mode $m = 1$, (b) meridional mode $n = 1$ and vertical mode $m = 1$, and (c) meridional mode $n = 1$ and vertical mode $m = 4$. From Žagar (2017), Fig. 11.

During the forecasts the percentage of the Rossby-mode spread increases as seen in Fig. 3.16 for the ECMWF ensemble data in 2015. After 15 days there is 55% and 75% of the spread in scales $k < 15$ and $k < 30$, respectively (integration of the black line in Fig. 3.16b). At this time, a crossing point between the Rossby-mode and IG spread can be seen close to zonal wavenumber 35 that corresponds to 400 km in the midlatitudes and about 600 km in the tropics (Fig. 3.16b). This scale and variance ratio approximately corresponds to the climatological variance.

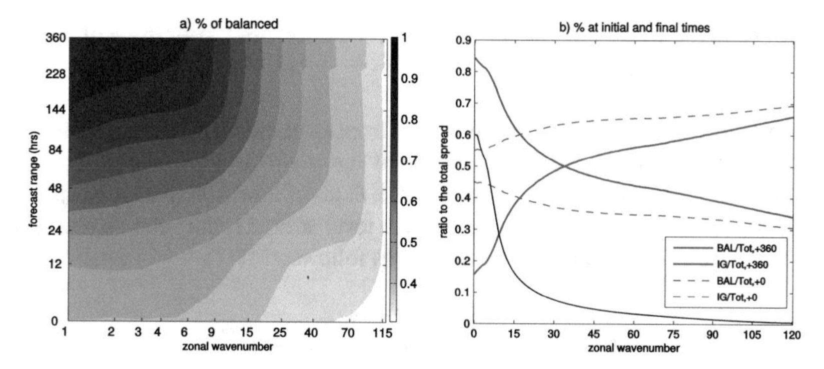

Fig. 3.16 (a) Balanced ensemble spread as a function of the zonal wavenumber normalized by the total spread at the same forecast range and zonal wavenumber, (b) Portion of the global spread at initial time and 15-day range in balanced and IG components in each zonal wavenumber. Different colour lines are specified in legend. Black line corresponds to the portion of the total spread in each zonal wavenumber at 2-week range multiplied by 10. From Žagar (2017), Fig. 14.

3.5 Tropical Data Assimilation Using the Normal Modes on the Equatorial β−Plane

Previous sections demonstrated a strong impact of the equatorial Kelvin wave and mixed Rossby-gravity wave on the global analysis and forecast error variances suggesting their major role for the mass-wind coupling in the tropics. Aspects of data assimilation in the tropics have been explored in variational data assimilation and in the Kalman filter, both based on using the normal modes on the equatorial β plane for the representation of the background error covariance matrix. Daley (1993) presented some special solution for the Kalman filter in the tropics. Žagar et al. (2004b) introduced the equatorial IG waves in the background error covariance model for tropical 4D-Var. Harlim and Majda (2013) discussed moist aspects of tropical data assimilation for various analysis methods.

3.5.1 Tropical Background Error Variance Model

The normal modes on the equatorial β−plane have been used for the representation of the background error variances by Daley (1993) and Žagar et al. (2004b). The former paper used the Kalman filter whereas the latter applied the 4D-Var framework. Prior to Daley (1993), there was "almost nothing in the literature on tropical error covariances" apart from the above discussed paper by Parrish (1988).

3.5.1.1 Representation of the Tropical Forecast Errors Using the Equatorial Eigenmodes

The equatorial eigenmodes are used for the transformation between an assimilation increment defined in physical space $\delta\mathbf{x}$ and the multivariate assimilation control variable χ. If $\delta\mathbf{x} = |\delta h\ \delta u\ \delta v|^T$ represents the fields of forecast errors in geopotential height and wind variables at a single vertical level at time t, they can be represented by a parabolic cylinder function expansion as follows:

$$\delta\mathbf{x}(\lambda, \phi, t) = \sum_{k=-N_k}^{N_k} \sum_{n=0}^{3\times N_n} \chi_n^k(t)\, \mathbf{X}_n^k(y)\, e^{ikx}, \tag{3.58}$$

where $\mathbf{X}_n^k(y)$ represents the vector with the height, zonal wind and meridional wind for the parabolic cylinder functions for the wind components and the height error variable fields

$$|h_\nu(k, y)\ u_\nu(k, y)\ v_\nu(k, y)|^T, \tag{3.59}$$

with $\nu = \nu(k, n)$ the modal index for a single equatorial eigenmode.

The parabolic cylinder functions play the role of the Hough functions on the equatorial β–plane. For example, the time-dependent solution of the linearized shallow-water equations on the equatorial β–plane for the meridional wind is the real part of

$$v(x, y, t) = D_n\left(\sqrt{\frac{\beta}{c}}y\right) e^{i(kx-\omega t)}, \tag{3.60}$$

where the parabolic cylinder functions D_n is defined as

$$D_n\left(\frac{y}{a_e}\right) = 2^{-n/2}\exp\left[-\left(\frac{y}{2a_e}\right)^2\right] H_n\left(\frac{y}{\sqrt{2}a_e}\right). \tag{3.61}$$

Here, H_n is a Hermite polynomial of degree n, ω is the wave frequency and a_e is the equatorial radius of deformation given by $a_e = (c/2\beta)^{1/2}$ where c is the phase speed of a pure gravity wave without rotation. The full expression for h_ν, u_ν and v_ν can be derived following, for example, Gill (1982). Factor 3 in (3.58) stands for the three types of equatorial eigen modes; i.e., the summation over n goes from 0 to N_n for each kind of the mode species: the equatorial Rossby (ER), equatorial eastward inertio-gravity (EEIG) and westward inertio-gravity (EWIG) modes. In comparison with the global expansion of 3D errors in Section 3.3.2, equation (3.58) applies for a single equivalent depth, i.e. for a horizontal layer. As earlier, the same expansion coefficients $\chi_\nu(t)$ are used for u, v and h because the expansion (3.58) is complete as (N_k, N_n) goes to infinity.

Of largest importance for the mass-wind coupling is the choice as what modes to be included in the application of (3.58). One way to select eigenmodes included in (3.58) could be the choice of a "critical frequency" ω_c, derived from the dispersion relation for the equatorial waves. For example, Daley (1993) used a critical frequency $\omega_c = \sqrt{\beta c/2}$ or $\omega_c = \sqrt{\beta c}$ for his Kalman filter application of (3.58). With this choice, the MRG and some Kelvin waves were included in the background covariance matrix, so that Daley (1993) confirmed earlier results of Parrish (1988) regarding the importance of these modes for the mass-wind coupling close to the equator. Figure 3.17 from his paper shows a dramatically changed coupling between the zonal wind and geopotential height when the Kelvin wave in considered among the slow modes.

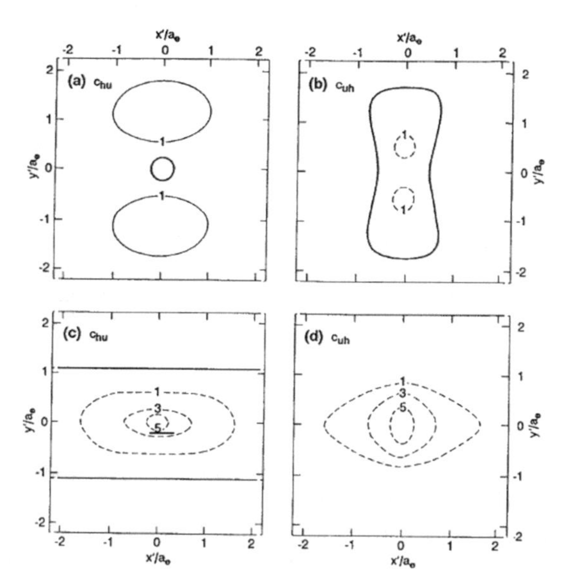

Fig. 3.17 Correlations (a) c_{hu} and (b) c_{uh} including the slow Kelvin mode, and (c) c_{hu} and (d) c_{uh} without the slow Kelvin mode. Contours are ±0.9, ±0.7, ±0.3, ±0.1 and 0.0. The heavy line is 0.0 and negative contours are dashed. Correlations are plotted as a function of distance from the equator normalised by the equatorial Rossby radius. From Daley (1993), Fig. 5. ©Taylor & Francis. Used with permission.

Žagar et al. (2004b) relaxed the frequency criterion by keeping both the eastward IG and westward IG modes for several lowest n, along with the Kelvin and MRG modes. A further realism was achieved by applying (3.58) to the simulated tropical forecast error data of ECMWF using the ensemble method. The result presented in Fig. 3.18 shows that the projection of the tropical forecast errors on equatorial normal modes can explain on average 60-70% of the error variance in the tropical free atmosphere. The largest part of the explained variance is represented by the equatorial Rossby (ER) modes, and a significant percentage pertains to the equatorial inertio-gravity (EIG) modes. The equatorial Rossby modes dominate the spectra at all scales except in the stratosphere. Here, all eastward-propagating modes have a larger variance than the Rossby waves for $k > 10$, with Kelvin waves becoming relatively more important at long scales; the latter feature is in agreement with observational and theoretical studies of the role of Kelvin waves in the equatorial stratosphere.

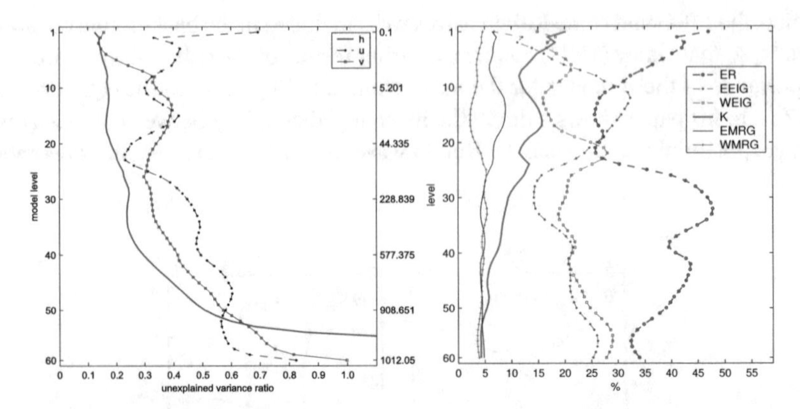

Fig. 3.18 Left: Vertical distribution of the unexplained variance ratio (area mean), representing an error of the equatorial wave approach to modelling the height and wind field errors at model levels. See text for definition. The right y-axis shows average pressure of model levels marked on the left y-axis. Right: Vertical distribution of the variance among various equatorial eigenmodes: equatorial Rossby (ER), eastward-propagating (EEIG) and westward-propagating (WEIG) equatorial inertio-gravity modes, Kelvin (K), eastward- (EMRG) and westward-propagating (WMRG) mixed Rossby-gravity waves. From Žagar et al. (2004), Fig. 2,4.

In could be noticed in Fig. 3.18 that eastward-propagating equatorial IG modes have maximum variance in the stratosphere, where the short-wave variance in westward-moving waves is particularly small. This feature was shown to be related to the phase of the quasi-biennial oscillation (QBO) in the study period Žagar et al. (2007), thus suggesting that there could be a significant temporal variation in longer-term time series of such statistics. The difference between vertical distributions of error variance in the two QBO phases is shown in the right panel of 3.18; the difference is significant only in the stratosphere and it extends all the way up to the stratopause. In the positive QBO phase there is no marked minima for the westward-propagating equatorial IG and MRG modes. Instead, the variance for these waves increases above the tropopause and the profile for westward IG modes follows that of equatorial Rossby waves until the stratopause. In the same QBO phase some differences in the stratosphere were related to the period of the descending wind shear zones in the easterly phase from which the data came. Žagar et al. (2007) discussed that the relative increase of variance in the westward IG modes in the positive phase of QBO, seen in Fig. 3.19, occurs at the expense of Kelvin, eastward MRG and eastward equatorial IG waves.

3.5.1.2 Formulation of Variational Data Assimilation in the Tropics

Žagar et al. (2004b) introduced other IG modes, in addition to the Kelvin and MRG modes, as strong constraints in the background-error covariance model for variational assimilation and systematically evaluated the role of various tropical modes in the

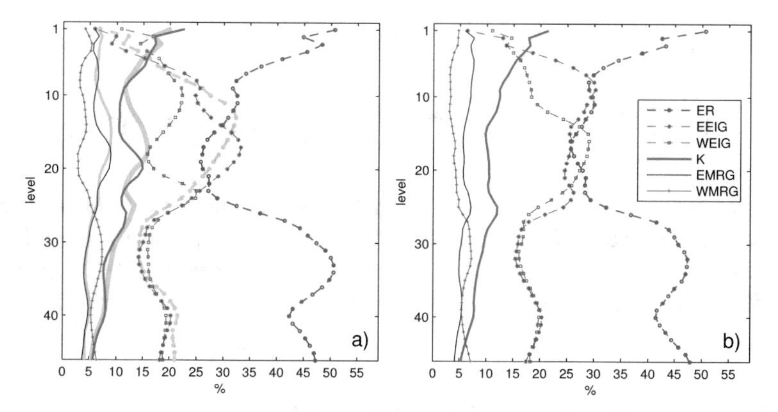

Fig. 3.19 As in Fig. 3.18b, but for datasets from the two phases of the quasi-biennial oscillation (QBO). Left: negative QBO phase, Right: positive QBO phase. Thick grey lines in a) correspond to the statistics shown in Fig. 3.18b for the same modes. From Žagar et al. (2007), Fig. 4.

mass-wind coupling near the equator. The tropical variational data assimilation was formulated using (3.58) to represent the **B** matrix. The development was performed using a non-linear shallow-water model on the equatorial β–plane, discretized in the spectral space using bi-periodic Fourier series. The transformation operator **L** between the subspace of the control variable χ and the grid-point space of the analysis increments $\delta\mathbf{x}$ is written as

$$\chi = \mathbf{L}\delta\mathbf{x}\ . \tag{3.62}$$

The transformation **L** is defined as

$$\mathbf{L} = \mathbf{D}\,\mathbf{P}_y\,\mathbf{F}_x\,\mathbf{F}^{-1}\ . \tag{3.63}$$

Here $\mathbf{F}^{-1}$ is the inverse Fourier transform to obtain assimilation increments in grid point space, and $\mathbf{F}_x$ is the direct Fourier transform in the zonal direction. The projection on the meridionally dependent part of the eigen modes in grid point space is denoted by $\mathbf{P}_y$, while **D** stands for the normalization by the spectral variance density.

It is the pseudo-inverse of the transformation (3.63) which is actually used during the assimilation, when moving from the space of the control vector χ to the model space spanned by the increment $\delta\mathbf{x}$:

$$\delta\mathbf{x} = \mathbf{L}^{-1}\tilde{\mathbf{T}}^{-1}\chi = \mathbf{F}\,\mathbf{F}_x^{-1}\,\mathbf{P}_y^{-1}\,\mathbf{D}^{-1}\,\tilde{\mathbf{T}}^{-1}\chi\ . \tag{3.64}$$

The truncation operator $\tilde{\mathbf{T}}$ summarizes in symbolic form the means of applied truncations, i.e. Fourier truncation, elliptic truncation and frequency cut-off. We call (3.64) the "pseudo-inverse" because the notation $(\mathbf{TL})^{-1}$ is not mathematically correct (i.e. **TL** is not an invertible operator).

The variational cost function $J(\delta\mathbf{x}) = J_b + J_o$ is given by

$$J(\chi) = J_b + J_o = \frac{1}{2}\chi^T\chi + \\ + \frac{1}{2}\sum_{n=1}^{K}\left(\mathbf{y} - \mathbf{H}(\mathbf{x}_b + \mathbf{L}^{-1}\tilde{\mathbf{T}}^{-1}\chi)\right)_n^T \mathbf{R}^{-1}\left(\mathbf{y} - \mathbf{H}(\mathbf{x}_b + \mathbf{L}^{-1}\tilde{\mathbf{T}}^{-1}\chi)\right)_n . \quad (3.65)$$

Here, the background-error covariance matrix **B** is replaced by an identity matrix and the pseudo-inverse $\tilde{\mathbf{T}}^{-1}$ of the truncation operator T consists of filling in zeros for truncated components. The model was used in Žagar et al. (2004a) and Žagar et al. (2005) to discuss mass-wind coupling near the equator as described next.

3.5.2 Mass-Wind Coupling Near the Equator

Assuming an analytical background error variance spectrum for the Kelvin, MRG and $n = 1, 2$ westward-propagating IG modes, Žagar et al. (2004b) demonstrated the role of equatorial IG mode on trapping the analysis increments close to the equator, in addition to their role in tropical mass-wind coupling. This is illustrated starting with Fig. 3.20 that shows analysis increments due to a single geopotential height, zonal wind and meridional wind observation at the equator. The background error variance in this case is split among the modes so that the equatorial Rossby modes contain about 50% of the total variance, the Kelvin modes, the westward MRG about 15% each, while the eastward MRG and the westward IG modes were assigned about 10% of the variance field.

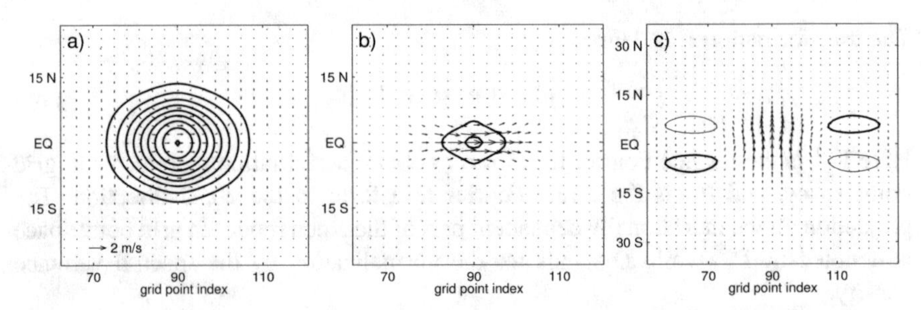

Fig. 3.20 Analysis increments for a single height observation (a) zonal wind observation (b) and a meridional wind observation (c) at the equator. Presented are the resulting assimilation increments for the height field and the wind vectors. Thick lines correspond to positive values, while thin lines are used for negative values. The equatorial Rossby, WIG, EIG modes including the MRG and Kelvin modes are included in the **B** matrix. From Žagar et al. (2004b), Fig 6.

The analysis increment for a height observation has the structure of a Kelvin wave (Fig. 3.20a). However, it is not a balanced Kelvin wave, because the balanced increment in the wind field is small. A zonal wind observation produces a height increment at the equator, which has a small positive amplitude, similar to the case of a southerly wind observation (Fig. 3.20c).

The increments based on the analytically modelled background variance spectra in Fig. 3.20 can be compared with the impact of the variance spectra derived from the ECMWF forecasts First, Fig. 3.21 shows analysis increments due to a single height observation at three vertical levels; the most striking feature is that the horizontal scale does not increase significantly with altitude. A disagreement with previous theoretical studies and global forecast-error statistics that suggested a significant increase with altitude of the horizontal correlation scale for the height field, is because of the focus of earlier studies on global, quasi-geostrophic statistics.

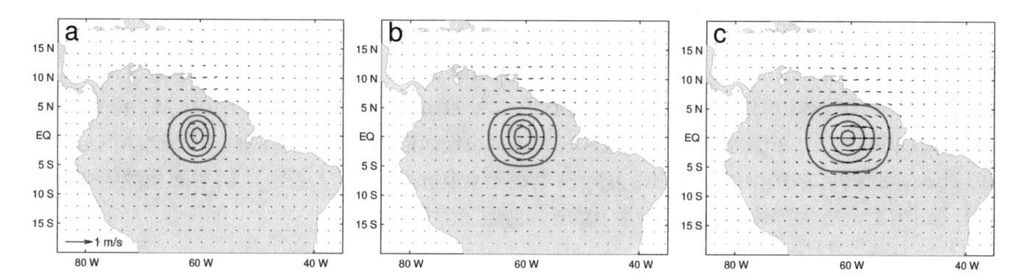

Fig. 3.21 Horizontal structure of the analysis increments due to a single northerly wind observation located (a-c) at the equator. The background error variance was derived from the ECMWF forecasts at model levels (a) approximately 654 hPa, (b) approximately 500 hPa, (c) approximately 12 hPa. From Žagar et al. (2005), Fig. 9.

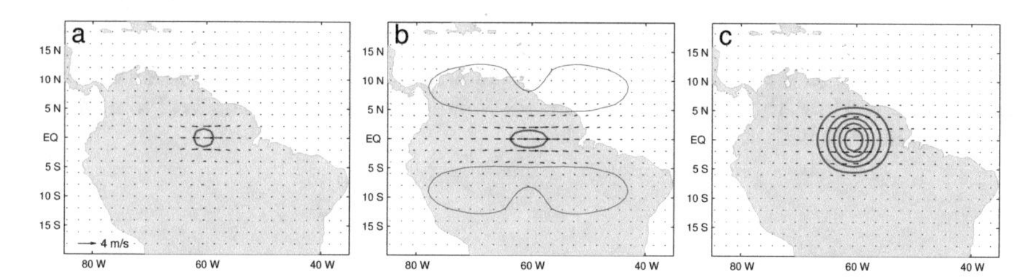

Fig. 3.22 As in Fig. 3.21, but for a single westerly wind observation. From Žagar et al. (2005), Fig. 10.

Figure 3.21 suggests that tropical mass-wind coupling, although weak, increases with altitude. The mass-wind balance is suggestive of a Kelvin wave although the fraction of the total error variance associated with Kelvin waves is comparatively small. The reason why the Kelvin-wave coupling is so effective near the equator is that it is the only mode with a significant amplitude in height at the equator coinciding with the strongest zonal wind. Away from the equator the importance of Kelvin waves diminishes, at least in the troposphere, and the equatorial Rossby type of mass-wind balance becomes more pronounced.

Contrary to what holds true for the height data, assimilation of wind observations results in increments with horizontal scales increasing with altitude (Figs. 3.22-

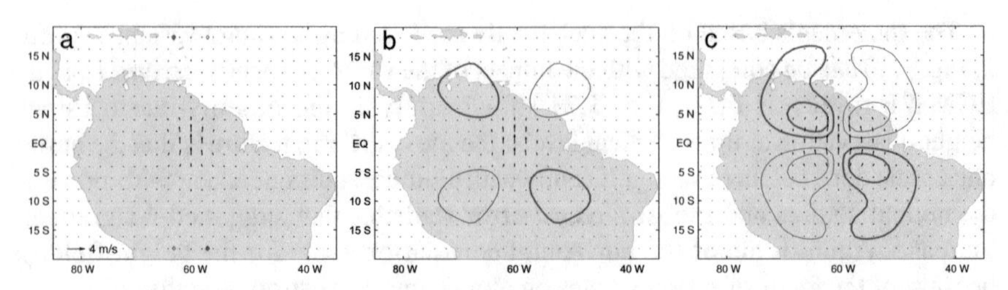

Fig. 3.23 As in Fig. 3.21, but for a single northerly wind observation. From Žagar et al. (2005), Fig. 11.

3.23). In addition, the equatorial wave balance gives rise to a coupling with the height field, especially in the stratosphere. A zonal wind observation at the equator is associated with a positive height-field increment through most of the troposphere and the weakest coupling in the upper troposphere (Fig. 3.22b). In the stratosphere, both height and wind increments have larger scales.

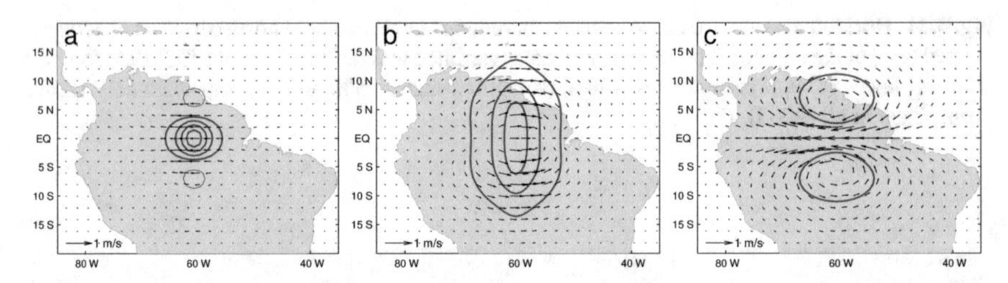

Fig. 3.24 As in Fig. 3.21, but for a height observation at model level near 500 hPa. (a) without Kelvin wave variance, (b) without IG variance, (c) without Kelvin and IG variance. From Žagar et al. (2005), Fig. 12a,b,c.

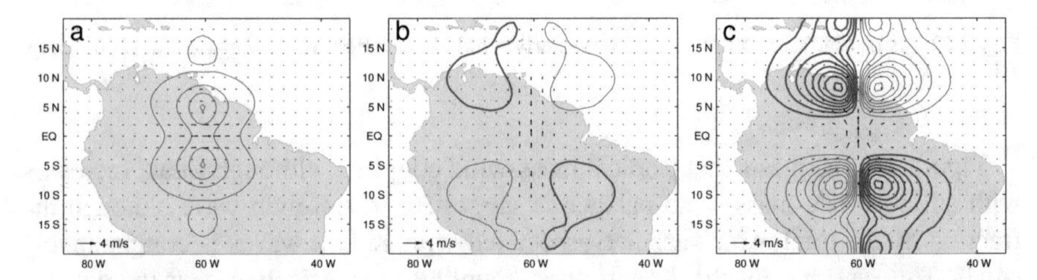

Fig. 3.25 As in Fig. 3.21, but for (a) zonal wind and (b-c) meridional wind at model level near 500 hPa. (a) without Kelvin mode variance, (b) without MRG mode variance, (c) without IG variance. From Žagar et al. (2005), Fig. 12d and Fig. 13.

The increase of the amplitude of the balanced height increments with altitude is largest for a meridional wind observation (Fig. 3.23). For a northerly wind observation centred at the equator, the shape of the increments resembles an MRG wave.

In spite of its relatively small contribution to the tropospheric variance, the Kelvin-wave coupling plays a decisive role for determining the characteristics of the horizontal correlation near the equator. An important role for the tropical mass-wind coupling is also played by the equatorial IG modes; these waves have a major impact by reducing the meridional correlation scales and the magnitudes of the balanced height-field increments. This is confirmed by sensitivity experiments shown in Figs. 3.24-3.25.

When Kelvin waves are removed from the spectrum, the increments at model level near 500 hPa (Fig. 3.24a, to be compared with Fig. 3.21b) are still centred on the equator, but the balanced winds have changed sign. A new configuration comprises 'wings' of negative correlations south and north of the equator and the whole structure is most reminiscent of an $n=1$ WEIG wave. The EEIG waves, to a similar degree present in the variance spectrum, would be associated with the opposite height-zonal wind balance at the equator. However, correlations due to equatorial Rossby modes act in the same direction as due to WEIG modes. Removal of the forecast variance associated with MRG waves does not produce dramatic changes similar to those for the Kelvin-mode variance for zonal winds. However, the equatorial IG modes have a major impact by reducing the meridional correlation scales and the magnitudes of the balanced height-field increments (Fig. 3.25b-c).

3.6 Concluding Remarks

Presented applications of normal mode functions provide understanding of balance issues of relevance for data assimilation beyond the non-linear normal-mode initialization. The normal-mode decomposition serves particularly well the tropics, where various inertio-gravity waves, each characterized by their own coupling between the mass field and the wind field, play an insufficiently well understood role in coupling between convection and horizontal motions. There is no such a problem in the midlatitudes where the quasi-geostrophic theory provides the basic framework for understanding dynamics on large scales.

We discussed how the information content of observations in data assimilation is computed in the normal-mode framework to compare uncertainty reduction in balanced and IG modes across scales. In the perfect-model framework, it was shown that data assimilation is about one-third less efficient in reducing the large-scale forecast uncertainties associated with unbalanced motions than with the balanced motions. However, this result was sensitive to the applied radius of the covariance localization in the ensemble Kalman filter data assimilation. Among the inertio-gravity modes, the Kelvin mode was initialized the best which together with its largest variance in dynamical fields, suggests its multivariate properties in the background error covariance matrix could be beneficial in NWP.

The relative impact of different equatorially-trapped wave solutions on forecast-error variance spectra in the tropics was illuminated using the variational data assimilation with a simplified model. A realism of this simplified framework was provided by the background error variance spectra from the simulated tropical forecast errors of an NWP system. It was shown that the Kelvin wave coupling plays a decisive role for determining the characteristics of the horizontal mass-zonal wind correlations near the equator, in spite of its relatively small contribution to the tropospheric variance. An important role for the tropical mass-wind coupling is also played by other equatorial IG modes; in particular, IG modes are crucial for the reduction of the meridional correlation scales and the magnitudes of the balanced height-field increments when the wind observations are assimilated. A mixture of equatorial waves contributing to the mass-wind coupling in the tropics may explain the inefficiency of data assimilation to extract information about the tropical wind field from the mass-field observations in 3D-Var; the time-averaged information about the variance of tropical waves, built into the background error covariance matrix results in analysis increments which appear nearly univariate even though they result from the advanced multivariate assimilation methodology.

A more realistic representation of tropical mass-wind couplings requires quantification of forecast-error variances in relation to flow, for example from the ensemble of 4D-Var analyses. Internal model dynamics in 4D-Var can provide significant increments in wind and temperature fields from moisture and aerosol observations in the tropics (Zaplotnik et al., 2018). Details of mass-wind couplings and coupling between the convection and circulation relevant for data assimilation can be understood by using intermediate complexity models alongside operational NWP systems, as demonstrated here in the normal-mode framework.

References

Baer, F. and Tribbia, J. (1977). On complete filtering of gravity modes through nonlinear initialization. *Mon. Wea. Rev.*, 105:1536–1539.

Brier, M. (1950). Verification of forecasts expressed in terms of probability. *Mon. Wea. Rev.*, 78:1–3.

Buizza, R. (2014). The TIGGE global, medium-range ensembles. ECMWF Research Department Technical Memorandum n. 739, pp. 53. Available from http://www.ecmwf.int/en/research/publications.

Callaghan, P., Fusco, A., Francis, G., and Salby, M. (1999). Hough spectral model for three-dimensional studies of the middle atmosphere. *J. Atmos. Sci.*, 56:1461–1480.

Candille, G. and Talagrand, O. (2005). Evaluation of probabilistic prediction systems for a scalar variable. *Q. J. R. Meteorol. Soc.*, 131:2131–2150.

Cardinali, C., Žagar, N., Radnoti, G., and Buizza, R. (2014). Representing model error in ensemble data assimilation. *Nonlin. Processes Geophys*, 21:971–985.

Cats, G. J. and Wergen, W. (1982). Analysis of large scale normal modes by the ECMWF analysis scheme. Proc. ECMWF Workshop on "Current problems in data assimilation", Reading, U.K., 8-10 November 1982, 343-372.

Courtier, P., Andersson, E., Heckley, W., Pailleux, J., Vasiljević, D., Hamrud, M., Hollingsworth, A., Rabier, F., and Fisher, M. (1998). The ECMWF implementation of three-dimensional variational assimilation (3D-Var). I: Formulation. *Q. J. R. Meteorol. Soc.*, 124:1783–1807.

Daley, R. (1991). Atmospheric data analysis. *Cambridge University Press, Cambridge, UK*, page 460 pp.

Daley, R. (1993). Atmospheric data analysis on the equatorial beta plane. *Atmos.-Ocean*, 31:421–450.

Dee, D. P. (2005). Bias and data assimilation. *Q. J. R. Meteorol. Soc.*, 131:3323–3343.

Dee, D. P. and Silva, A. D. (1986). Using Hough harmonics to validate and assess nonlinear shallow-water models. *Mon. Wea. Rev.*, 114:2191–2196.

Derber, J. C. and Bouttier, F. (1999). Formulation of the background error covariances in the ECMWF global data assimilation system. *Tellus*, 51A:195–221.

Fisher, M. (2003). Estimation of entropy reduction and degrees of freedom for signal for large variational analysis systems. ECMWF Research Department Technical Memorandum n. 39. Available from http://www.ecmwf.int/en/research/publications.

Flattery, T. W. (1970). Spectral models for global analysis and forecasting. U.S. Naval Academy, Air Weather Service Tech. Rep., 242, 42-53.

Gustafsson, N. (1992). Use of a digital filter as weak constraint in variational data assimilation. *Proc. ECMWF Workshop on "Variational assimilation with special emphasis three-dimensional aspects", Reading, U.K., 9-12 November 1992*, pages 327–338.

Hagedorn, R., Buizza, R., Hamill, T., Leutbecher, M., and Palmer, T. N. (2012). Comparing TIGGE multi-model forecasts with re-forecast calibrated ECMWF ensemble forecasts. *Q. J. R. Meteorol. Soc.*, 138:1814–1827.

Halberstam, I. and Tung, S. L. (1984). Objective analysis using Hough vectors evaluated at irregularly spaced locations. *Mon. Wea. Rev.*, 112:1804–1817.

Harlim, J. and Majda, A. J. (2013). Test models for filtering and prediction of moisture-coupled tropical waves. *Q. J. R. Meteorol. Soc*, 139:119–136.

Heckley, W., Courtier, P., Pailleux, J., and Andersson, E. (1993). The ECMWF variational analysis: General formulation and use of background information. *Proc. ECMWF Workshop on "Variational Assimilation, with Special Emphasis on Three Dimensional Aspects", Reading, 9-12 November 1992*, pages 49–93.

Kasahara, A. (1977). Numerical integration of the global barotropic primitive equations with Hough harmonic expansions. *J. Atmos. Sci.*, 34:687–701.

Kasahara, A. (1978). Further studies on a spectral model of the global barotropic primitive equations with Hough harmonic expansions. *J. Atmos. Sci.*, 35:2043–2051.

Lynch, P. and Huang, X. Y. (1992). Initialization of the HIRLAM model using a digital filter. *Mon. Wea. Rev.*, 120:1019–1034.

Machenhauer, B. (1977). On the dynamics of gravity oscillations in a shallow water model, with applications to normal mode initialization. *Contrib. Atmos. Phys.*, 50:253–271.

Parrish, D. (1988). The introduction of Hough functions into optimum interpolation. *Proc. Eighth Conf. on NWP, Am. Meteorol. Soc., Boston, Mass.*, pages 191–196.

Parrish, D. F. and Derber, J. C. (1992). The National Meteorological Center's spectral statistical-interpolation analysis system. *Mon. Wea. Rev.*, 120:1747–1763.

Phillips, N. (1986). The spatial statistics of random geostrophic modes and first-guess errors. *Tellus*, 38A:314–322.

Rabier, F., McNally, A. P., Andersson, E., Courtier, P., Undén, P., Eyre, J., Hollingsworth, A., and Bouttier, F. (1998). The ECMWF implementation of three-dimensional variational assimilation (3D-Var). II: Structure functions. *Q. J. R. Meteorol. Soc.*, 124:1809–1829.

Reynolds, C. A., Webster, P. J., and Kalnay, E. (1994). Random error growth in NMC's global forecasts. *Mon. Wea. Rev.*, 122:1281–1305.

Rogers, C. D. (2000). *Inverse Methods for Atmospheric Sounding: Theory and Practice*. Series on Atmospheric, Oceanic and Planetary Physics, World Scientific Publ., Singapore.

Saetra, O., Bidlot, J. R., Hersbach, H., and Richardson, D. (2002). Effects of observations errors on the statistics for ensemble spread and reliability. ECMWF Research Department Technical Memorandum n. 393, pp. 12. Available from http://www.ecmwf.int/en/research/publications.

Satterfield, E. and Szunyogh, I. (2011). Assessing the performance of an ensemble forecast system in predicting the magnitude and the spectrum of analysis and forecast uncertainties. *Mon. Wea. Rev.*, 139:1207–1223.

Singh, K., Sandu, A., Jardak, M., Bowman, K. W., and Lee, M. (2013). A practical method to estimate information content in the context of 4D-Var data assimilation. *SIAM/ASA Journal on Uncertainty Quantification*, 1:106–138.

Swarztrauber, P. N. and Kasahara, A. (1985). The vector harmonic analysis of Laplace tidal equations. *SIAM J. Stat. Comput.*, 6:464–491.

Tribbia, J. (1984). A simple scheme for high-order nonlinear normal mode initialization. *Mon. Wea. Rev.*, 112:278–284.

Žagar, N. (2017). A global perspective of the limits of prediction skill of NWP models. *Tellus A*, 69:1317573.

Žagar, N., Anderson, J. L., Collins, N., Hoar, T., Raeder, K., Lei, L., and Tribbia, J. (2016). Scale-dependent representation of the information content of observations in the global ensemble Kalman filter data assimilation. *Mon. Wea. Rev.*, 144:2927–2945.

Žagar, N., Andersson, E., and Fisher, M. (2005). Balanced tropical data assimilation based on a study of equatorial waves in ECMWF short-range forecast errors. *Q. J. R. Meteorol. Soc.*, 131:987–1011.

Žagar, N., Andersson, E., Fisher, M., and Untch, A. (2007). Influence of the quasi-biennial oscillation on the ECMWF model short-range forecast errors in the tropical stratosphere. *Q. J. R. Meteorol. Soc.*, 133:1843–1853.

Žagar, N., Buizza, R., and Tribbia, J. (2015). A three-dimensional multivariate modal analysis of atmospheric predictability with application to the ECMWF ensemble. *J. Atmos. Sci.*, 72:4423–4444.

Žagar, N., Gustafsson, N., and Källén, E. (2004a). Dynamical response of equatorial waves in four-dimensional variational data assimilation. *Tellus*, 56A:29–46.

Žagar, N., Gustafsson, N., and Källén, E. (2004b). Variational data assimilation in the tropics: the impact of a background error constraint. *Q. J. R. Meteorol. Soc.*, 130:103–125.

Žagar, N., Isaksen, L., Tan, D., and Tribbia, J. (2013). Balance and flow-dependency of background-error variances in the ECMWF 4D-Var ensemble. *Q. J. R. Meteorol. Soc.*, 139:1229–1238.

Žagar, N., Tribbia, J., Anderson, J. L., and Raeder, K. (2011). Balance of the background-error variances in the ensemble assimilation system DART/CAM. *Mon. Wea. Rev.*, 139:2061–2079.

Žagar, N., Tribbia, J., Anderson, J. L., Raeder, K., and Kleist, D. T. (2010). Diagnosis of systematic analysis increments by using normal modes. *Q. J. R. Meteorol. Soc.*, 136:61–76.

Wilks, D. (2011). *Statistical Methods in the Atmospheric Sciences*. Academic Press, ISBN: 978-0-12-385022-5.

Zaplotnik, v., Žagar, N., and Gustafsson, N. (2018). An intermediate-complexity model for four-dimensional variational data assimilation including moist processes. *Q. J. R. Meteorol. Soc.*, 144:1772–1787.

Chapter 4
3D Modal Variability and Energy Transformations on the Sphere

Hiroshi L. Tanaka and Nedjeljka Žagar

Abstract The three-dimensional (3D) normal-mode function (NMF) decomposition is derived in the system with pressure as the vertical coordinate followed by the applications of NMFs to atmospheric variability and energetics. Energy generation and transfers are analyzed in modal space. It is demonstrated that 3D NMF decomposition filters inertio-gravity waves even outside the tropics. Energy decomposition into the Rossby and inertio-gravity components shows that modern analysis data have the Rossby wave energy spectrum obeying a k^{-3} law, where k is the zonal wavenumber. The energy spectrum of inertio-gravity modes follows a $k^{-5/3}$ law at smaller synoptic scales which are well resolved by analyses whereas the analyses still lack variability at mesoscale. The energy spectrum of the Rossby modes obeys the 2 power of the eigenfrequency for the barotropic mode, as expected from the Rossby wave saturation theory. A clear energy peak at the spherical Rhines scale separates the nonlinear turbulent regime from the linear wave regime. Scale-dependent diagnostics of climate model biases is derived in relation to spatio-temporal variability of the models in comparison with reanalysis data.

4.1 Introduction

This Chapter derives the normal-mode functions in the system with pressure as the vertical coordinate followed by the presentations of 3D spectral energetics and energy transformations. An analytical form of the vertical structure functions and the computation of the vertical velocity are also developed for the system with pressure vertical coordinate. An overview of variability studies in modal space using the pressure system is complemented by recent results of the spatio-temporal variability and model biases using the NMFs in the sigma coordinate system derived in chapter 1.

The original studies on which much of the material presented in this chapter is based use a different notation for the zonal wavenumber, meridional and vertical mode indices than the first three chapters. In this and the following two chapters the

N. Žagar and J. Tribbia (eds.), *Modal View of Atmospheric Variability*,
Mathematics of Planet Earth 8, https://doi.org/10.1007/978-3-030-60963-4_4

notation is unified in order to present both the pressure and σ systems together: the zonal wavenumber is denoted k, the meridional mode index n and the vertical mode index m.

4.2 Derivation of the Normal-Mode Functions in Pressure System

4.2.1 Background and Motivation

Since the atmospheric energy flow was discussed by Lorenz (1955) using the concept of available potential energy, the energetical role of the atmospheric eddies has been extensively investigated. Saltzman (1957) expanded the energy equations into the wavenumber domain and showed that the kinetic energy of the cyclone-scale waves is transformed into both the planetary waves and the short waves in terms of nonlinear wave-wave interactions. The study by Saltzman was followed by Kao (1968) and Hayashi (1980) who extended this approach to the wavenumber-frequency domain making use of the two-dimensional Fourier expansion or the space-time spectral method. The energy decomposition was further pursued in the meridional wavenumber domain using spherical harmonics (Eliasen and Machenhauer, 1965) and in the vertical wavenumber domain using empirical orthogonal functions (Holmström, 1963).

Kasahara (1976), on the other hand, presented a computational scheme using Hough functions (called horizontal normal-mode functions) in the barotropic atmosphere. He applied the Hough functions as an orthonormal basis for the energy decomposition in the meridional wavenumber domain.

Since Kasahara and Puri (1981) first obtained orthonormal eigensolutions to the vertical structure equation, it became possible to expand the 3D atmospheric data into the three-dimensional harmonics of the eigensolutions. The Hough functions derived for a motionless atmosphere would be distorted by the presence of the realistic basic state (Kasahara, 1980; Salby, 1981). However, the distortion is small for the lowest-order Hough modes (Ahlquist, 1982).

The normal mode approach is useful especially for the research of tropical waves because Kelvin waves or mixed Rossby-gravity waves are obtained as the normal modes of the atmospheric oscillations. Furthermore, some oscillations in the middle latitude atmosphere, such as 5-day waves or 16-day waves, are identified with external Rossby waves of (1,1) and (1,3) modes, respectively (see Madden, 1978). According to Lindzen et al. (1984), the observed planetary scale oscillations are identified with the low-order external Rossby waves which are expected by theoretical research.

It is still uncertain, however, how these modes are created, amplified or dissipated. This problem is closely related to an amplification of planetary waves in conjunction with the blocking phenomena in the troposphere and the sudden warming in the stratosphere. Garcia and Geisler (1981) suggested that the waves are created by stochastic noise. Regularly oscillating zonal wind (Hirota, 1971) may be one of the explanations for the variations of the atmospheric normal mode. Although several researchers investigated the statistics (Kasahara, 1976) or time variations (Lindzen et al., 1984) of the normal modes by projecting the energy onto the Hough functions, the energy flow among these modes has not yet been thoroughly investigated. A brief review of NMF applications was recently given by Žagar et al. (2016), and a useful software package of normal-mode functions MODES is open to the public (Žagar et al., 2015).

In this section we develop a diagnostic energetics scheme which describes the energy spectrum and energy flow among the different normal modes in the atmosphere. Hereafter, we will call such a scheme a normal-mode energetics scheme analogous to the spectral energetics scheme by Saltzman. In order to develop the normal mode energetics scheme, we have applied the three-dimensional expansion onto normal-mode functions developed by Kasahara and Puri (1981). The scheme can provide useful information concerning the energetics for particular normal modes such as 5-day waves or 16-day waves. By summing the normal mode energetic terms within the same physical categories, for example, barotropic mode, baroclinic mode, Rossby mode or gravity mode, it is also possible to investigate the energy interactions among them.

By applying the scheme to reanalysis data and model simulations, we investigate the energy distributions and the energy interactions as functions of zonal wavenumber, vertical mode or meridional mode index. Since each normal mode is associated with a unique eigenfrequency, it is also attempted to present the energy distributions in the eigenfrequency domain.

4.2.2 NMFs in Pressure System

A set of primitive equations on the sphere with pressure as the vertical coordinate may be written as

$$\frac{\partial u}{\partial t} - 2\Omega \sin\theta v + \frac{1}{a\cos\theta}\frac{\partial \phi}{\partial \lambda} = -\mathbf{V}\cdot\nabla u - \omega\frac{\partial u}{\partial p} + \frac{\tan\theta}{a}uv + F_u, \tag{4.1}$$

$$\frac{\partial v}{\partial t} + 2\Omega \sin\theta u + \frac{1}{a}\frac{\partial \phi}{\partial \theta} = -\mathbf{V}\cdot\nabla v - \omega\frac{\partial v}{\partial p} - \frac{\tan\theta}{a}uu + F_v, \tag{4.2}$$

$$\frac{\partial}{\partial t}\left[-\left(\frac{\partial}{\partial p}\frac{p^2}{R\gamma}\frac{\partial}{\partial p}\right)\phi\right] + \nabla\cdot\mathbf{V} = \frac{\partial}{\partial p}\left[\frac{p}{\gamma}\left(-\mathbf{V}\cdot\nabla T - \omega\frac{\partial T}{\partial p}\right)\right] + \frac{\partial}{\partial p}\frac{pQ}{c_p\gamma}, \tag{4.3}$$

where

$$\gamma = \frac{RT_0}{C_p} - p\frac{dT_0}{dp}. \tag{4.4}$$

The symbols used in the equations are customary and summarized in Table 4.1. In order to obtain the energy conservation law, one term has been neglected in (4.3) assuming that the perturbation temperature, T, is negligible compared with the basic state temperature, T_0. The static stability parameter, γ, which is determined by the basic state temperature is a function of p only. The right hand sides of (4.1)-(4.3) involve nonlinear terms, frictional forces and a diabetic heat source.

By a separation of variables for the linearized version of (4.1)-(4.3), we obtain a vertical structure equation which constitutes an eigenvalue problem to obtain equivalent heights, h_m,

$$-\left(\frac{d}{dp}\frac{p^2}{R\gamma}\frac{d}{dp}\right)G_m(p) = \frac{1}{gh_m}G_m(p)\,. \tag{4.5}$$

Applied to the proper boundary conditions, Eq. (4.5) is solved by the finite difference method. The m-th vertical eigenvectors, $G_m(p)$, satisfy the orthonormal condition:

$$\frac{1}{p_s}\int_0^{p_s} G_m(p)G_j(p)\,dp = \delta_{mj}\,, \tag{4.6}$$

where the subscript j refers to a different eigenvector, and δ_{mj} denotes Kronecker delta. Global mean surface pressure, p_s, is substituted for (4.6). Refer to Kasahara and Puri (1981) and Kasahara (1984) for discrete and continuous formula to get a set of orthonormal eigenvectors. Typical values of the basic state temperature, T_0, stability parameter, γ, and equivalent height, h_m, based on the FGGE data are listed in Table 4.2.

Table 4.1 List of Symbols

t	: time (s)
λ	: longitude (rad)
θ	: latitude (rad)
p	: pressure (hPa)
u	: zonal wind (m/s)
v	: meridional wind (m/s)
ϕ	: geopotential (m^2/s^2)
T	: temperature (K)
ω	: vertical p-velocity (Pa/s)
F_u	: zonal component of frictional force
F_v	: meridional component of frictional force
Q	: diabatic heating rate
Ω	: angular speed of earth's rotation (7.29×10^{-5} rad/s)
a	: radius of the earth (6371.22 km)
g	: gravity of the earth (9.806 m/s^2)
c_p	: specific heat at constant pressure (1004 J/K/kg)
R	: specific gas constant of dry air (287.04 J/K/kg)

Hereafter, the original paper Tanaka (1985) is referred to as T1985. For example, T1985 presented the orthonormal vertical eigenvectors obtained by the finite difference scheme with 12 vertical levels. The vertical mode $m = 0$ is called a barotropic mode because the values of the mode are approximately constant and have no node in the vertical. The vertical mode $m = 1$ has one node in the vertical, $m = 2$ has two nodes and so on. The vertical mode number is defined in this section so that the mode number corresponds to the number of nodes in the vertical. The modes $m \geq 1$ are regarded as baroclinic modes. The vertical structure for the higher modes may depend on the selection of vertical levels for finite difference method.

Using the orthonormality condition (4.6), we can construct a set of vertical transforms:

$$f(p) = \sum_{m=0}^{\infty} f_m G_m(p), \tag{4.7}$$

$$f_m = \frac{1}{p_s} \int_0^{p_s} f(p) G_m(p) dp, \tag{4.8}$$

where $f(p)$ ia an arbitrary function of pressure. By applying the vertical transforms to (4.1)-(4.3), we obtain a dimensionless equation in a vector form:

$$\frac{\partial}{\partial \tau} \mathbf{W}_m + \mathbf{L}\mathbf{W}_m = \mathbf{B}_m + \mathbf{C}_m + \mathbf{D}_m, \tag{4.9}$$

where

$$\mathbf{W}_m = \begin{pmatrix} u_m \\ v_m \\ \phi_m \end{pmatrix} = \begin{pmatrix} \sqrt{gh_m} & 0 & 0 \\ 0 & \sqrt{gh_m} & 0 \\ 0 & 0 & gh_m \end{pmatrix}^{-1} \begin{pmatrix} u \\ v \\ \phi \end{pmatrix}_m, \tag{4.10}$$

$$\mathbf{B}_m = \begin{pmatrix} 2\Omega\sqrt{gh_m} & 0 & 0 \\ 0 & 2\Omega\sqrt{gh_m} & 0 \\ 0 & 0 & 2\Omega \end{pmatrix}^{-1} \begin{pmatrix} -\mathbf{V}\cdot\nabla u - \omega\frac{\partial u}{\partial p} + \frac{\tan\theta}{a} uv, \\ -\mathbf{V}\cdot\nabla v - \omega\frac{\partial v}{\partial p} - \frac{\tan\theta}{a} uu, \\ 0 \end{pmatrix}_m, \tag{4.11}$$

$$\mathbf{C}_m = \begin{pmatrix} 2\Omega\sqrt{gh_m} & 0 & 0 \\ 0 & 2\Omega\sqrt{gh_m} & 0 \\ 0 & 0 & 2\Omega \end{pmatrix}^{-1} \begin{pmatrix} 0 \\ 0 \\ \frac{\partial}{\partial p}(\frac{p}{\gamma}[-\mathbf{V}\cdot\nabla T - \omega\frac{\partial T}{\partial p}]) \end{pmatrix}_m, \tag{4.12}$$

$$\mathbf{D}_m = \begin{pmatrix} 2\Omega\sqrt{gh_m} & 0 & 0 \\ 0 & 2\Omega\sqrt{gh_m} & 0 \\ 0 & 0 & 2\Omega \end{pmatrix}^{-1} \begin{pmatrix} F_u \\ F_v \\ \frac{\partial}{\partial p}\left(\frac{pQ}{c_p\gamma}\right) \end{pmatrix}_m . \tag{4.13}$$

The subscript m denotes the m-th component of the vertical transform. After obtaining the m-th vertical component, we make the vectors dimensionless by a scaling matrix involving the equivalent height h_m and 2Ω for time. The dimensionless time is defined as $\tau = 2\Omega t$. The linear operator, $\mathbf{L}$, is given by

$$\mathbf{L} = \begin{pmatrix} 0 & -\sin\theta & \frac{\alpha_m}{\cos\theta}\frac{\partial}{\partial\lambda} \\ \sin\theta & 0 & \alpha_m\frac{\partial}{\partial\theta} \\ \frac{\alpha_m}{\cos\theta}\frac{\partial}{\partial\lambda} & \frac{\alpha_m}{\cos\theta}\frac{\partial()\cos\theta}{\partial\theta} & 0 \end{pmatrix}, \tag{4.14}$$

where the dimensionless coefficient, α_m, is called Kasahara's number and is defined as

$$\alpha_m = \frac{\sqrt{gh_m}}{2\Omega a} . \tag{4.15}$$

The linearized equation (4.9) substituted by zero for the right hand side is called a horizontal structure equation (equivalent to the Laplace's tidal equation), and the solutions are called Hough harmonics, $\mathbf{H}_{knm}$. The Hough harmonics are obtained as an eigenvalue problem with eigenfrequencies for the free waves:

$$\mathbf{L}\cdot\mathbf{H}_{knm} = i\,\sigma_{knm}\mathbf{H}_{knm} , \tag{4.16}$$

where

$$\mathbf{H}_{knm}(\lambda,\theta) = \mathbf{\Theta}_{knm}(\theta)\exp(ik\lambda), \tag{4.17}$$

and the Hough vector functions, $\mathbf{\Theta}_{knm}$, are given by

$$\Theta_{knm} = \begin{pmatrix} U \\ iV \\ Z \end{pmatrix}_{knm} (\theta) . \tag{4.18}$$

Refer to Longuet-Higgins (1968) and Kasahara (1976) for details concerning the Hough vector functions. The subscripts k and n denote zonal wavenumber and meridional mode number, respectively. The meridional mode number is defined as a sequence of the three distinct modes. The westward propagating Rossby mode is specified by n_R. The other two are westward and eastward propagating gravity modes, n_W and n_E. The Hough harmonics satisfy the orthonormal condition in the following sense:

$$\frac{1}{2\pi}\int_{-\pi/2}^{\pi/2}\int_0^{2\pi} \mathbf{H}^*_{knm}\cdot\mathbf{H}_{k'n'm}\cos\theta d\lambda d\theta = \delta_{kk'}\delta_{nn'} , \tag{4.19}$$

where the asterisk denotes a complex conjugate and the primes refer to another Hough harmonics.

Table 4.2 Global mean temperature T_0(K), static stability parameter γ(K) at pressure p(hPa), equivalent height h_m (m), and h_m^{-1}(m^{-1}) for the vertical index m. From Tanaka (1985), Table 1.

p	T_0	γ	m	h_m	h_m^{-1}
30	215.56	66.84	0	9623.9	1.0×10^{-4}
50	212.72	67.91	1	2297.1	4.4×10^{-4}
100	205.49	57.62	2	475.9	2.1×10^{-3}
150	211.44	39.55	3	272.0	3.7×10^{-3}
200	219.17	33.77	4	150.0	6.7×10^{-3}
250	225.94	28.65	5	79.5	1.3×10^{-2}
300	233.47	22.51	6	42.4	2.4×10^{-2}
400	247.63	20.23	7	26.3	3.8×10^{-2}
500	258.97	24.19	8	21.6	4.6×10^{-2}
700	274.38	34.36	9	13.4	7.5×10^{-2}
850	282.48	37.07	10	9.4	1.1×10^{-1}
1000	289.82	33.84	11	9.0	1.1×10^{-1}

Using the orthonormal condition, we can construct a set of Fourier-Hough transforms:

$$\mathbf{W}_m(\lambda, \theta, t) = \sum_{k=-\infty}^{\infty} \sum_{n=0}^{\infty} w_{knm}(t)\mathbf{H}_{knm}(\lambda, \theta), \tag{4.20}$$

$$w_{knm}(t) = \frac{1}{2\pi} \int_{-\pi/2}^{\pi/2} \int_0^{2\pi} \mathbf{H}_{knm}^* \cdot \mathbf{W}_m \cos\theta d\lambda d\theta \,. \tag{4.21}$$

By applying the Fourier-Hough transforms to (4.9), we obtain

$$\frac{d}{d\tau} w_{knm} + i\,\sigma_{knm} w_{knm} = b_{knm} + c_{knm} + d_{knm}\,, \tag{4.22}$$

where the complex variables w_{knm}, b_{knm}, c_{knm}, and d_{knm} are the Fourier-Hough transforms of the vectors of (4.10)-(4.13), respectively. According to (4.22), the time change of the complex expansion coefficient of a normal mode, w_{knm}, is caused by four terms, i.e., a linear term related with phase change of the wave, nonlinear terms due to wind field and mass field, and a diabatic process. Since the eigenfrequency σ_{knm} is always real, the linear term contributes only to the phase change of the wave, but not to the amplitude change.

On the other hand, the summation of kinetic energy, K, and available potential energy, A, is conserved provided that $F_u = F_v = 0$:

$$\frac{d}{dt}\left[\frac{1}{gS}\int_S \int_0^{p_s} E\,dp + \frac{1}{2}\frac{p_s}{RT_s}\phi_s^2\,dS\right] = 0, \tag{4.23}$$

where $E = K + A$ is total energy and

$$K = \frac{1}{2}\left(u^2 + v^2\right), \tag{4.24}$$

$$A = \frac{1}{2}\frac{p^2}{R\gamma}\left(\frac{\partial\phi}{\partial p}\right)^2, \tag{4.25}$$

$$\frac{d}{dt}\left[\frac{1}{gS}\int_S \frac{1}{2}\frac{p_s}{RT_s}\phi_s^2\, dS\right] = \frac{1}{gS}\int_S \phi_s\omega_s\, dS, \tag{4.26}$$

and the subscripts s for variables and coefficients of the basic state denote surface values. It is found that the second term in (4.23) represents the geopotential flux across the surface. By expanding the dependent variables in (4.23) into the vertical normal modes using (4.7), the equation of energy conservation is reduced to a sum of squares of dimensionless variables u_m, v_m, and ϕ_m:

$$\frac{d}{dt}\left[\sum_{m=0}^{\infty}\frac{p_s h_m}{2S}\int_S\left(u_m^2 + v_m^2 + \phi_m^2\right) dS\right] = 0. \tag{4.27}$$

Moreover, by expanding them into the Hough harmonics using (4.20), we finally obtain the equation of energy conservation in terms of a summation of energies associated with each mode:

$$\sum_{m=0}^{\infty}\sum_{n=0}^{\infty}\sum_{k=0}^{\infty}\frac{d}{dt}E_{knm} = 0, \tag{4.28}$$

where

$$E_{0nm} = \frac{1}{4}p_s h_m |w_{0nm}|^2, \tag{4.29}$$

$$E_{knm} = \frac{1}{2}p_s h_m |w_{knm}|^2. \tag{4.30}$$

The energy of a normal mode is defined as the square of the absolute value of the complex expansion coefficient, multiplied by a dimensional factor chosen so that the energy is expressed in Jm^{-2}. The kinetic energy of zonal and meridional components, K_u and K_v, and the available potential energy, A, for each mode, may be approximated by E_{knm} through multiplication by coefficients, β_u, β_v, β_z which represent energy rations of U, V, and Z of the normalized Hough vector functions:

$$\begin{pmatrix} K_u \\ K_v \\ A \end{pmatrix}_{knm} = E_{knm}\begin{pmatrix} \beta_u \\ \beta_v \\ \beta_z \end{pmatrix}_{knm} = E_{knm}\int_{-\pi/2}^{\pi/2}\begin{pmatrix} U^2 \\ V^2 \\ Z^2 \end{pmatrix}_{knm}\cos\theta\, d\theta\,. \tag{4.31}$$

In general, this separation of energy is not correct because U, V, Z are not orthogonal to one another. However, the energy separation in K and A so obtained is comparable to the physical separation by the established spectral energetics by Saltzman. This fact suggests that the present energy separation is a good approximation to the reality.

In order to obtain energy balance equations for the normal modes, Eqs. (4.29) and (4.30) are differentiated with respect to time, t. Substituting (4.22) into the time derivatives of w_{knm}, we obtain finally:

$$\frac{d}{dt}E_{knm} = B_{knm} + C_{knm} + D_{knm}, \tag{4.32}$$

where

$$B_{knm} = p_s\Omega h_m[w^*_{knm}b_{knm} + w_{knm}b^*_{knm}], \tag{4.33}$$

$$C_{knm} = p_s\Omega h_m[w^*_{knm}c_{knm} + w_{knm}c^*_{knm}], \tag{4.34}$$

$$D_{knm} = p_s\Omega h_m[w^*_{knm}d_{knm} + w_{knm}d^*_{knm}]. \tag{4.35}$$

According to (4.32), the time change of E_{knm} is caused by the three terms which appear in the right hand side of (4.32). The terms B_{knm} and C_{knm} are respectively associated with nonlinear mode-mode interaction of kinetic and available potential energies, and D_{knm} represents an energy source or sink due to the diabatic process and dissipation. The linear term in (4.22) does not appear in the energy balance equation because this term does not contribute to the time change of the magnitude of w_{knm}. Equations (4.33)-(4.35) should be multiplied by 0.5 for $k = 0$ as in (4.29).

By means of the inverse transforms of the vertical and Fourier-Hough transforms, it can be shown that the summations of all nonlinear mode-mode interactions, B_{knm} and C_{knm}, are zero because they represent global integrals of the flux convergences of the kinetic and the available potential energies, respectively:

$$\sum_{m=0}^{\infty}\sum_{n=0}^{\infty}\sum_{k=0}^{\infty} B_{knm} = \frac{1}{gS}\int_S\int_0^{p_s}\left[-\nabla\cdot K\mathbf{V} - \frac{\partial K\omega}{\partial p}\right] dpdS = 0, \tag{4.36}$$

$$\sum_{m=0}^{\infty}\sum_{n=0}^{\infty}\sum_{k=0}^{\infty} C_{knm} = \frac{1}{gS}\int_S\int_0^{p_s}\left[-\nabla\cdot A\mathbf{V} - \frac{\partial A\omega}{\partial p}\right] dpdS = 0. \tag{4.37}$$

$$\sum_{m=0}^{\infty}\sum_{n=0}^{\infty}\sum_{k=0}^{\infty} D_{knm} = \frac{1}{gS}\int_S\int_0^{p_s}\left(uF_u + vF_v + \frac{RTQ}{c_p\gamma}\right) dp + \frac{p_s\phi_s Q_s}{c_p\gamma} dS. \tag{4.38}$$

The vertical change of γ is assumed to be negligible, and the vertical geopotential flux is also assumed to be negligible for the surface integral at p_s, so as to obtain the relation (4.37). The second term in (4.23) represents the vertical geopotential flux at the lower surface. This term has been considered of the secondary importance for the global energetics analysis.

4.3 Normal Modes of the Zonally-Averaged Geostrophic Motions

4.3.1 Background

Small-amplitude motions of a thin, uniform layer of fluid over a rotating sphere are governed by Laplace's tidal equations (Lamb, 1932). Historically, eigensolutions of Laplace's tidal equations have been used to solve atmospheric tidal problems (Chapman and Lindzen, 1970). The eigensolutions of free oscillations described by Laplace's tidal equations, referred to as the normal modes, have been applied to the problem of data initialization (Errico, 1989), to the numerical integrations of the global shallow-water equations (Kasahara, 1977; Salby et al., 1990), and to the diagnosis of global atmospheric energetics (Kasahara and Puri, 1981; Tanaka, 1985; Tanaka et al., 1986; Tanaka and Kung, 1988).

The characteristics of the normal modes of Laplace's tidal equations have been discussed, for example, by Longuet-Higgins (1968). In general, there are two kinds of solutions. One, called oscillations of the first kind, consists of the gravity-inertia waves which propagate eastwards and westwards. The other, called oscillations of the second kind, consists of the westward propagating rotational waves, often referred to as Rossby-Haurwitz waves. Because Hough (1898) used spherical harmonics to solve the normal mode problem, we refer to the eigenfunctions of Laplace's tidal equations, by tradition, as Hough harmonics.

For non-zonal motions with zonal wavenumber greater than zero, Hough harmonics are discrete and orthogonal (Kasahara, 1976, 1977). The orthogonality supports the completeness, at least for a finite dimensional Hilbert space. However, the case of zonal wavenumber zero is special in that the frequencies of gravity modes (first kind) appear as pairs of positive and negative values of the same magnitudes, and the frequencies of the rotational modes (second kind) are all zero (Longuet-Higgins, 1968). Therefore, the rotational modes corresponding to zonal wavenumber $k = 0$ are not unique. It is necessary to have a complete set of the eigenfunctions of zonal rotational motions in order to expand atmospheric data in terms of a series of Hough harmonics.

In the case of $k = 0$, Laplace's tidal equations for the rotational motions degenerate to the linear balance equation for zonal flows on the sphere. A similar equation is discussed by Merilees (1966). Since the balance equation is a generalized form of the geostrophic equation, we shall refer to the Hough harmonics of the rotational modes for $k = 0$ as geostrophic modes. Kasahara (1978) constructed a set of meridional functions corresponding to the geostrophic modes by a series of Legendre polynomials, and applied the Gram-Schmidt procedure to obtain an orthonormal set. These will be referred to as the K-modes. Tribbia (1979) adopted a similar procedure to construct geostrophic modes for his study of data initialization using the equatorial beta-plane shallow-water model.

For nonzero wavenumbers, the normal modes of linearized equatorial beta-plane shallow-water system form a complete and orthogonal set (Matsuno, 1966). The eigenfrequencies of the rotational modes vanish in the case of zonal wavenumber zero and, therefore, the proof of the orthogonality of eigenfunctions fails in much the same way as the case of Laplace's tidal equations. To fulfill the need of orthogonal expansion functions for the rotational motions, Silva Dias and Schubert (1979) constructed an orthogonal set of geostrophic modes by taking the limit of the eigenfunctions of the rotational modes as the zonal wavenumber approaches to zero, and applying L'Hôpital's rule to the eigenfunctions.

Shigehisa (1983) obtained the geostrophic modes of Laplace's tidal equations as the limit of rotational modes for $k \to 0$. This is essentially the same as Silva Dias and Schubert's approach for the case of the equatorial beta-plane model. However, unlike the equatorial beta-plane model, in which the zonal wavenumber is a real number, the zonal wavenumber k for a spherical domain becomes an integer. The limit of eigensolutions of the rotational modes is calculated by considering k to be a continuous parameter and the ratio between k and the corresponding eigenfrequency σ to be finite. The latter condition ensures the phase speed $c = \sigma/k$ to be continuous with respect to k. By this approach, Shigehisa (1983) obtained orthogonal geostrophic modes, which are referred to as the S-modes. They have similar characteristics with the rotational modes for $k > 0$. In a software package developed by Swarztrauber and Kasahara (1985), the S-modes are calculated instead of the K-modes.

Since two sets of the geostrophic modes have been proposed, it is meaningful to examine the difference in the properties of these two sets. We are particularly interested in the spectral characteristics of the observed zonal mean atmospheric states in terms of the two different sets of the geostrophic modes. On one hand, it is desirable to have a series to converge faster to the observed values in atmospheric data expansion. For example, a spectral model in terms of the Hough harmonic expansion prefers fewer basis functions to represent the data. On the other hand, the spectral slope of the energy spectrum is an important information in understanding the atmospheric disturbances. In this case, the meridional expansion functions should be defined using the governing dynamical equations in a way consistent with Hough harmonics for $k > 0$. Since our knowledge on the meridional spectral characteristics of the observed zonal mean atmospheric states is insufficient, it is worthwhile to examine the characteristics of geostrophic modes as expansion basis functions of atmospheric variables.

The purpose of this section is to derive normal modes for the zonal man state and to compare the spectral characteristics of the K-modes and S-modes as described in Tanaka and Kasahara (1992) (hereafter TK1992). The comparison based on the real data is included in the next section.

4.3.2 Derivation of the Normal Modes for Zonal Wavenumber Zero

We describe briefly the derivation of the normal modes for wavenumber $k = 0$ following Swarztrauber and Kasahara (1985) (hereafter referred to as SK). A system of linearized shallow water equations in spherical coordinates of longitude λ and latitude θ for a resting basic state may be written in the matrix form of (4.16). The eigensolutions $\mathbf{H}(\lambda, \theta)$ of (4.16) are referred to as Hough harmonics of wavenumber k after multiplication by $e^{ik\lambda}$:

$$\mathbf{H}(\lambda, \theta) = \begin{bmatrix} U \\ -iV \\ Z \end{bmatrix} (\theta)\, e^{ik\lambda}, \tag{4.39}$$

where U, V, and Z represent the dimensionless longitudinal and meridional velocity components, scaled by $\sqrt{gh}$, and the dimensionless geopotential, scaled by gh, respectively. Also, σ is the dimensionless frequency, scaled by 2Ω.

In order to determine the Hough vector functions $(U, iV, Z)^T$, we assume a series solution for $\mathbf{H}_k(\lambda, \theta)$, in terms of spherical vector harmonics $y^k_{n,1}$, $y^k_{n,2}$, and $y^k_{n,3}$ with expansion coefficients A^k_n, B^k_n, and C^k_n:

$$\mathbf{H}_k(\lambda, \theta) = \sum_{n=0}^{\infty} (iA^k_n y^k_{n,1} + B^k_n y^k_{n,2} - C^k_n y^k_{n,3})\,. \tag{4.40}$$

We refer the reader to SK for the description of the spherical vector harmonics which form a complete set of vector functions defined on the sphere under a suitable inner product. Substituting (4.40) into (4.16) and collecting the expansion coefficients of the same spherical vector harmonics, we obtain

$$(\sigma + \frac{k}{n(n+1)})A^k_n = r_n C^k_n + p^k_n B^k_{n-1} + p^k_{n+1} B^k_{n+1}\,, \tag{4.41}$$

$$(\sigma + \frac{k}{n(n+1)})B^k_n = p^k_n A^k_{n-1} + p^k_{n+1} A^k_{n+1}\,, \tag{4.42}$$

$$\sigma C^k_n = r_n A^k_n\,, \tag{4.43}$$

where

$$p^k_n = \sqrt{\frac{(n-1)(n+1)(n-k)(n+k)}{n^2(2n-1)(2n+1)}}\,, \tag{4.44}$$

$$r_n = \alpha\sqrt{n(n+1)}\,. \tag{4.45}$$

We shall now discuss the solutions of (4.16) for $k = 0$. In this case, the system is reduced to

$$\sigma A_n = r_n C_n + p_n B_{n-1} + p_{n+1} B_{n+1}\,, \tag{4.46}$$

$$\sigma B_n = p_n A_{n-1} + p_{n+1} A_{n+1}\,, \tag{4.47}$$

$$\sigma C_n = r_n A_n\,, \tag{4.48}$$

where the superscript k is eliminated from the coefficients and p_n reduces to

$$p_n = \sqrt{\frac{(n-1)(n+1)}{(2n-1)(2n+1)}}\,. \tag{4.49}$$

It is known that the first-kind solutions, called gravity modes, appear as a pair of positive and negative frequencies σ and that the second-kind solutions, called rotational modes, have a frequency whose value is zero for $k = 0$. In this case, the eigenfunctions associated with the same eigenvalues are not necessarily linearly independent.

We first describe the first-kind solutions, gravity modes. By eliminating B_n and C_n from (4.47), we obtain

$$p_{n-1}p_n A_{n-2} + (r_n^2 + p_n^2 + p_{n+1}^2 - \sigma^2)A_n + p_{n+1}p_{n+2}A_{n+2} = 0\,. \tag{4.50}$$

This system can be separated into symmetric solutions with respect to the equator for n=0, 2, 4, ..., and antisymmetric solutions for n=1, 3, 5, ... Each of the two distinct tridiagonal symmetric systems constitutes an eigenvalue problem for σ^2. The eigenfrequencies are calculated as plus and minus the square root of the eigenvalues. The corresponding Hough vector functions in (4.39) appear as a complex conjugate pair. Conventionally, the meridional index may be assigned as n_E=0, 1, 2, ... for positive σ and n_W=0, 1, 2, ... for negative σ on the analogy of eastward and westward propagating nonzonal gravity modes. Note that the eigenfrequencies for $n_E = 0$ and $n_W = 0$, which correspond to the modes for $n = 0$ in (4.50), are zero and the associated normal modes are identically zero. Refer to SK for its explanation.

Next, we describe the second-kind solutions, rotational modes. The frequencies of the rotational modes are identically zero for $k = 0$. Hence, A_n becomes zero according to (4.48). This results in $\mathbf{V} = 0$, a strictly zonal flow. Also, from (4.47), we have

$$r_n C_n + p_n B_{n-1} + p_{n+1} B_{n+1} = 0\,. \tag{4.51}$$

The required relation to be satisfied by the rotational modes is a geostrophic balance between U and Z. We thus refer to the rotational modes for $k = 0$ as geostrophic modes. The meridional index may be assigned as n_R =0, 1, 2, ... by the analogy of the nonzonal rotational modes. Because the geostrophic modes are not unique, their functional forms are not discussed in, for example, Longuet-Higgins (1968). However, it is necessary to have a complete set of the geostrophic modes for the expansion of atmospheric data.

Kasahara (1978) observed that any combination of U and Z satisfying the geostrophic relation (4.51) can be a basis function of the rotational modes for $k = 0$.

One such a set of U and Z is constructed by specifying

$$B_n = 1, \quad \text{for} \quad n = l_R, \tag{4.52}$$

$$B_n = 0, \quad \text{for all other } n, \tag{4.53}$$

where l_R=1, 2, 3, ... We then calculate C_n from (4.51) as

$$C_{n-1} = -p_n/r_{n-1}, \quad \text{for} \quad C_{n+1} = -p_{n+1}/r_{n+1}, \quad \text{for} \quad n = n_R, \tag{4.54}$$

$$C_n = 0, \quad \text{for all other } n. \tag{4.55}$$

It is clear from (4.55) that the magnitude of nonzero C_n increases while the nonzero B_n remains unity as $h \to 0$. The mode corresponding to $n_R = 0$ is that both U and V are identically zero and Z is a nonzero constant. The resulting modes are not orthogonal, so they are orthogonalized using the Gram-Schmidt process. Hereafter, we refer to these as the K-modes.

Shigehisa (1983) proposed an alternative derivation of geostrophic modes by assuming that the following limits exist: $\sigma/k \to c$ and $A_n/k \to \hat{A}_n$ as $k \to 0$. The quantity c has the analogy of dimensionless phase speed. In this case, from (4.48) and (4.48), we obtain

$$[c + \frac{1}{n(n+1)}]B_n = p_n\hat{A}_{n-1} + p_{n+1}\hat{A}_{n+1}, \tag{4.56}$$

$$c\,C_n = r_n\hat{A}_n. \tag{4.57}$$

By eliminating $\hat{A}_n$ and C_n from (4.51), (4.56), and (4.57), we obtain

$$d_{n-1}e_{n-1}\hat{B}_{n-2} + [n(n+1) + e_{n-1}^2 + d_{n+1}^2 + c^{-1}]\hat{B}_n + d_{n+1}e_{n+1}\hat{B}_{n+2} = 0, \tag{4.58}$$

where we have replaced B_n by $\hat{B}_n$ as

$$\hat{B}_n = \frac{B_n}{\sqrt{n(n+1)}}, \tag{4.59}$$

and

$$d_n = \frac{(n-1)}{\alpha\sqrt{(2n-1)(2n+1)}}, \tag{4.60}$$

$$e_n = \frac{(n+2)}{\alpha\sqrt{(2n+1)(2n+3)}}. \tag{4.61}$$

In terms of the variable $\hat{B}_n$, the balance condition (4.51) is expressed by

$$C_n = -d_n\hat{B}_{n-1} - e_n\hat{B}_{n+1}. \tag{4.62}$$

Equation (4.58) can be separated into two independent systems of tridiagonal equations for $\hat{B}_n$ with the eigenvalue c^{-1} from which the normal modes are constructed. One system gives symmetric solutions with odd subscripts n=-1, 1, 3, ... The other gives antisymmetric solutions with even subscripts n= 2, 4, 6, ... For the symmetric solutions, let

$$\mathbf{S} = (-i\hat{B}_{-1}, \hat{B}_1, \hat{B}_3, ...)^T \,, \tag{4.63}$$

and

$$\mathbf{E} = \begin{bmatrix} -\alpha^{-2} & 2\alpha^{-2}/\sqrt{3} & & & \\ -2\alpha^{-2}/\sqrt{3} & 2 + e_0^2 + d_2^2 & d_2 e_2 & & \\ & d_2 e_2 & 12 + e_2^2 + d_4^2 & d_4 e_4 & \\ & & d_4 e_4 & \cdots & \end{bmatrix} \tag{4.64}$$

then (4.58) is reduced to an eigenvalue problem:

$$\mathbf{E} \cdot \mathbf{S} = -c^{-1} \mathbf{S} \,. \tag{4.65}$$

Note that the real matrix (4.64) is a similarity transform of a complex matrix originally discussed in SK. The scalar c^{-1} and vector $\mathbf{S}$ are determined as the eigenpairs of (4.65). Those eigenpairs are found to be real-valued, thus $\hat{B}_{-1}$ in (4.63) is strictly imaginary. Since d_0 is also strictly imaginary, C_0 is a real-valued coefficient. Although $\hat{B}_{-1}$ is complex-valued, $B_{-1} = 0$ because of (4.59). Also, all eigenvalues c^{-1} are found to be negative except one, which is positive for the lowest symmetric mode corresponding to $n = -1$. Therefore, we designate the lowest symmetric mode to be $n_R = -1$, as done by Shigehisa (1983). This mode has a property analogous to the eastward propagating Kelvin mode.

Once $\hat{B}_n$ is determined, B_n can be obtained using (4.59) and C_n from (4.62). The Hough vector functions $(U, -iV, Z)^T$ in (4.39) are evaluated using B_n and C_n as a series of (4.40). For the antisymmetric solutions, the procedure is similar to the symmetric solutions, but more straightforward. Refer to SK for the detail of antisymmetric solutions. We should comment that the lowest antisymmetric mode corresponds to $n = 2$, and we designate it to be $n_R = 2$. Hence, the index of the geostrophic modes runs like $n_R = -1, 1, 2, 3, ...$ Hereafter we refer to these geostrophic modes as the S-modes. The proof of the orthogonality of S-modes has been discussed by Shigehisa (1983) and SK.

Finally, the Hough vector functions so obtained must be normalized. We use the following normalization:

$$\int_{-\pi/2}^{\pi/2} \left(U^2 + V^2 + Z^2\right) \cos\theta d\theta = 1. \tag{4.66}$$

The meridional integration is performed by the Gaussian quadrature in a consistent manner with the truncation of series (4.40).

We define the components of kinetic energy and the potential energy associated with the normalized normal modes as in Kasahara (1976):

$$\begin{bmatrix} K_u \\ K_v \\ A \end{bmatrix} = \frac{1}{2} \int_{-\pi/2}^{\pi/2} \begin{bmatrix} U^2 \\ V^2 \\ Z^2 \end{bmatrix} \cos\theta d\theta \,. \tag{4.67}$$

By definition of (4.67), $K_u + K_v + A = 1/2$, and $K_v = 0$ for the geostrophic modes.

Figure 4.1 illustrates the meridional structures of symmetric K-modes and S-modes for h_m=150 m. The structures of the K-modes show large amplitudes of Z compared with those of U and have a global extent with large amplitudes in higher latitudes. The larger amplitudes in high latitudes resemble the characteristics of forced Hough modes. The basic meridional geopotential distribution of the atmosphere, having warmer tropics and colder polar regions, are effectively projected onto the symmetric $n_R = 1$ K-mode (it is denoted by l in the figure). In contrast, the structures of S-modes show comparable amplitudes of Z and U and have large amplitudes in low latitudes. The characteristics of the equatorially trapped internal modes resemble the free-mode characteristics for nonzero wavenumbers. The asymptotic form of the normal modes for $h \rightarrow 0$ resembles the free solutions of the equatorial beta-plane shallow water equations which are equatorially trapped (Longuet-Higgins, 1968).

TK1992 compared the normalized energy levels of K (= K_u) and A for the K-modes and S-modes, respectively, as functions of the inverse of equivalent height h. They showed that the K-modes have the accelerated tendency of kinetic energy and potential energy to approach the values close to 1/2 and zero, respectively, for increasing meridional index n_R at $h = 10^4$ m. This implies that the variance of U dominates over that of Z in the modal structures. As h^{-1} increases, the K/A energy ratio decreases and eventually approaches zero as $h \rightarrow 0$. At small h, such as $h = 10$ m, potential energy dominates over kinetic energy, i.e., Z dominates U in their variance. It is shown from (4.55) that K tends to zero and A tends to 1/2 as $h \rightarrow 0$. Both symmetric and antisymmetric modes show similar characteristics. The energy levels with respect to meridional index n_R do not line up in order between the symmetric and antisymmetric modes due to the application of the Gram-Schmidt procedure for orthogonalization.

In contrast, the S-modes show the tendency of concentrated kinetic and potential energy values near the top and the bottom of the figures, respectively, at $h = 10^4$ m (Fig. 1 of TK1992). As h^{-1} increases, the K/A energy ratio becomes smaller resulting from decreasing K and increasing A as in the case of K-modes. However, there is a clear turning point such that the K/A ratios approach to unity. It is shown by Shigehisa (1983) that both K and A tend to 1/4 as $h \rightarrow 0$. This tendency agrees with that of nonzonal rotational modes. The S-modes share common characteristics with the nonzonal rotational modes, but the K-modes do not have this feature. Particularly, the characteristics of S-modes are quite different from K-modes for small h.

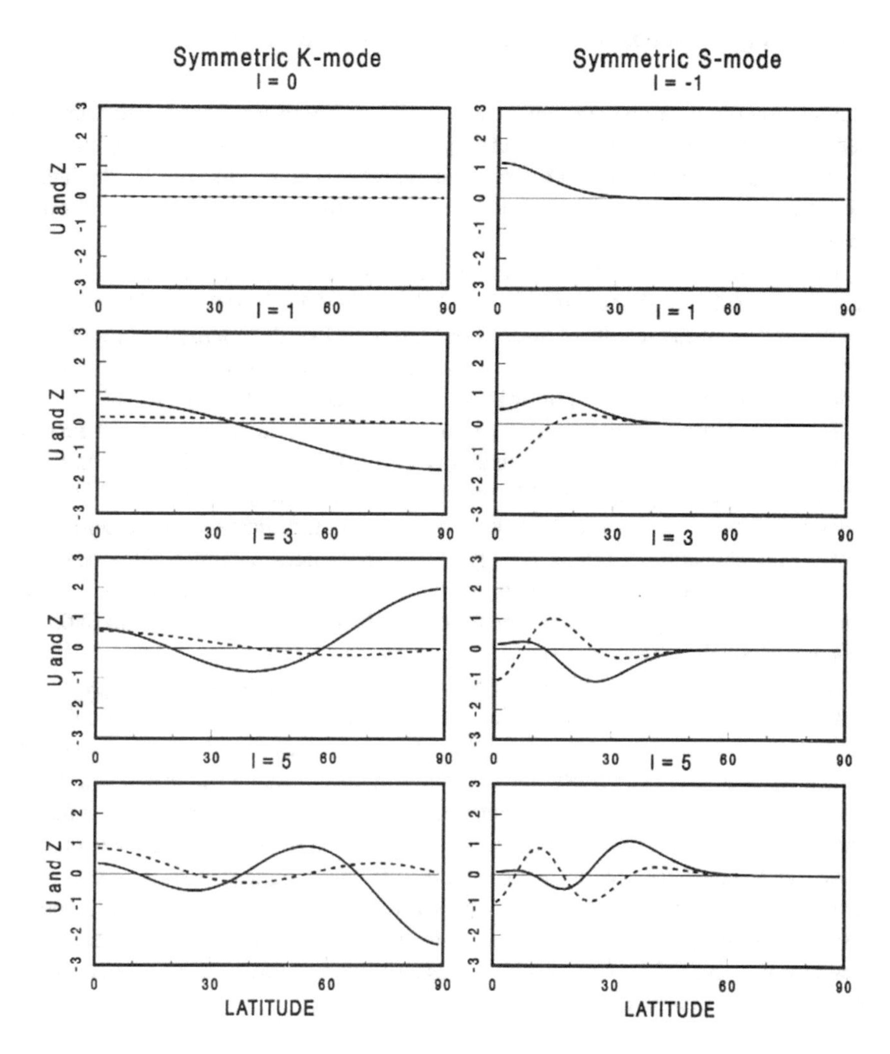

Fig. 4.1 Meridional structures of symmetric K-modes (left) and S-modes (right) for the mode with equivalent depth h_m=150 m. Solid lines denote the dimensionless geopotential Z, and dashed lines the dimensionless zonal wind U. Note that the first meridional index (denoted l_R) is zero for the K-modes -1 for the S-modes. From Tanaka and Kasahara (1992), Fig. 6.

4.4 Energy Distribution and Energy Interactions

This section provides an overview of application of NMFs to atmospheric energetics starting from early studies in pressure system followed by a more detailed analysis of energy spectra using recent operational analyses. The results support the idea of inertio-gravity waves associated with unbalanced dynamics behind a $k^{-5/3}$ power law on the mesoscale.

The shallowing of the kinetic energy power spectrum at scales below 1000 km has been the subject of much research over the years since Nastrom and Gage (1985) pointed out from the results of aircraft observations that the energy spectra of zonal wind, meridional wind, and potential temperature obey the k^{-3} power law in the synoptic scale and the $k^{-5/3}$ power law in the mesoscale, respectively.

The idea that shallower energy spectrum at mesoscale results from a forward energy cascade exists for several decades (Dewan, 1979; Van Zandt, 1982; Bartello, 1995; Lindborg and Cho, 2001; Lindborg, 2006). In particular, Dewan (1979) associated a shallower energy spectrum at mesoscale with the quasi-linear inertio-gravity waves. This idea is supported by energy spectra from high-resolution numerical simulations that diagnose inertio-gravity waves (Kitamura and Matsuda, 2010; Terasaki et al., 2011) or use divergence as a proxy of inertio-gravity waves in the midlatitudes (Waite and Snyder, 2009, 2013).

Kitamura and Matsuda (2006) investigated the characteristics of the energy spectra of the vortical and gravitational modes using their model which assumes a nonhydrostatic, incompressible Boussinesq fluid on f−plane. They found that the downscale energy cascade associated with the divergent modes plays an important role in the formation of the mesoscale $k^{-5/3}$ spectrum. Kitamura and Matsuda (2010) conducted a notable numerical experiments in relation to the mesoscale energy cascade process in the free atmosphere. According to their results, the nonlinear interactions between geostrophic and gravitational modes play a crucial role in the downscale energy cascades for forming a $k^{-5/3}$ spectrum. The modal decomposition presented in this section reinforces the inertio-gravity waves and Rossby-gravity wave interactions as the mechanism behind a $k^{-5/3}$ power law at mesoscale.

4.4.1 Expansion of Atmospheric Data in Terms of the Normal-Mode Functions

In order to expand the three-dimensional atmospheric variables of longitudinal u and meridional v wind components and deviation of the geopotential ϕ from a global mean reference state, we extend the Hough harmonics to three-dimensional normal mode functions $\mathbf{\Pi}_{knm}(\lambda, \theta, \sigma)$ by the additional tensor product with the vertical structure functions:

$$\mathbf{\Pi}_{knm}(\lambda, \theta, \sigma) = \mathbf{H}_{knm}(\lambda, \theta)\, \mathbf{G}_m(\sigma)\,. \tag{4.68}$$

Here the zonal wavenumber is denoted k, meridional mode index is n and the vertical mode is m. We can show that $\mathbf{\Pi}_{knm}(\lambda, \theta, \sigma)$ are mutually orthonormal under a natural inner product:

$$\frac{1}{2\pi}\int_{-\pi/2}^{\pi/2}\int_{0}^{2\pi}\int_{0}^{1} \mathbf{\Pi}^{*}_{knm} \cdot \mathbf{\Pi}_{k'n'm'} \, cos\theta d\sigma d\lambda d\theta$$
$$=< \mathbf{\Pi}_{knm}, \mathbf{\Pi}_{k'n'm'} >= \delta_{kk'}\,\delta_{nn'}\,\delta_{mm'}, \quad (4.69)$$

where the asterisk denotes complex conjugation, and δ is the Kronecker delta. A series of $\mathbf{\Pi}_{knm}$ forms a complete set of expansion basis functions for arbitrary vectors on the sphere. Based on this property, we expand the atmospheric variables $\mathbf{W}(\lambda, \theta, \sigma) = (u, v, \phi)^T$ by a series of $\mathbf{\Pi}_{knm}(\lambda, \theta, \sigma)$:

$$\mathbf{W}(\lambda, \theta, \sigma) = \sum_{k,n,m} w_{knm}\, \mathbf{X}_m\, \mathbf{\Pi}_{knm}(\lambda, \theta, \sigma), \quad (4.70)$$

where the scaling matrix $\mathbf{X}_m$ is defined for each vertical index m:

$$\mathbf{X}_m = diag\left(\sqrt{g h_m}, \sqrt{g h_m}, g h_m\right). \quad (4.71)$$

The expansion coefficients w_{knm} can be determined by the orthonormality condition (4.69) as follows:

$$w_{knm} =< \mathbf{X}_m^{-1}\, \mathbf{\Pi}_{knm}, \mathbf{W} > . \quad (4.72)$$

Once w_{knm} are obtained, an energy element E_{knm} in a dimensional form for a particular basis function is evaluated from

$$E_{knm} = \frac{1}{4} p_s\, h_m\, |w_{knm}|^2 \quad \text{for} \quad k = 0, \quad (4.73)$$
$$= \frac{1}{2} p_s\, h_s\, |w_{knm}|^2 \quad \text{for} \quad k > 0. \quad (4.74)$$

4.4.2 Energetics of the Zonal Mean State

For the mean zonal state ($k = 0$), TK1992 showed that the energy spectra for K- and S-modes agree well for large h_m. This suggests that the number of functions used in the K- and S-mode expansions is sufficient to represent the atmospheric data for large h_m. However, the S-mode expansion contains less energy than the K-mode expansion in small h_m. On the other hand, the S-modes share the properties of the rotational modes for nonzero zonal wavenumbers. This is not the case of the K-modes.

For the external component (m = 0), the meridional spectra of K- and S-modes are very close to each other, especially in large n_R. Nevertheless, there are minor differences in small n_R. The K-mode energy spectrum shows the maximum at n_R=1, whereas the S-mode spectrum shows the maximum at n_R=3. The gravity mode spectrum shows the energy maximum at n_R=1.

For the internal components (m >0), the energy spectra of K- and S-modes show marked differences with respect to meridional indices. For example, the second internal mode analyzed by TK1992 has the energy levels of K-modes decreasing almost monotonically as n_R increases starting from the maximum at n_R=1, while the values of S-modes do not decrease markedly until n_R reaches 10. This means that several meridional S-modes are necessary to capture the majority of energy, while only the first two K-modes are sufficient. The tendency for the data projection onto the S-modes to require many meridional functions becomes more evident for higher internal modes. The different spectral characteristics between K- and S-modes are related to difference in the energy ratio between U and Z.

Observed atmospheric zonal flows possess a large amount of available potential energy compared with kinetic energy and are nearly in geostrophic balance. Available potential energy is concentrated in a zonal internal component near h_m=150 m, whereas kinetic energy is concentrated in the zonal external component. Available potential energy dominates kinetic energy in the observed internal components (Tanaka and Kung, 1988). The K-modes have a property of large variance in Z for small h_m, because of their globally extended structures. This property is effective in describing the distribution of atmospheric available potential energy. The S-modes, in contrast, are suitable to represent evenly partitioned kinetic and potential energies for small h_k. We find that this partitioning is not observed in the atmosphere. The atmospheric zonal fields are fundamentally forced by differential heating. The S-modes, that have the property of free modes, are not effective to describe predominantly forced motions. Therefore, the K-mode series converge faster than the S-mode's to represent observed zonal fields.

Although the eigenfrequency of geostrophic modes vanishes, the S-modes have finite values of c as the limit of σ/k when both the frequency σ and the wavenumber k approach zero. Therefore, the phase velocity c can be used as an intrinsic index, in place of the meridional scale index n_R, against which the energy spectra of S-modes are plotted. In fact, Tanaka (1985) presented the energy spectra of rotational modes plotted against the phase speed $|c| = |\sigma|/k$ for k >0 and showed that the slope of the external-rotational mode energy spectra in the range of $|c| < 1.5 \times 10^{-2}$ (This dimensionless phase speed corresponds to approximately 14 m/s at the equator) follows approximately the 2-power of $|c|$. Moreover, the normal-mode energy distributions are dependent only on c and approximately independent of k.

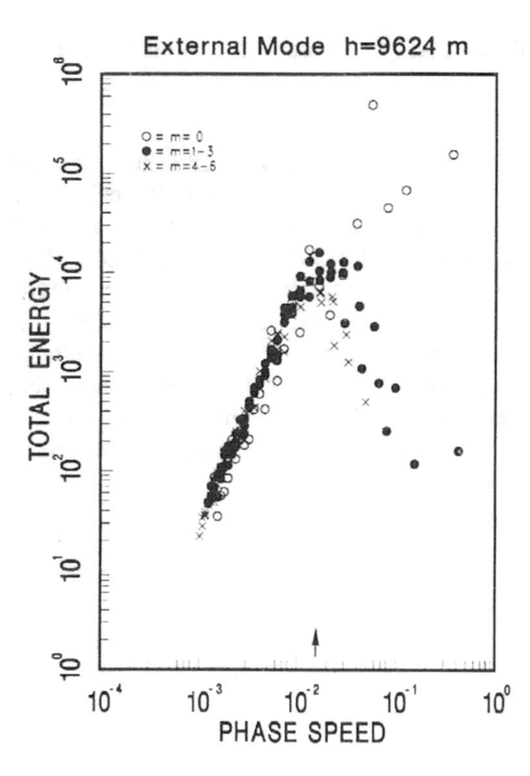

Fig. 4.2 Spectral distributions of atmospheric total energy of barotropic (or external) rotational modes as functions of the dimensionless phase speed $|c|$. Units are J m^{-2}. Circles denote for S-modes of zonal wavenumber $k = 0$, dots are for rotational modes of $k = 1 - 3$, and crosses are for those of $k = 4 - 6$. Notice that zonal wavenumber is labeled m (not k) in the figure. From Tanaka and Kasahara (1992), Fig. 7.

Since the K- and S- mode expansions basically produce similar energy spectra for the external-mode motions when plotted against n_R, it is of interest to examine the dependence of the external-mode energy spectra on the phase speed. Figure 4.2 illustrates the external-mode energy spectra for the S-modes, indicated by open circles, as a function of $|c|$. In addition, the energy spectra of the rotational disturbances for k=1 to 6 are shown by using solid dots (k=1 to 3) and crosses (k=4 to 6). We see clearly a 2-power range extending over $|c| < 1.5 \times 10^{-2}$. In this range, we see that the S-mode energy values fall right on other energy values for $k > 0$. The energy levels of S-modes increase for $|c| > 1.5 \times 10^{-2}$, whereas those of the nonzonal rotational modes decrease toward the -5/3 power regime (Tanaka and Sun, 1990). The present result demonstrates that the S-mode expansion reveals a property of the zonal atmospheric motions that is observed in the external energy spectrum of the nonzonal atmospheric disturbances.

4.4.3 Energy Spectra in Modal Space: Early Studies

Here we discuss energy distribution in terms of the zonal and meridional modes based on early results of Tanaka (1985). His analysis was based on the GFDL version of the FGGE IIIb data for 1 December 1978 though 30 November 1979 for 12 vertical levels of 1000, 850, 700, 500, 400, 300, 250, 200, 150, 100, 50, 30 hPa. These FGGE data are the same as used by Kung and Tanaka (1983, 1984) for the spectral energetics analysis of the global circulation (refer to these papers for details). Twice daily data were obtained on a 1.875°× 1.875°latitude longitude grid but interpolated to a 4°× 5°grid with 46 latitudes The Hough vector functions are truncated at 26 Rossby modes (n_R=0-25) and 12 gravity modes (n_W=0-11, n_E=0-11). Energetic terms are computed for each observation time and averaged during the data period.

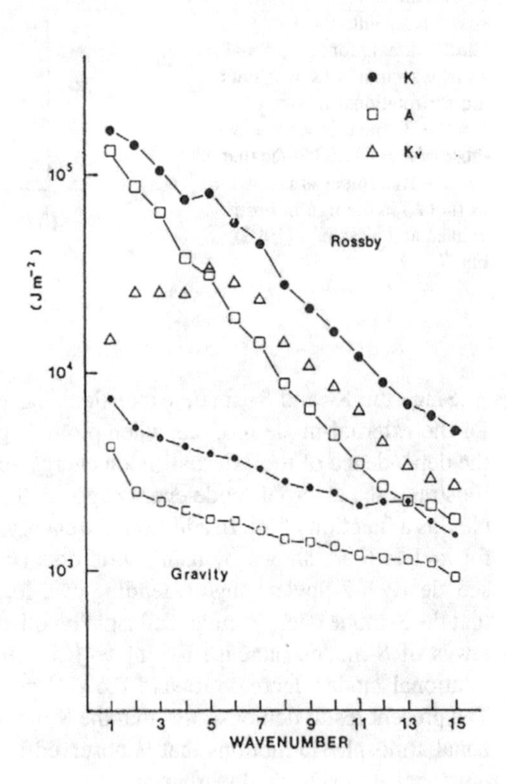

Fig. 4.3 Energy distributions in the zonal wavenumber domain. K: kinetic energy, A: available potential energy, K_v: v-component of K. From Tanaka (1985), Fig. 2.

The distributions of kinetic and available potential energies for Rossby and gravity modes are illustrated in Fig. 4.3 as a function of zonal wavenumber. The meridional components of the kinetic energy for the Rossby modes are also illustrated in the figure. As noted before, this separation is an approximation by the non-orthogonality of the variables. The kinetic energy spectrum for the Rossby modes follows approximately the -3 power law for $k \geq 7$ (Leith, 1971). The available potential energy spectrum also follows the -3 power law (Chen and Wiin-Neilsen, 1978). This range is regarded as an inertial subrange for two dimensional isotropic turbulence in the atmosphere. The kinetic energy level is lower than that expected by the -3 power law for k=1 to 6. This deflection of the energy distribution is attributable to the meridional component of the kinetic energy. The energy distributions for the gravity modes is more shallow which is associated with with the three-dimensional isotropic turbulence in the atmosphere. The kinetic energy spectra in the meridional wavenumber domain are illustrated in Fig. 4.4 for $k = 1$ through 6 and $m = 0$ and $m = 4$.

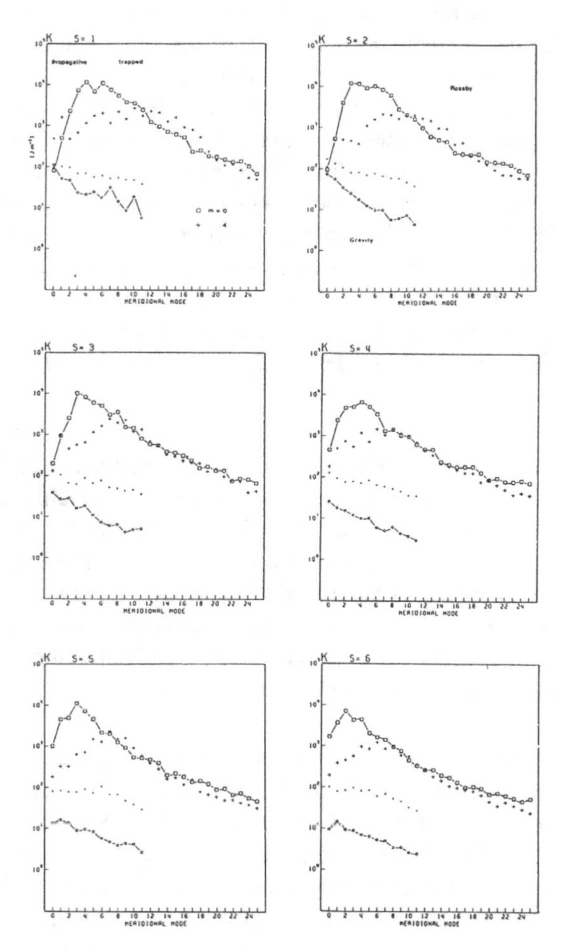

Fig. 4.4 Kinetic energy distributions in the meridional mode domain for vertical modes $m = 0$ and $m = 4$ for the wavenumber $k = 1$ through 6 (denoted s in the figure). From (Tanaka, 1985), Fig. 4.

The kinetic energy spectrum for $k = 0$ (not shown) indicates the energy peaks at the first two symmetric modes of the barotropic mode. The distribution of available potential energy for k=0 shows an energy peak at the first symmetric mode of m=4. The kinetic energy distribution for the Rossby modes of $k = 1$, $m = 0$ shows the energy peaks at $n_R = 4$ and 6. The distribution approximately follows the -3 power of the meridional mode number at the range of large meridional modes. A similar result was found by Kasahara and Puri (1981). There is an apparent cut-off of energy at the range of $n_R \leq 4$. It is discussed in Tanaka (1984) that for $k = 1$, $m = 0$, a transition of energy peaks was observed during January 1979 form $n_R = 8$ via 6 to 4. The wave energy started to propagate vertically when the energy peak reached $n_R = 4$ or 3 which is the critical meridional scale for the vertical propagation (Dickinson, 1968). It is found by the intermediate results of the present study that the energy of $m = 0$ is transformed to $m = 1$ while the vertical propagation occurred. From this results, the meridional index with the energy peak is considered as the critical meridional scale for the vertical propagation of wave energy. The peaks also correspond to the scale where westward propagating Rossby mode becomes stationary within the westerly jet. The range where the -3 power law is applicable is regarded as pertaining to trapped mode. Conversely the range of smallest meridional index represents the propagative mode. The energy peaks are seen at $n_R = 3$ for $k = 2, 3, 5$ and at $n_R = 2$ for $k = 6$. The kinetic energy spectra for $k = 4$ show energy peaks at $n_R = 6, 7$ for $k = 5$ and 6. These energy peaks are associated with the characteristic meridional scale of the cyclone-scale waves. The energy peaks for $m = 4$ are flattened in $k = 1$ or $k = 2$.

Synthesizing the energies with respect to all the zonal wavenumbers and meridional indices for Rossby and gravity modes separately, we obtain the energy spectra in the vertical wavenumber domain. Fig. 4.5 illustrates the energy distributions of eddy kinetic and eddy available potential energies ($k = 1 - 15$) for Rossby and gravity modes, respectively. A large amount of kinetic energy of the Rossby mode is in the barotropic mode ($m = 0$) with another peak at $m = 4$. These peaks in the spectra have been associated with the shapes of the eigenvectors for these vertical modes in the upper troposphere near the jets maxima.

A Hough function is associated with an eigenfrequency which is determined by the horizontal scale of the wave. Using the eigenfrequency as a coordinate, we can investigate the energy spectra in the frequency domain. In Fig. 4.6 the kinetic energy distribution of barotropic mode is plotted as a function of dimensionless frequency σ. The westward propagating Rossby modes (large symbols) and gravity modes (small symbols) are plotted in the left half of the figure, whereas the small symbols in the right half indicate the eastward propagating gravity modes. Since the energy levels of the gravity modes are low and are accommodated with high frequencies, the energy distributions for the gravity modes are positioned in the lower left and lower right corners of the figure. The distributions show clear energy peaks at the frequency (period) σ=0.03 (16 day) for $k = 1$, and $\sigma = 0.07$ (7day) for $k = 6$. The

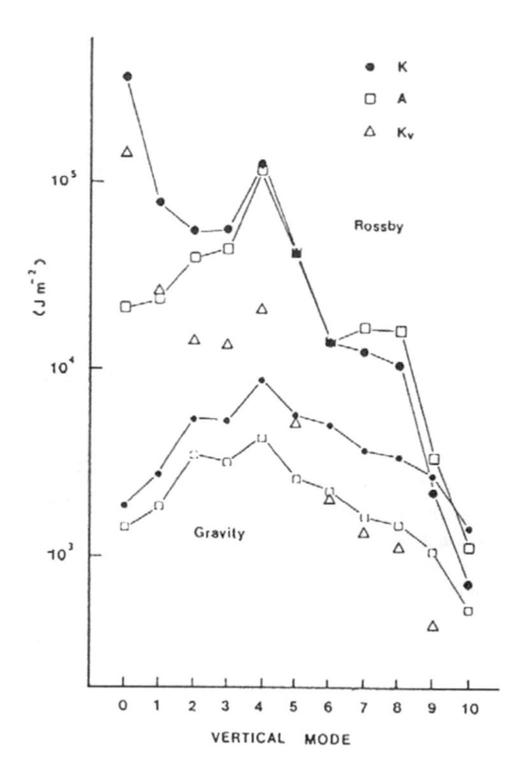

Fig. 4.5 Energy distributions in the vertical mode domain. From Tanaka (1985), Fig. 3.

energy peak is a function of the wavenumber. These energy maxima correspond to those in the meridional mode domain (Fig. 4.4).

The frequency for which the energy peak is observed, corresponds to the critical meridional scale for the vertical propagation where the westward propagating Rossby wave becomes stationary within the westerly jet. The energy spectra follow approximately the 2 power of the frequency at the low frequency range. On the other hand, the spectrum follows approximately the -5/3 power at the high frequency range of the gravity modes. As is seen in Fig. 4.3 and 4.4, the energy spectra of the gravity modes seem to follow the -5/3 power of the zonal wavenumber and meridional index. Because the phase velocity of the gravity mode is a function of the depth of the fluid (equivalent height), and taking into account the relation $c = \sigma/k$, we thus have an analogy with the -5/3 power law for the frequency domain. By using the dispersion relationship of the Rossby-Haurwitz waves, the same argument gives us an analogy of the 2 power law for the low frequency range. The most interesting features of this result are that the energy distribution of the largest-scale Rossby modes not only seems to follow the -5/3 power law, but also merges continuously with the distribution of the gravity mode. The mixed Rossby-gravity modes are positioned

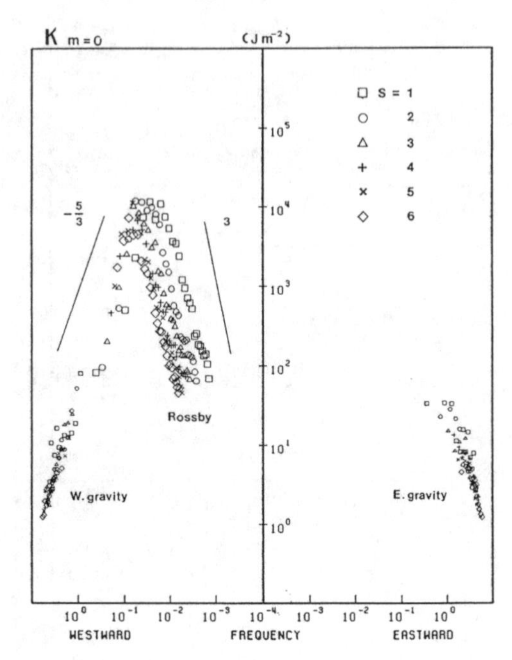

Fig. 4.6 Kinetic energy distributions in the dimensionless frequency domain for barotropic mode ($m = 0$). Zonal wavenumber k is in the figure denoted s. From Tanaka (1985), Fig. 5.

between the two types of modes. The frequency in the abscissa is determined by the theory for free waves in the motionless atmosphere. Nevertheless, the energy peaks in the frequency domain may be significantly related to forced and free stationary waves in the westerlies. Supposing that the energy peaks correspond to the stationary waves, the spectral distribution turns out to be similar to that found in previous research concerning the space-time spectra of progressive and retrogressive waves (e.g. Hayashi, 1982).

T1985 showed also the energy distributions for Kelvin modes and mixed Rossby-gravity modes. Most of the Kelvin wave energy was included in the planetary waves whereas the mixed Rossby-gravity modes indicate an energy peak in the cyclone-scale waves. The energy distributions as a function of vertical mode obtained by a summation over all the wavenumbers showed that most of the energy is included in the barotropic mode for the mixed Rossby-gravity modes whereas the Kelvin modes contained energy maximum in the range of $m = 2-4$ and the very low energy level in the barotropic mode.

4.4.4 Zonal Energy Spectra in Modern Analyses

In recent years, the global analyses have reached the resolution at which the IG waves across many scales are being resolved. Resolved IG waves may be contributed both by increased model resolution and improved parameterizations that provide more realistic (background) forecasts (e.g. Bechtold et al., 2008) as well as by advances in data assimilation methods and an increase in assimilated satellite data. Referring to Chapter 1 for the derivation of normal modes and modal space energetics in σ system, we apply the NMF decomposition to ECMWF (re)analyses and present their total energy spectra, the sum of kinetic and available potential energies, as a function of zonal scale and motion type.

The partition of total energy into the kinetic and available potential energies is calculated from the complex coefficients $\chi_n^k(m)$ as the following scalar product, denoted $I_n^k(m)$:

$$I_n^k(m) = \frac{1}{2} g D_m \, \chi_n^k(m) \left[\chi_n^k(m)\right]^* , \tag{4.75}$$

where $\left[\chi_n^k\right]^*$ is the complex conjugate χ_n^k. Depending on the value of the index n, (4.75) refers to the energy component of the Rossby or IG spectrum. The global energy product of the mth vertical mode is defined as

$$I_m = I_{m,R} + I_{m,IG} = \frac{1}{2} g D_m \sum_{n=1}^{R} \sum_{k=-K}^{K} \chi_n^k(m) \left[\chi_n^k(m)\right]^* . \tag{4.76}$$

The summation over n includes both Rossby and IG waves and the two components are obtained by splitting the summation between $n = 1$ to $n = N_R$ and $n = 1$ to $n = N_G$. In Chapter 1 it was shown that that the scalar product defined by (4.76) is equal

$$I_m = \frac{1}{2} g D_m \int_0^{2\pi} \int_{-1}^{1} \left(\tilde{u}_m^2 + \tilde{v}_m^2 + \tilde{h}_m^2\right) \mathrm{d}\lambda \mathrm{d}\mu = \int_0^{2\pi} \int_{-1}^{1} (K_m + P_m) \, \mathrm{d}\mu \mathrm{d}\lambda , \tag{4.77}$$

where

$$K_m = \frac{1}{2} \left(u_m^2 + v_m^2\right) \quad \text{and} \quad P_m = \frac{1}{2} \frac{g}{D_m} h_m^2 \tag{4.78}$$

denote the specific kinetic energy and available potential energy, respectively, of the mth vertical mode. Likewise, the energy spectrum with respect to the kth zonal wavenumber can be calculated as

$$I_k = I_{k,R} + I_{k,IG} = \frac{1}{2}\sum_{m=1}^{M} gD_m \sum_{n=1}^{R} \chi_n^k(m)\left[\chi_n^k(m)\right]^* . \quad (4.79)$$

The 2D horizontal energy spectra can be computed for a vertical mode m or as a vertically integrated quantity representing the average horizontal energy distribution as

$$I_n^k = \frac{1}{2}\sum_{m=1}^{M} gD_m\, \chi_n^k(m)\left[\chi_n^k(m)\right]^* . \quad (4.80)$$

The maximal number of vertical modes is denoted by M and it can be smaller or equal to the number of levels with data.

4.4.4.1 Filtering Inertio-Gravity Waves: A Case Study

It is worth presenting a case study of inertio-gravity waves in midlatitudes to demonstrate that NMF derived using the reference state at rest can be filtered. The case study was published in Žagar et al. (2017) and the event took place in July 2014, the period of the DEEPWAVE experiment that took place over New Zealand (Fritts and Coauthors, 2016). DEEPWAVE measured the IG waves in the region and collected a large observation dataset that can be used to estimate the gravity wave fluxes (Fritts and Coauthors, 2016). We demonstrate how our decomposition method can be used for the validation of the global weather and climate models with such data.

We present the IG wave packet in the ECMWF analyses 4 July. Figures 4.7-4.8 show the IG wave packet associated with the westerly jet south of Australia in terms of the temperature perturbations superposed on the background wind. The IG temperature perturbations are obtained by filtering the gravity modes with the cut-off scale $K_1 = 15$ (around 950 km at 45°N) that has usually been used as a cut-off scale between the synoptic range and mesoscale range. The highest wavenumber is $K_2 = 300$. First, Fig. 4.7 compares the wave features in the forecasts and analyses at 3 time steps on 5 July 2015. The IG wave packet is seen propagating nearly zonally along 40°S. Such IG packets are known to be generated by a strong lateral shear of the jet. In Fig. 4.8, we show the same fields but in the vertical cross-sections along the latitude 40°S every 6 hours on 5 July. The comparison of the four consecutive analysis times shows a gradual increase of the jet strength close to 200 hPa along with its eastward movement. At the leading edge of the jet the vertically tilted phase lines of the IG waves are clearly visible with the wave amplitude increasing during the day. The wave is propagating upward with its amplitude increasing. During the day the core of the jet has moved some 15-20 degrees eastward with an average speed of around 20 m/s.

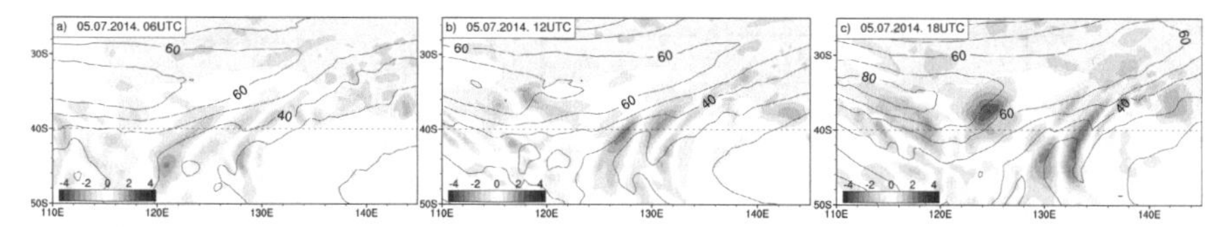

Fig. 4.7 Temperature perturbations associated with IG waves and the horizontal background wind speed at 200 hPa on 5 July 2014, at 06, 12 and 18 UTC in ECMWF analyses, The wind speed is shown with contours (in m/s) and the temperature perturbations with colours (the colour scale is in Kelvins). From Žagar et al. (2017). ©American Meteorological Society. Used with permission.

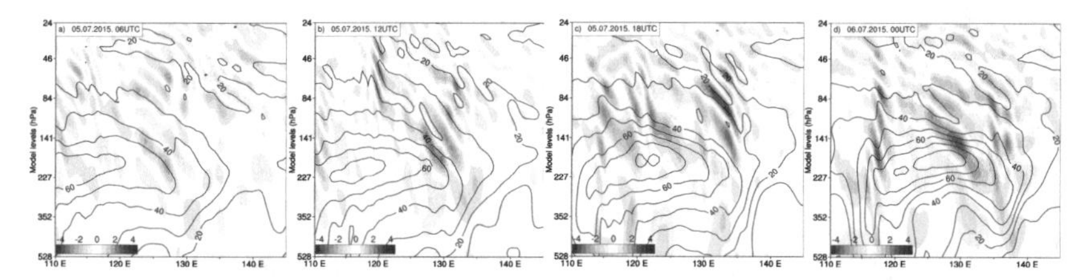

Fig. 4.8 Vertical cross-sections along between 110°E and 145°E along the latitude circle 40°S of temperature perturbations associated with inertio-gravity (IG) waves in ECMWF analyses. Figures are valid on 5 July 2014, 06, 12, 18 UTC and on 6 July 2014, 00 UTC. From Žagar et al. (2017). ©American Meteorological Society. Used with permission.

Further properties of the IG wave packet are provided in Figs. 4.9. Figure 4.9 (left) shows the vertical profile of the zonal wind perturbation in the upper troposphere and the lower stratosphere at four subsequent hours of the forecast. Shown are forecast winds at step +30, +31, +32 and +33 hours of the forecast started at 00 UTC, 4 July 2014. The location of the profile at 130°E, 40°S. The figure shows that the phase of the wave is moving downward which is associated with the upward energy propagation. In Fig. 4.9 (right) we show the IG wave winds at 12 UTC on 5 July at nearly the same location (129°E, 40°S). Two different presentations of the wave are shown. One is a classical hodograph (e.g. Hamilton, 1991) showing winds at different vertical levels between level 76 (around 216 hPa) and level 59 (around 93 hPa). As one moves upward, a counterclockwise rotating wind vector produces the elliptically-shaped hodograph shown in Fig. 4.9 (right a). Such hodographs have been shown very useful to analyze observations of single vertical profiles (e.g. rockets) (e.g. Gubenko et al., 2008; Fritts and Alexander, 2003). It is compared with the simulated evolution of the IG wave shown in Fig. 4.9 (right b). Different points in this figure correspond to the hourly values of the forecasted winds starting from 6 UTC on 5 July (30-hour forecast).

Without the temporal information about the IG wave evolution, the hodograph in Fig. 4.9 (right top panel) and the dispersion and polarization relationships for IG waves provide estimates of the intrinsic wave properties such as the frequency, phase and group speeds. From the polarization relation, it follows that the ratio of the major and the minor semi-axes of the polarization ellipse equals f/ω, where ω

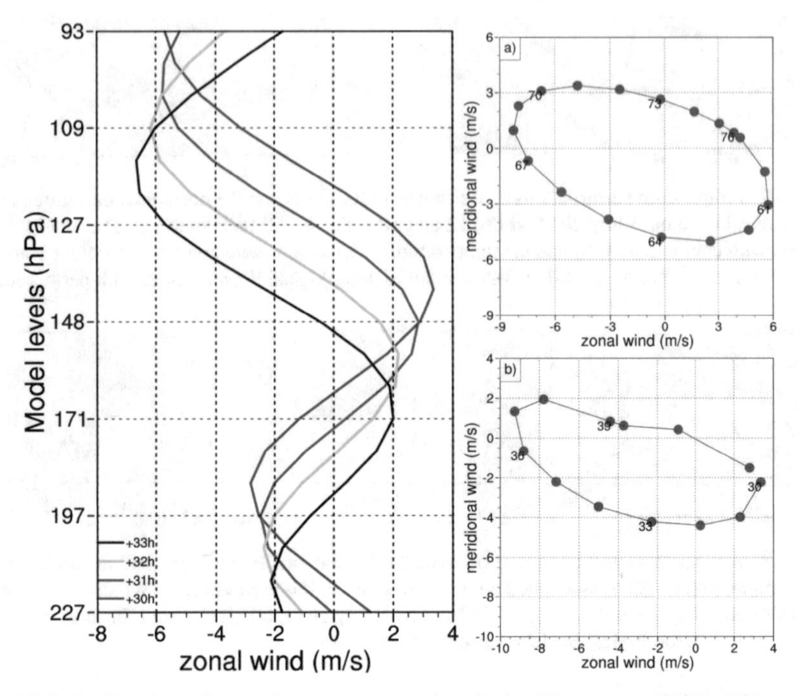

Fig. 4.9 Left: Zonal wind perturbations associated with the IG wave at 130°E, 40°S at four subsequent forecast times in the forecast initialized at 00 UTC, 4 July 2014. Right: (a) Hodograph of IG winds in ECMWF analysis valid on 5 July 2014, 12 UTC at 130°E, 40°S. Each dot presents a pair of (u_{IG}, v_{IG}) values at the denoted model level, starting from model level 76 (around 216h Pa) up to level 59 (93 hPa). (b) Evolution of IG wind at model level 69 (around 150 hPa) in point 130°E, 40°S. Each dot presents a pair of (u_{IG}, v_{IG}) values at the denoted hour of the forecast initialized at 00 UTC on 4 July 2014. The first point is the forecast length 30 hours and the last dot belongs to the 42-hour long forecast. From Žagar et al. (2017). ©American Meteorological Society. Used with permission.

is the intrinsic wave frequency. The vector lengths of the major and minor axes are estimated at 7.1 ms^{-1} and 3.7 ms^{-1} respectively, which results in $\omega = 1.8 \times 10^{-4}$ rad s^{-1}, and a wave period of 9.7 hours. The wave period suggested by Fig. 4.9 (right bottom panel) is somewhat larger, around 12hours. The zonal and vertical group speed estimated from the hodograph are 5.93 ms^{-1} and 0.13 ms^{-1} respectively.

In summary, presented case demonstrates that NMFs can be used the scale-dependent quantification of temperature and wind perturbations associated with the inertio-gravity waves in extratropics in spite of the derivation of NMFs with respect to the state of rest. We showed that the linear wave features readily persist in time despite of wave frequency-independent filtering of circulation data.

4.4.4.2 Slopes of Energy Spectra

A systematical investigation of the slopes of Rossby and IG energy spectra in the most recent operational analyses of ECMWF compared to ERA-Interim reanalyses was carried out by Žagar et al. (2017). ERA-Interim reanalysis data was produced by a somewhat older version of the ECMWF model at lower resolution (Dee and Coauthors, 2011). The operational analyses were evaluated on different horizontal resolutions such as N128 grid (512×256 grid points) with $K = 200$ waves (70 km scale in the midlatitudes) and on N200 grid that consists of 800×400 grid points equivalent to a grid spacing of about 50 km at the equator i.e about 35 km in the midlatitudes. All 137 model levels were considered and such spectra were available on a daily basis since November 2014 for 00 UTC analyses and high-resolution 10-day forecasts. Since December 2016 the decomposition excluded the top three model levels where some numerical artefacts may occasionally contaminate the IG wave signal. This dataset is denoted L134 and represents the state-of-the-art global Rossby and IG spectra. Operational analysis spectra are compared with the average spectra from the ERA Interim reanalyses over 35 years between 1981 and 2015 analyzed once per day on the grid N64 including 256×128 data in the zonal and meridional directions respectively. The expansion included levels up to about 0.8 hPa (denoted L57) and is denoted L57 ERAI. Some features of the ERA Interim dataset have previously been presented in Žagar et al. (2015) that also presented the vertical structure functions for the ERA Interim vertical discretization. All results are presented for the spectra that sum up energy of all vertical modes.

Table 4.3 Slopes of energy spectra in the three ranges of the zonal wavenumbers k in ECMWF analyses

	Inertia-gravity			Rossby			Total		
Dataset	$1 \leq k \leq 6$	$7 \leq k \leq 25$	$31 \leq k \leq 50$	$1 \leq k \leq 6$	$7 \leq k \leq 25$	$31 \leq k \leq 50$	$1 \leq k \leq 6$	$7 \leq k \leq 25$	$31 \leq k \leq 50$
L57 ERAI	-1.0	-1.9	-2.5	-1.2	-3.2	-3.7	-1.2	-3.0	-3.2
L137 2015-2016	-1.0	-1.5	-2.1	-1.1	-3.1	-3.1	-1.1	-2.6	-2.4
L134 DJF 2016/17	-1.0	-1.6	-2.0	-1.2	-3.1	-3.0	-1.2	-2.8	-2.5

The global energy spectra for the Rossby and IG modes during 2016/2017 NH winter season as a function of the zonal wavenumber is shown Fig. 4.10 (left). They can be compared with the average ERA-Interim spectra in Fig. 4.10 (right). Both spectra have a slope close to -3 for a wide range of the Rossby waves between zonal wavenumber $k = 7$ and the smallest resolvable scale. However, even a visual inspection of Fig. 4.10 suggests that the Rossby wave part of the spectrum is steeper than -3. The IG wave part of the spectrum in Fig. 4.10 continuously follows a more shallow line then the Rossby spectrum. When the two spectra are summed up, the total global energy spectrum is somewhat more shallow than the -3 slope. Overall the IG spectrum appears more shallow than $-5/3$ in planetary scales and steeper than $-5/3$ on small scales. In between there is a range of scales for which a $-5/3$

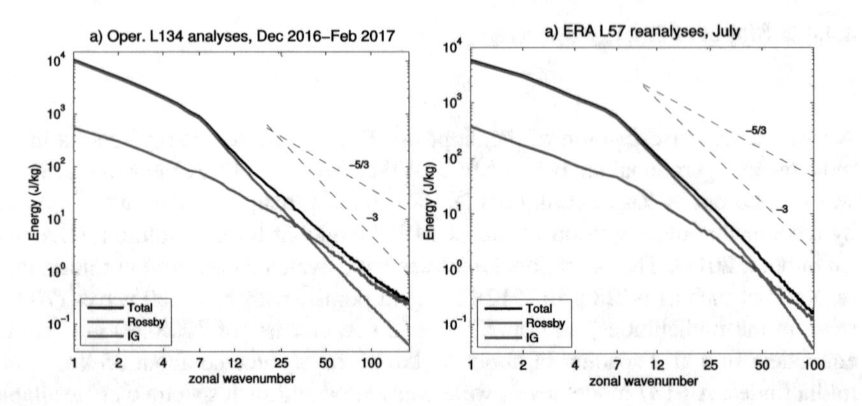

Fig. 4.10 Energy distribution as a function of the zonal wavenumber k in ECMWF analyses. Left: DJF period of 2016/2017 in operational analyses on 134 levels were under 5 Pa, Right: ERA Interim reanalyses for July using 57 model levels under 0.8 hPa. All meridional and vertical modes are summed up. From Žagar et al. (2017). ©American Meteorological Society. Used with permission.

slope fits the data relatively well. The best linear estimates of the slope for various ranges of the zonal wavenumbers and two datasets are collected in Table 4.3.

From the Rossby and IG spectra in Fig. 4.10 we can deduce a crossing wavenumber k_c defined as a wavenumber at which the IG wave energy becomes dominant over the Rossby wave energy. For example, July 2014 L137 analyses have crossing at wavenumber $k_c = 35$ (about 570 km at the equator, and 400 km in the midlatitudes). Throughout the 2015-2016 period, the value k_c deviates less than 100 km from the average depending on the flow, season and the analyzed model depth. The vertical model depth (i.e. analyzed layer) is otherwise an important factor for the location of the crossing point as illustrated in Fig. 4.11 for the 3-day period in July 2014. The number of levels reduces from 134 (top at 6 Pa) to 123, 108 and 89 levels (with top level at 53 hPa); the corresponding crossing scale k_c increases from 39 to 41, 52 and 58, respectively. The change in k_c occurs due to the IG spectrum since there is no variability in Rossby waves in these wavenumbers in the stratosphere and higher. In addition to the mean zonal state ($k = 0$) which is not discussed, only the planetary waves $1 \leq k \leq 3$ of the Rossby spectrum are affected when the lower mesosphere and the upper stratosphere are excluded from the consideration, whereas the IG spectra are affected at all scales. Notice that these larger values of k_c than in Fig. 4.10 are due to a 3-day period in NH summer.

In general, both ERA Interim and operational analysis data suggest that k_c is marginally greater in July than in December, likely associated with more intense unbalanced circulation during NH summers. The ERA Interim data, that are based on a decade older forecast model with a lower resolution, less advanced physical parametrisations and a lower model top, are expected to be less reliable for the IG spectrum and to contain less variability in small scales. Correspondingly, the

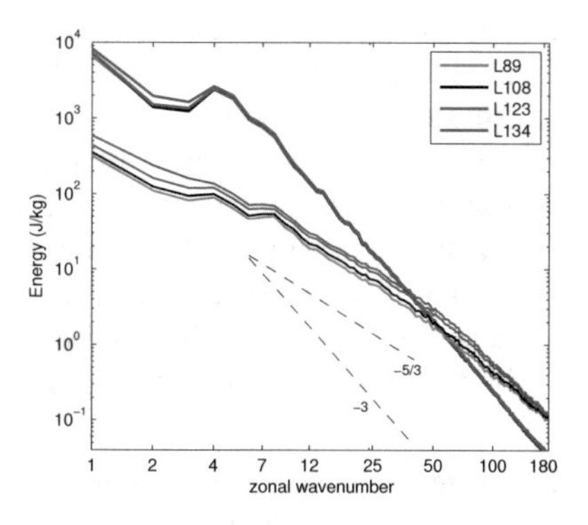

Fig. 4.11 As in Fig. 4.10 but averaged over 12 analysis times during 4-6 July 2014 for different model depths: 134 levels with the top level around 6 Pa, 123 levels with the top level around 1.2 hPa, 108 levels with the top level around 11 hPa, and 89 levels with the top level around 53 hPa. From Žagar et al. (2017). ©American Meteorological Society. Used with permission.

crossing scale in Fig. 4.10 (right) is found at larger wavenumbers around $k_c = 50$ which corresponds to zonal scales around 400 km.

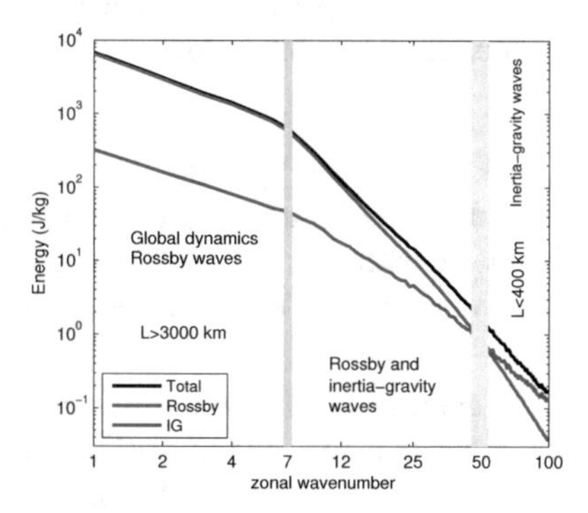

Fig. 4.12 Global energy distribution as a function of the zonal wavenumber in ERA Interim reanalysis data. The spectra are based on data once per day during period 35-year period 1981-2015. From Žagar et al. (2017). ©American Meteorological Society. Used with permission.

The global energy spectrum based on the L57 ERA Interim data can be considered representative for the current state-of-the-art renalysis products spanning both troposphere and stratosphere. Presented in Fig. 4.12, the climatological spectrum for the Rossby and IG waves in divided in three distinctive dynamical regimes:

1. the large-scale regime dominated by the Rossby waves (balanced dynamics). In this regime, the upscale flow of energy feeds the global mean zonal circulation. The unbalanced part of spectrum associated with the IG modes in this regime is mainly a projection of the tropical large-scale circulation features such as the Hadley and Walker circulation. In the midlatitudes, the gradient wind balance within the stratospheric polar vortex as well as stationary orographic waves due to large scale orography of Antarctica and Greenland also contribute to the large-scale IG spectrum (figures not shown). The evidence of these properties in physical space was provided in Žagar et al. (2015). The slope of energy spectra in this regime is in ERA Interim data close to -1.

2. the synoptic-scale weather regime is found in ERA reanalyses between the scales around 3000 km and about 400 km. Dynamics in this regime involves both the Rossby and IG waves. The regime starts at $k = 7$ where the Rossby waves dominate and a majority of the unbalanced circulation is contributed by the tropics. As we move downscale, the IG waves become relatively more energetic leading eventually to the crossing of the Rossby and IG wave spectra at $k_c = 47$ ($\sim$ 430–km scale). The scale vary with season and the analyzed vertical model depth. The slope of the total energy spectrum in ERA Interim is -3 with the balanced component steeper and IG part more shallow than -3.

3. the mesoscale regime beyond the crossing scale of the Rossby and IG waves. In this regime, the temperature and wind perturbations are primarily associated with IG waves. In other words, circulation is dominated by flow divergence and deformation. The average amount of global energy in this regime is very small, not reaching above 1 J/kg energy at scale of 400 km. The average slope of mesoscale energy spectrum in ERA Interim is somewhat steeper than -3 and it is an average between steeper than -3 spectrum of Rossby waves and the spectrum of the IG modes with a slope around -2.5.

The portion of energy in Rossby and IG modes as a function of scale is characterized by a nearly constant ratio of energy in the two types of modes in wavenumbers $k = 1-6$ with around 95% of energy being associated with balanced dynamics (not shown). From $k = 7$ the percentage of Rossby wave energy steadily reduces.

In contrast to ERA Interim analyses, produced using a 10 year older version of the forecast model, the operational analyses maintain the -3 slope for the Rossby modes for all $k > 6$ (Fig. 4.10). Moreover, for any particular month in the period since 2014 the balanced spectrum has a slope -3 ± 0.05 (not shown). For large scales ($k \leq 6$), the slope of balanced spectrum is -1.1 to -1.2 i.e. similar to L57 reanalysis in spite of their different vertical depth.

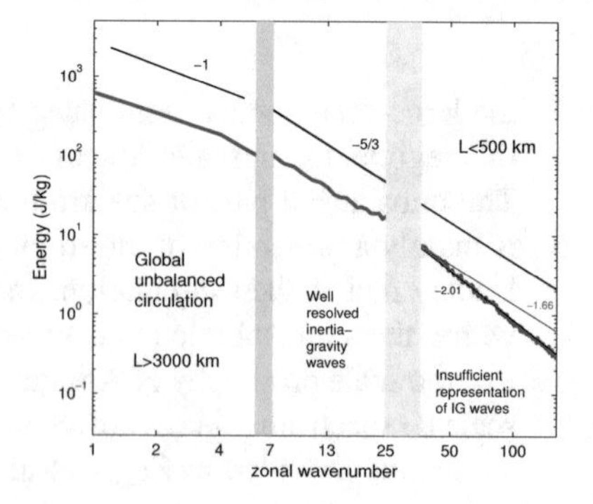

Fig. 4.13 Three ranges of the energy associated with the global large-scale unbalanced circulation and with the inertio-gravity waves. The IG spectrum is derived from the operational ECMWF analysis data in 2015-2016. From Žagar et al. (2017). ©American Meteorological Society. Used with permission.

The global IG spectrum derived from 2015-2016 ECMWF analyses is shown in Fig. 4.13. In comparison to the IG spectrum for the ERA Interim data, the spectrum in Fig. 4.13 comes from a more advanced model and data assimilation system and encompasses a deeper layer of the atmosphere (up to 1 Pa at about 80 km). Correspondingly, amplitudes of IG energy in J/kg are greater and the spectrum is more shallow. Similar to Fig. 4.12, we divide the global IG spectrum in three parts with their slopes listed in Table 4.3. The most shallow part of the IG spectrum corresponds to scales greater than about 3000 km. As discussed above and in Žagar et al. (2015), large-scale unbalanced circulation is associated primarily with tropical circulations and to the role of orography and stratospheric vortex in the extratropics. The estimated k_c is smaller than in ERA-Interim reanalysis and the difference in scale is 100-200 km. Based on the L137 dataset and sensitivity to the vertical depth, we define the crossing scale L_c for the current whole atmosphere ECMWF analyses to be at 500-600 km. Then the synoptic-scale regime is defined between about 3000 km ($k = 7$) and 500-600 km ($k \approx 35$). As the slope of the IG spectrum in this regime is -1.5 to -1.6, we can consider unbalanced dynamics in this range of scales well resolved by the current ECMWF analyses and forecasts.

According to Fig. 4.13 the globally integrated IG wave energy in the synoptic scales varies between a few J/kg up to several tens J/kg or even 100 J/kg. Such estimates will depend on the analyzed depth of the global atmosphere and vary by season and latitude. These orders of magnitudes of IG energy agree with the early estimates by Van Zandt (1982). More recent observational studies are mostly based on temperature observations and thus provide the estimates of IG wave potential energy (e.g. Tsuda et al., 2000). They can not be directly compared to the spectra derived using the NMF method that contain both potential and kinetic energy integrated over the atmosphere depth.

4.4.5 Zonal Energy Spectrum in High-Resolution Simulations

Here we examine whether a high-resolution model can reproduce the $k^{-5/3}$ power spectra. The model is called NICAM (Nonhydrostatic ICosahedral Atmospheric Model) and it is developed by CCSR (the Center for Climate System Research), University of Tokyo, and Frontier Research Center for Global Change/Japan Agency for Marine-Earth Science and Technology (Satoh et al., 2008). Icosahedral grid system with quasi-homogeneous grids over the sphere is used to overcome the pole problem. The results of the global model experiments with 3.5 km horizontal mesh are reported by Tomita et al. (2005); Nasuno et al. (2007); Miura et al. (2007). Tomita et al. (2005) succeeded in demonstrating a Madden- Julian Oscillation (MJO)-like intraseasonal oscillation, and diurnal precipitation cycles on the aqua planet experiment with glevel-10. Nasuno et al. (2007) showed that super cloud cluster propagated eastward; these have a Kelvin wave structure in dynamical fields. Miura et al. (2007) further showed that the MJO-event is realistically simulated with multi-scale structures of tropical convective systems. It is important to examine the characteristics of NICAM from various points of view for improving the new model. The results presented in this section come from Terasaki et al. (2009) and Terasaki et al. (2011).

The terrain-following vertical coordinate ζ is adopted, which has a relation with the height z from the sea level,

$$\zeta = \frac{z_T(z - z_s)}{(z_T - z_s)} \tag{4.81}$$

where z_T is the top of the model domain, which is set to 40 km in this study, and z_s is the height of topography. The output data are interpolated from the icosahedral grids to equally spaced horizontal grids, and from ζ coordinate to 17 mandatory vertical levels from 1000 to 10 hPa. In this study, the energy spectrum of vertical wind speed w from 40°N to 50°N is compared first with that of zonal wind speed u and meridional wind speed v. The kinetic energy (K) is calculated as follows,

$$K(k) = \frac{1}{2}(|U_k|^2 + |V_k|^2 + |W_k|^2), \tag{4.82}$$

$$\approx \frac{1}{2}(|U_k|^2 + |V_k|^2) \tag{4.83}$$

where k is a zonal wavenumber, and U, V and W are the coefficients of Fourier transform of u, v, and w, respectively. The truncation wavenumbers of each global experiment were set to 80, 160, 320, 640, 1280, 2560, and 5120, respectively. As will be shown in Fig. 4.14, kinetic energy of the vertical component is negligible compared with that of zonal and meridional components. Therefore, the kinetic energy is evaluated using only the horizontal components.

We first compared the kinetic energy of vertical component with that of zonal and meridional components in Fig. 4.14. While the kinetic energy spectra of the zonal and meridional components become a red noise spectrum at small scales, the kinetic energy of the vertical component is supposed to be a blue noise spectrum (the energy is included mostly in small scales) (Kevin Trenberth, personal communication). However, it is found in this study that kinetic energy of the vertical component becomes a white noise spectrum, in which the vertical kinetic energy has almost the same magnitude for all wavenumbers less than 1000. The horizontal wind energy spectrum follows the $k^{-5/3}$ power up to $k = 500$. The energy spectrum drops down in the region of spectral tail. However, the behavior of the spectral tail depends on the numerical diffusion (Tomita et al. 2008), although the k^{-3} and $k^{-5/3}$ power laws are produced by the atmospheric internal dynamics (Takahashi et al. (2006)).

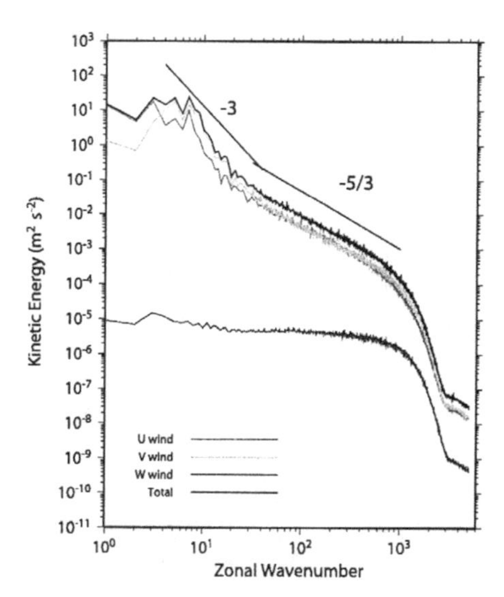

Fig. 4.14 Kinetic energy spectra for zonal (top), meridional (middle) and vertical (bottom) winds in the zonal wavenumber domain at 200 hPa level averaged for 40-50°N, produced by NICAM. The total kinetic energy is drawn with black line in units of m^2 s^{-2}. The k^{-3} and $k^{-5/3}$ power lines are drawn in the figure. From Terasaki et al. (2009), Fig. 1.

The modal spectra are obtained using NMF decomposition in pressure system of the simulation performed by way of "Meteorological Research Consortium", a framework for research cooperation between JMA and MSJ (Meteorological Society of Japan). The resolution of the model is TL959-L60 (the model top is 0.1 hPa), corresponding roughly to a 20 km horizontal grid spacing. The data contain meteorological variables of horizontal wind u, v, geopotential ϕ, specific humidity q, and air temperature T. The data period used in this study is from 00Z 1 September 2009 to 18Z 7 September 2009. The data are interpolated on 60 gaussian vertical levels in the -log (p/p_s) coordinate by cubic spline method, the top of the gaussian vertical level is 0.4 hPa. The 900 Hough modes are divided into 3 parts; Rossby modes, eastward gravity modes, and westward gravity modes.

Figure 4.15 shows the energy spectrum in the zonal wavenumber domain. The energy spectra of Rossby and gravity modes are obtained by summing up all vertical and meridional modes in Hough modes, respectively. The energy spectrum of Rossby modes follows -3 power law in the synoptic scale (k = 6 to 300). The slope of a linear regression for k = 6 to 300 is -2.84. It is found that the energy spectrum of gravity mode exactly obeys the -5/3 power law in both synoptic and mesoscales. The slope of a linear regression for k = 6 to 350 is -1.66. The spectra for Rossby and gravity modes become less steeper than k^{-3} and $k^{-5/3}$ slopes from around k = 300. The spectrum behind the wavenumber k =600 drops rapidly, because of the truncation and the viscosity and diffusion. It can also be seen in Fig. 4.15 that the spectra for Rossby and gravity energy cross each other around a zonal wavenumber 80, although the total energy spectrum does not clearly show this shift. This scale corresponds to about 350 km on 45°circle.

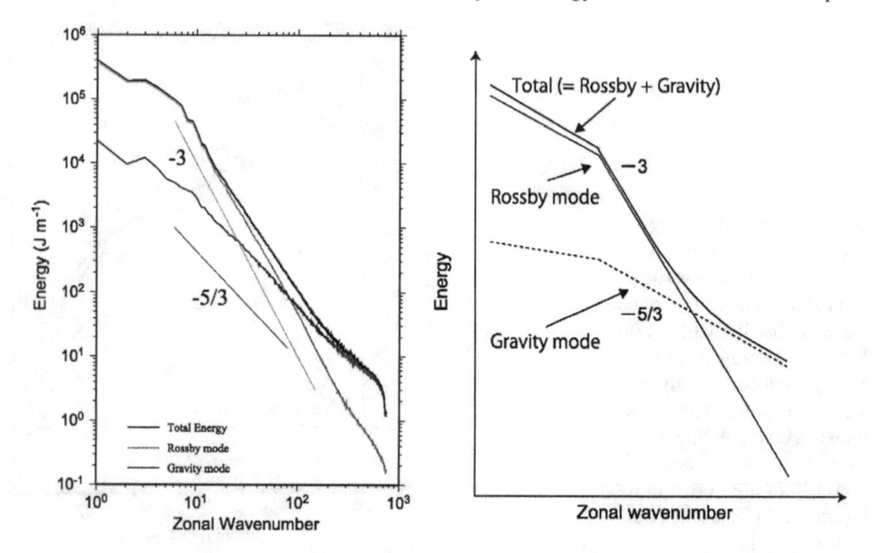

Fig. 4.15 Left: The energy spectra for total, Rossby modes, and gravity modes in the zonal wavenumber domain for JMA data. The units are J m^{-2} The Rossby modes obey k^{-3} and gravity modes obey $k^{-5/3}$ power law. Right: Schematic diagram of energy spectrum for baroclinic atmosphere. The dotted and dashed lines show the energy spectra for Rossby and gravity modes, respectively. The solid line shows the total of them. From Terasaki et al. (2011), Figs. 1 and 4.

Finally, the schematic diagram of energy spectrum is included in Fig. 4.15.

4.4.6 Energy Interactions

Tanaka and Kung (1988, hereafter TK1988) examined energy interactions in the zonal wavenumber domain by synthesizing all the vertical and meridional modes and comparing it to the standard spectral energetics result by Kung (1988). The results showed a reasonable agreement is confirmed for the two energetics schemes. The positive values of C and $(S + R)$ at all wavenumbers indicate the transformation of zonal mean available potential energy to eddy available potential energy, whereas the negative values of B and $(L + M)$ show the transformation of eddy kinetic energy to the zoanl mean component. Most of these eddy energy transformations are associated with the transient motion of the general circulation.

It was found that the positive values of C is mostly contained in the vertical mode $m = 4$, centerred at the zonal wavenumber $k = 6$ in synoptic eddies. It implies that the available potentail energy in zonal baroclinic component is transformed to eddy baroclinic component by the mode-mode interactions. Approximately same amount of kinetic energy is generated at the vertical mode m=4, centered at the zonal wavenumber $k = 6$ as indicated by negative values of B. It is clearly shown that the kinetic energy is generated by the baroclinic conversion from availabel potential energy at this vertical mode in the synoptic eddies. It is important to note that the vertical mode separation for B shows opposite sign for the barotropic mode $m = 0$ for all zonal waves. The result implies that the baroclinic kinetic energy is transformed to barotropic kinetic energy. This process is called barotropization.

This energy transformation can be regarded as a characteristic of the atmospheric baroclinic instability (Charney, 1947), by which the synoptic waves gain energy from the zonal available potential energy. The baroclinic conversion from available potential energy to kinetic energy appears in this study as the compensation of the negative B_T with the positive C_T in the baroclinic component. The accumulated barotropic energy is then transformed from synoptic eddies to planetary waves and zonal motions by the up-scale energy cascade characterized by the two-dimensional fluid mechanics in the barotropic component. This result appears reasonable, and agree with existing studies (e.g. Wiin-Nielsen, 1967).

An overall flow of normal mode energy during the FGGE year is shown in a box diagram in Fig. 4 of TK1988. The four boxes represent the energy levels of baroclinic (upper boxes) and barotropic (lower boxes) components of zonal (left boxes) and eddy (right boxes) energies. For the barotropic energy it may be shown that the kinetic energy of the vertical mean flow dominates. The kinetic energy of the vertical shear flow is generally associated with the exsitence of available potential energy through the thermal wind relation, with their sums seen as the total baroclinic energy. The flow pattern of normal mode energy, as shown in Fig. 4 of TK1988, is the net energy input and output of processes B and C. The process D is evaluated as the residual of the energy balance.

There is a net generation of zonal baroclinic energy by the differential heating, with this energy being the initial energy input into the system. The zonal baroclinic energy is first transformed into eddy baroclinic energy through the transformation of zoanl available potential energy to eddy available potential energy, which is further converted into eddy kinetic energy of the vertical shear flow within the box for eddy baroclinic energy. The eddy baroclinic energy is transformed to eddy barotropic energy as the shear flow kinetic energy is transformed to mean flow kinetic energy. The accumulated eddy barotropic energy in synoptic scale is transformed to zonal motions by the up-scale energy cascade. This process is called zonalization. Finally, the net energy dissipations take place with zonal and eddy barotropic components. Radiative cooling should be contained in the dissipation at the eddy baroclinic component. This box diagram can be rearranged to represent an energy flow from eddy baroclinic to eddy barotropic components and an energy flow from eddy barotropic to zonal barotropic components.

In summary, early studies by the author H. Tanaka and collaborators using the FGGE data suggested that available potential energy peak in the vertical mode $m = 4$, and kinetic energy in the barotropic mode $m = 0$. The kinetic energy spectrum shows a secondary energy peak at $m = 4$. The available potential energy generated at the zonal baroclinic components (especially $m = 4$) is transformed to the eddy available potential energy of baroclinic components. This result is sensitive to the treatment of the kinematical surface wind. In Tanaka and Kung (1988) we assume the vanishing lower surface wind for the data analysis. With this assumption, available potential energy is transformed from zonal to eddy of the baroclinic component. The available potential energy is converted to kinetic energy by the baroclinic conversion at eddy baroclinic component. On the other hand, the kinetic energy of the cyclone-scale baroclinic mode ($m = 2 - 4$) is transformed to eddy kinetic energies of the barotropic mode by barotropization characterized by the removal of baroclinicity due to the baroclinic instability. The accumulated eddy barotropic kinetic energy is transformed to zonal barotropic kinetic energy by the upscale energy cascade of the two dimensional fluid mechanics. Finally the barotropic kinetic energy is dissipated by surface friction. Consequently, the atmospheric energy flow is characterized by the energy interaction from zonal baroclinic mode to eddy baroclinic mode, to eddy barotropic mode, and finally to zoanl barotropic mode.

Parameterizing the horizontal scale of waves by their eigenfrequencies, we find in the frequency domain that the kinetic energy spectra for the barotropic mode indicate clear energy peaks at the frequency σ=0.03 (16 day) for the wavenumber 1 and σ=0.07 (7 day) for the wavenumber 6. This frequency corresponds to the stationary Rossby waves, beyond which the Rossby waves propagate westward. This frequency (scale) is referred to as a shperical Rhines scale. It is confirmed that the dominant nonliniarity produces geostrophic turbulance at the low-frequency range, whereas the dominant linear term produces normal mode Rossby waves at the high-frequency range separated by the spherical Rhines scale. The spectral characteristics are distinguished by the 3-D scale of the waves. The energy spectra follow approximately the 2 power of the frequency for the smaller-scale waves in the low frequency range. However, for the largest-scale Rossby modes in the high frequency range, the spectrum obeys the -5/3 power law and merges continuously with the spectrum of the gravity modes. It is suggested from these results that the energy spectrum of the planetary Rossby waves is connected to that of gravity waves for the global scale waves.

4.5 Vertical Mode Energy Spectrum Derived Using Analytical Vertical Structure Functions

4.5.1 Motivation

Since Kasahara and Puri (1981) first obtained orthonormal eigensolutions to the vertical structure equation, it became possible to expand the atmospheric data into the three-dimensional harmonics of the eigensolutions. Tanaka (1985) and Tanaka and Kung (1988) studied the atmospheric energy spectrum and interactions expanding the atmospheric data to the three-dimensional normal mode functions. The vertical structure functions in these studies were obtained by solving the vertical structure equation with a finite difference method. The numerical vertical structure functions may have large aliasing for higher order vertical modes indicating largest amplitudes near the sea level despite that the analytical solutions indicate the largest amplitudes always in the upper atmosphere (Sasaki and Chang, 1985).

The energy spectrum and interactions, which are calculated by using the numerical vertical structure functions, may be influenced by aliasing for high order modes. For this reason, it is desired to calculate the atmospheric energy spectrum and interactions by using the analytical vertical structure functions which have no aliasing.

In this section, we first obtain a set of analytical vertical structure functions by assuming a constant static stability parameter as described by Terasaki and Tanaka (2007, hereafter TT2007). The vertical energy spectrum and interactions are then analyzed using the analytical vertical structure functions for their 3D NMF analysis of global energetics in two reanalysis datasets.

4.5.2 Analytical Vertical Structure Functions

The governing equations used in this section are the 3D spectral primitive equations on a sphere (Tanaka, 1985; Tanaka and Terasaki, 2005). The basis functions in the vertical direction are the solutions of the vertical structure equation as follows:

$$-\frac{d}{d\sigma}\left(\sigma^2 \frac{dG_m}{d\sigma}\right) = \lambda_m G_m, \qquad \text{for} \qquad \epsilon < \sigma < 1, \tag{4.84}$$

$$\frac{dG_m}{d\sigma} = 0, \qquad \text{at} \qquad \sigma = \epsilon, \tag{4.85}$$

$$\frac{dG_m}{d\sigma} + \alpha G_m = 0, \qquad \text{at} \qquad \sigma = 1, \tag{4.86}$$

where G_m is the vertical structure function, $\sigma = p/p_s$, $\lambda_m = \frac{R\gamma}{gh_m}$, $\alpha = \gamma/T_s$, p_s (=1013.25 hPa) and T_s (=300 K) are surface pressure and surface temperature of the reference state, R is the gas constant of dry air, and γ (=30 K) is a static stability parameter which is assumed to be a constant in this section.

Since γ is a constant, the vertical structure equation becomes the so-called Euler equation. Applied to a rigid top boundary condition at $\sigma = \epsilon$, the problem is reduced to the regular boundary value problem of Sturm-Liouville type. In this study the top of the atmosphere is assumed at p =1 hPa.

Under this geometric configuration, we can solve the Euler equation as a series solution (William and Richard, 1977), and the infinite series of the vertical structure functions are represented as follows:

$$G_0(\sigma) = C_1\sigma^{r_1} + C_2\sigma^{r_2}, \tag{4.87}$$

$$G_m(\sigma) = \sigma^{-\frac{1}{2}}\{C_1\cos(\mu\ln\sigma) + C_2\sin(\mu\ln\sigma)\}, \tag{4.88}$$

$$r_1 = -\frac{1}{2} + \mu, \qquad r_2 = -\frac{1}{2} - \mu, \qquad \mu = \sqrt{\left|\frac{1}{4} - \lambda_m\right|}, \tag{4.89}$$

where the eigenvalues λ_m are obtained by solving the eigenvalue problem of (4.84), and the equivalent height h_m of each vertical mode is listed in Table 1 of TT2007. The equivalent heights h_m range from 9726.6 m to 8.8 m for vertical index m=0 to 22. C_1 and C_2 are obtained from the boundary conditions (4.85) and (4.86), and normalized as $C_1^2 + C_2^2 = 1$.

The energy balance equations for normal modes may be described by the following equation:

$$\frac{dE_{knm}}{dt} = B_{knm} + C_{knm} + D_{knm}, \tag{4.90}$$

where k, n, m are zonal, meridional, and vertical wavenumbers, and

$$B_{knm} = p_s\Omega h_m(w_{knm}^* b_{knm} + w_{knm} b_{knm}^*), \tag{4.91}$$

$$C_{knm} = p_s\Omega h_m(w_{knm}^* c_{knm} + w_{knm} c_{knm}^*), \tag{4.92}$$

$$D_{knm} = p_s\Omega h_m(w_{knm}^* d_{knm} + w_{knm} d_{knm}^*). \tag{4.93}$$

According to (4.90), the time change of the energy is caused by the three terms which appear in the right hand side of (4.90). B_{knm} and C_{knm} are respectively associated with the nonlinear mode-mode interactions of kinetic and available potential energies, and D_{knm} represents an energy source and sink due to the diabatic process and dissipation.

The nonlinear vertical energy flux can be obtained by summing up nonlinear interaction terms B and C over all zonal k and meridional modes n:

$$F_B(M) = \sum_{m=0}^{M} \left(\sum_k \sum_n B_{knm} \right), \tag{4.94}$$

$$F_C(M) = \sum_{m=0}^{M} \left(\sum_k \sum_n C_{knm} \right), \tag{4.95}$$

where F_B and F_C are the vertical kinetic energy flux and the vertical available potential energy flux, respectively.

4.5.3 Energy Spectrum and Interactions

We present the results of TT2007 who used the analytical vertical structures functions. The datasets are four-times daily (00, 06, 12, and 18 UTC) JRA-25 (Japanese Re-Analysis 25 years) (Onogi et al., 2007) and ERA-40 (ECMWF 40 years Re-Analysis) (Uppala et al., 2005) for December, January and February from 1979 to 2000. The analyzed variables are horizontal winds u, v, vertical p-velocity, temperature, and geopotential ϕ, defined at every 2.5° longitude by 2.5° latitude grid points over 23 mandatory pressure levels from 1000 to 0.4 hPa for JRA-25 and from 1000 to 1 hPa for ERA-40. The data are interpolated on the 46 Gaussian vertical levels in the log (p/p_s) coordinate by cubic spline method. The Hough vector functions are truncated by Rossby modes ($n_R = 0-25$) and 12 eastward and westward propagating gravity modes ($n_E = 0 - 11, n_W = 0 - 11$). The energetic terms are computed for each observational time and averaged during the data period.

Figure 4.16 illustrates the analytical vertical structure functions for $m = 0$ to 5 and $m = 17$ to 22, respectively. The envelope function of $\sigma^{-\frac{1}{2}}$ is superimposed in the figure. The analytical vertical structure functions for baroclinic modes are represented by trigonometric functions multiplied by the envelope function, so the profiles have larger amplitudes at the upper atmosphere. Compared with Fig. 1 of TT2007, it is shown that the vertical structure functions by the finite difference method are approximately correct for the lower order vertical modes. However, those for the higher

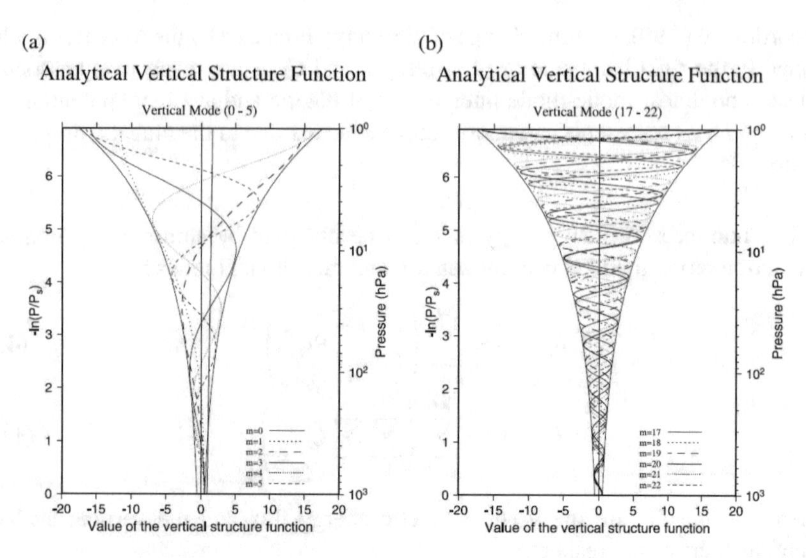

Fig. 4.16 The profiles of the analytical vertical structure functions for (a) m=0-5 and for (b) m=17-22. From Terasaki and Tanaka (2007), Fig. 2.

order vertical modes are totally different due to the finite difference approximation. The difference may have caused by an aliasing of the numerical solution. The vertical energy spectrum and the interactions in previous studies must be highly deformed by the aliasing effect, so it must be reexamined using the analytical solutions.

The kinetic energy spectrum using these analytical vertical structure functions obeys -3 power law when it is plotted against the geometrical vertical wavenumber μ in (4.89) instead of the inverse of equivalent height as shown in Fig. 4.17. There is a marked energy peak at the vertical wavenumber $m = 4$ for both K and A. This energy peak is caused by the vertical structure function for $m = 4$ having a maximum at about 200 hPa and the opposite sign at lower and higher troposphere. The tropospheric jet around upper troposphere may cause the secondary maximum of kinetic energy. The baroclinic structure of temperature deviation from the global mean, which has an opposite sign at lower and higher troposphere, may be reflected at $m = 4$. The total energy at this peak is mostly explained by the available potential energy, whereas the barotropic energy at $m = 0$ is mostly contained in kinetic energy. The higher order vertical modes are mostly explained by the available potential energy, and the contributions from kinetic energy are less than 1/10. Overall the vertical mode spectrum is significantly different than shown by Tanaka (1985) and Tanaka and Kung (1988), where the spectrum for large vertical mode indices appeared to have a zigzag shape possibly influenced by aliasing due to the numerical solutions.

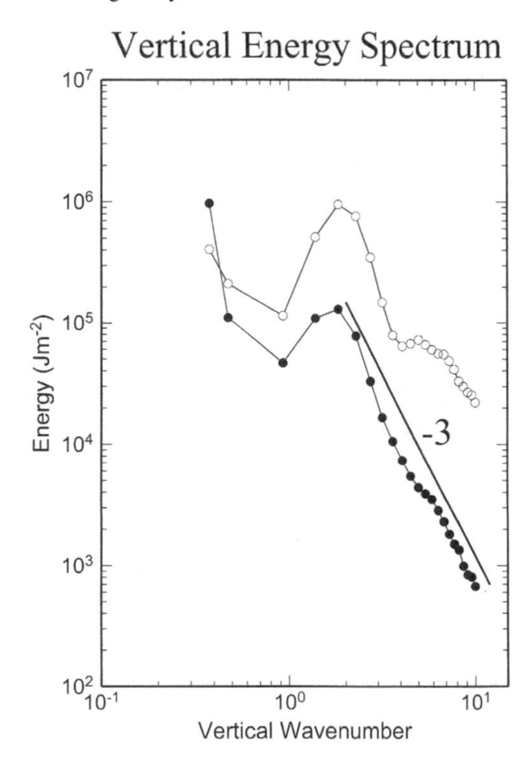

Fig. 4.17 Energy spectra of kinetic energy (black) and available potential energy (white) for the analytical vertical structure functions using a vertical wavenumber μ in the abscissa. Adapted from Terasaki and Tanaka (2007), Fig. 4.

TT2007 also computed energy interactions in the zonal wavenumber domain and in the vertical spectral domain are analyzed in this section based on (4.90) The effect of using the analytical vertical structure functions is found at the nonlinear interactions for the higher order vertical modes. Small but consistently negative values of the available potential energy interactions over $m > 6$ indicated that an energy source of the atmospheric general circulation exists in higher order vertical modes. There were some available potential energy sources at $m = 1$ and 3. The kinetic energy interactions are hardly seen in the higher order vertical modes.

Figure 4.18 shows the energy flux of kinetic energy (F_B), available potential energy (F_C), and total energy (F_N) in the vertical wavenumber domain. Negative and positive values indicate upscale and downscale cascades, respectively. It shows that energy flux basically has negative value, indicating dominant inverse energy cascade from smaller vertical scale to larger vertical scale motions. As a result, the atmospheric energy is transformed from baroclinic to barotropic components. The kinetic energy flux is the largest at the vertical wavenumber $m = 2$, and peak of the available potential energy flux is seen at $m = 7$. It is suggested from these analyses using the analytical vertical structure functions that the energy interactions are performed by relatively larger vertical scale motions for the kinetic energy, whereas the structures of the available potential energy interactions are complicated in the higher order vertical modes.

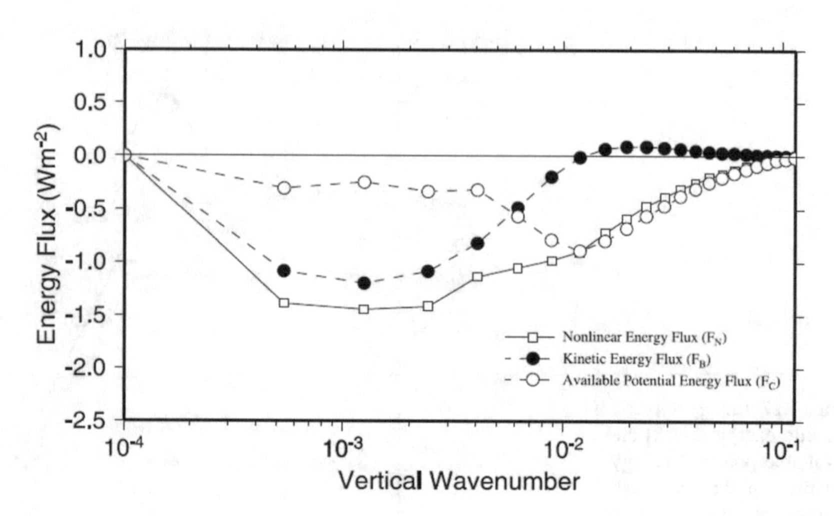

Fig. 4.18 Energy flux of kinetic energy, available poteintail energy, and total energy in the vertical mode domain as a function of the inverse of equivalent heights. From Terasaki and Tanaka (2007), Fig. 8.

According to the result for the energy interactions, energy flows are represented from the zonal baroclinic energy to eddy baroclinic energy to eddy barotropic energy, and finally to zonal barotropic energy, as is consistent with the result by Tanaka and Kung (1988). It is found in this study using the analytical vertical structure functions that there are small but consistently negative values of nonlinear interactions of available potential energy at zonal baroclinic components in the higher order vertical modes. The result suggests that the source of available potential energy in the zonal field is distributed in wide range of the vertical spectrum at large vertical wavenumbers. Energy injection by the solar radiation is mostly distributed in lower order vertical mode at $m = 4$. The energy may be injected by the solar radiation in higher order vertical modes, but further analysis is needed to ensure the dynamic interporetation for this. The analysis of vertical energy flux shows that the energy injected at the higher order baroclinic modes by the solar radiation is transformed to lower order vertical modes, ultimately to barotropic mode.

4.6 Vertical Motions of Rossby and Gravity Modes

4.6.1 Background and Motivation

Vertical motion is the key variable for any diagnostic study in the general circulation, especially for energy and water budgets and various kinds of material transfer in the 3-D atmosphere. Except for some point-wise observations by wind profilers (e.g. Sato, 1994), vertical motion is not an observable quantity in a global network. Therefore, we must evaluate it diagnostically from observed horizontal wind data by integrating a mass continuity equation with respect to the vertical coordinate. The vertical p-velocity, ω, can be estimated using the omega equation, assuming an inviscid and adiabatic flow. Modern data assimilation scheme provide vertical motions that is kinematically consistent within the model atmosphere. However, these fields may not always be available and it is difficult to separate the contributions from vertical motions in Rossby waves from those associated with unbalanced dynamics.

The purpose of this section is to compare the vertical motions evaluated by some classical kinematic methods in the global atmosphere with that obtained by the normal mode expansion method. Schemes to be compared include a central difference scheme, plane fitting method, and spherical harmonic expansion method. The presentation of this section follows Tanaka and Yatagai (2000) (hereafter TY2000).

4.6.2 Computation of Vertical Velocity

The vertical motion ω is evaluated from horizontal divergence, $\nabla \cdot \mathbf{v}$, in the pressure coordinate by integrating the continuity equation with respect to pressure.

$$\nabla \cdot \mathbf{V} + \frac{\partial \omega}{\partial p} = 0 \quad \rightarrow \quad \omega = -\int_0^p \nabla \cdot \mathbf{V}\, dp\,. \tag{4.96}$$

Here, the boundary condition should be ω=0 at the limit of $p \rightarrow 0$.

In the following, we present a finite difference method, plane fitting method, and spectral method for evaluating divergence. The divergence is then integrated with respect to the vertical from the top to bottom of the atmosphere using the boundary condition. It has been, however, integrated conventionally from the bottom to top, as in this study, by evaluating surface ω under a proper assumption. We use the tendency of geopotential at 1000 hPa surface for evaluating surface ω as described in Tanaka and Milkovich (1990). The resulting contradiction at the top boundary is then adjusted by O'Brien (1970) quadratic correction. It is well-known that the kinematically estimated surface ω can be erroneous over the mountainous region, such as Tibetan Plateau and Antarctic. Here, a normal mode method is introduced for the computation of ω, with the divergence integrated from the top to bottom of the atmosphere. It will be shown that the problem in the mountainous region is reduced in this method.

In the finite difference scheme, the horizontal divergence is approximated by a standard central finite difference method:

$$\nabla \cdot \mathbf{V} = \frac{u(x + \Delta x) - u(x - \Delta x)}{2\Delta x} + \frac{v(y + \Delta y) - v(y - \Delta y)}{2\Delta y}, \tag{4.97}$$

where Δx and Δy are the Cartesian grid intervals. The divergence is evaluated by the small difference in wind speed. Unfortunately, this scheme contains large error in small scales since the observed wind has, at least, 10 percent of error. Hence, the divergence sometimes contains more than 100 percent of error. For example, when two adjacent zonal winds are 10 and 11 m s^{-1}, the divergence becomes $1/\Delta x$. If the wind contains 10% error, the error in divergence becomes at most $2/\Delta x$, which appears to be 200% error. For this reason, the simple difference scheme is danger to use unless some additional care is taken.

The plane fitting method aimed to remove such an erroneous divergence in small scales by fitting wind data near the origin onto a plane using a lease square method (Kung, 1972).

$$\begin{aligned} u(x, y) &= u_0 + ax + by, \\ v(x, y) &= v_0 + cx + dy, \\ \nabla \cdot \mathbf{V} &= a + d, \end{aligned} \tag{4.98}$$

where the regression coefficients a, b, c, d are evaluated from a set of gridded data using the standard multiple regression code. For the square gridded analysis data, surrounding 9 points may be adequate to fit a plane. The deviations from the plane are considered mostly as an observational error. The divergence tends to be smoother if the number of the data points increase. This method has been widely used since it is applicable for randomly distributed observation stations. The optimal number of the data points for the fitting is determined empirically depending on the purpose of the study. The result of energy budget would, however, depend highly on the choice of that number.

The spherical harmonics method is straightforward to apply to global data. Here, the divergence δ is expanded in spherical harmonics and synthesized over the wavenumbers approximately a half of the Nyquist wavenumber.

$$\nabla \cdot \mathbf{v} = \delta = \sum_k \sum_n \delta_n^k P_n^k(\mu) e^{ik\lambda}, \tag{4.99}$$

$$\delta_n^k = \frac{1}{\left\|P_n^k\right\|^2} \int_{-1}^{1} \frac{ikU^k}{a(1-\mu^2)} P_n^k - \frac{V^k}{a(1-\mu^2)} H_n^k \, d\mu, \tag{4.100}$$

where a is the radius of the earth, U^m and V^m are zonal Fourier expansion coefficients of $u\cos\theta$ and $v\cos\theta$, λ and θ are longitude and latitude, $P_n^k(\mu)$ is the associated Legendre functions as a function of μ=sinθ, k and n are zonal and total wavenumbers, and $H_n^k(\mu) = (1-\mu^2)\frac{dP_n^k}{d\mu}$ is the derivative of P_n^m. The analytical expression is available for $H_n^k(\mu)$. The divergence δ is evaluated from integration of the observed wind rather than a finite differentiation of the wind.

4.6.3 Computation of ω Using the Normal-Mode Method

An alternative spectral method calculates ω by expanding the state variable in 3-D normal mode functions:

$$\mathbf{X}(\lambda, \theta, p, t) = \sum_{knm} w_{knm}(t) \, \mathbf{D}_m \, \mathbf{\Pi}_{knm}(\lambda, \theta, p), \tag{4.101}$$

$$w_{knm}(t) = < \mathbf{X}, \mathbf{D}_m^{-1} \mathbf{\Pi}_{knm} >, \tag{4.102}$$

where $\mathbf{X} = (u, v, \phi)^T$ is the state variable vector of horizontal wind velocity u, v and geopotential deviation from the reference state ϕ as functions of longitude λ, latitude θ, pressure p and time t. The 3-D Fourier expansion coefficient w_{knm} has triple subscripts of zonal, meridional, and vertical wavenumbers, respectively. The expansion basis function $\mathbf{\Pi}_{knm}$ is a tensor product of vertical structure functions $\mathbf{G}_m$ and Hough harmonics (Kasahara, 1984). The dimensional factor matrix $\mathbf{D}_m = diag(c_m, c_m, c_m^2)$ contains scale parameters $c_m = \sqrt{gh_m}$ involving the gravity g and the separation constant of the equivalent depth h_m. The equations in (4.101) and (4.102) construct a pair of Fourier transforms in the 3-D spectral domain with a proper inner product $<, >$ which satisfies an orthonormal condition for $\mathbf{\Pi}_{knm}$.

Once the expansion coefficient w_{knm} is obtained, the vertical motion ω may be calculated by the synthesis of the modes, corresponding to the inverse transform:

$$\omega = \sum_{knm} 2\Omega \frac{c_m^2 p^2}{R\gamma} \frac{dG_m}{dp} w_{knm} i\sigma_{knm} Z_{knm} e^{ik\lambda}, \tag{4.103}$$

where Ω is the angular speed of Earth's rotation, R the gas constant of dry air, γ the static stability parameter, σ_{knm} the eigenfrequency of Laplace's tidal equation, and Z_{knm} the geopotential component of the Hough function. Refer to Tanaka and Kung (1988) for the detail of the variables. The vertical integral from the top to bottom of the atmosphere has been accomplished analytically by integrating the vertical structure functions by parts:

$$\int_0^p G_m dp = -\frac{c_m^2 p^2}{R\gamma} \frac{dG_m}{dp}. \tag{4.104}$$

Since the analytical expression is available for the vertical derivative of G_m, the vertical motion is analytically obtained except for the truncations imposed on the synthesis. The 3-D modes should be synthesized over the wavenumbers approximately a half of the Nyquist wavenumber.

4.6.4 Comparison of ω Estimates by Different Methods

The four methods for the computation of ω were applied to NCEP data for a case of January 28, 1989, when an explosive cyclogenesis occurred at the Far East to yield a marked vertical motions (Hayasaki and Tanaka, 1999). The low pressure system over the Sea of Okhotsk has a barotropic structure since the surface cyclone and 500 hPa trough are located at the same geographical location. Another low-pressure system is seen over Alaska extending toward the north Pacific. The data were defined at every 2.5°longitude by 2.5°latitude grid point over 17 levels from 1000 to 10 hPa. The Hough mode truncations in this study are 26 Rossby modes and 24 gravity modes with 15 zonal wavenumbers. In the vertical, 7 modes are synthesized. Those truncation seems to have filtered most of the small-scale structures.

Since ω was provided in the NCEP and ECMWF reanalysis, ω fields from the two reanalyses could be compared to find to what extent these values agree with each other. The result by the normal mode method was smoother than spherical harmonics method, and no extreme values are found over Tibet, Greenland, and Antarctic.

Since the Hough functions can be partitioned in Rossby and gravity modes, the contributions from Rossby and gravity modes to the vertical motion can be examined separately. Figure 4.19 (left) compares the vertical motions associated with gravity modes and Rossby modes, respectively. The result clearly shows that the vertical motion is mostly contained in the gravity modes, despite the fact that the geopotential height is mostly represented by Rossby modes. According to Figs. 4.19, the difference between omega and geopotential fields is resulted from the weight of

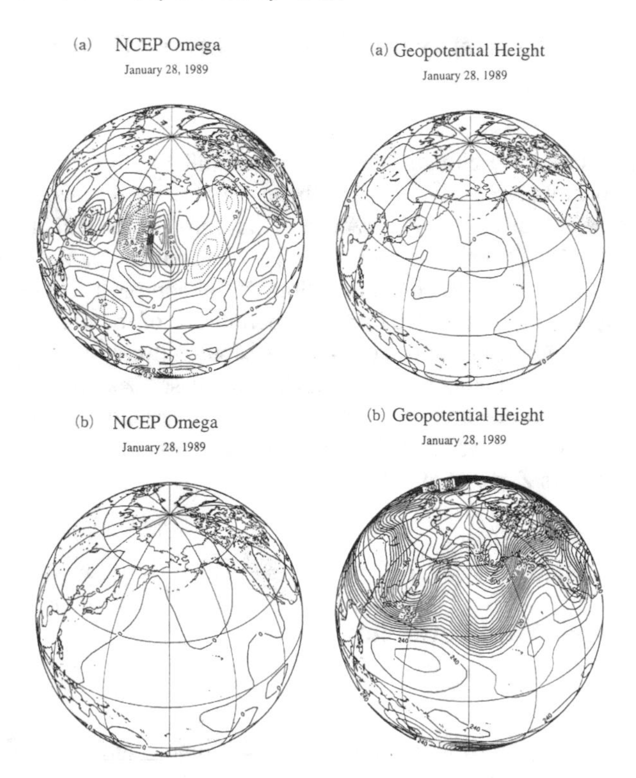

Fig. 4.19 Left: Distribution of ω (Pa s^{-1}) at the 500 hPa level at 00 UTC, January 28, 1989, represented by (a) gravity modes, and (b) Rossby modes, respectively. Right: As in Left, but for geopotential height represented by (a) gravity modes, and (b) Rossby modes, respectively. From Tanaka and Yatagai (2000), Figs. 3 and 4.

σ_{knm} since the rest of parameters are common to Rossby and gravity modes. High-frequency gravity modes have larger σ_{knm} compared with that of Rossby modes, resulting in larger contributions to the omega field.

The central difference scheme and plane fitting method provide a bunch of small-scale ω field as in ECMWF, but those are considered as noise. In contrast, the spherical harmonics method and normal mode method yield smooth ω field as in NCEP. The smooth result in the spectral method may be caused by smooth divergence field in NCEP. On the other hand, the smooth result in the normal mode method is caused by the severe truncations in the gravity modes. The difference in the magnitude for various estimators is substantial. However, the difference is within the difference range between ECMWF and NCEP reanalyses.

4.7 Temporal Variability and Biases in Modal Space

With reanalyses and the climate model outputs available in terms of k, n and m indices, all moments can be evaluated in the space of normal modes for the Rossby and inertio-gravity waves. This is attractive opportunity for the validation of climate models since spatio-temporal variability and biases in climate model thermodynamical fields and circulation are coupled, calling for an approach that considers them simultaneously. In this section we introduce definitions of temporal variability and bias in modal space using the σ coordinate system and the basic equations from chapter 1. We derive an analytical relationship that describes, for each mode, deficiencies in simulated mean circulation and temporal variance as a function of the bias amplitude and the covariance between the reanalysis mean state and bias. The presentation follows Žagar et al. (2020).

4.7.1 Spatio-temporal Variability

As in the previous sections, we work with the meridionally and vertically integrated energy which is computed from the NMF expansion in $\sigma-$system as

$$I_k = \frac{1}{2}\sum_{m=1}^{M} gD_m \sum_{n=1}^{R} \chi_n^k(m)\left[\chi_n^k(m)\right]^* . \qquad (4.105)$$

The spectrum I_k at a single time instant describes how energy is distributed as a function of the zonal wavenumber. Time-averaged energy spectrum is $\overline{I_k} = 1/N\sum_{t=1}^{N} I_k(t)$, with the length of the time series denoted N.

Notice that $\overline{I_k}$ is different from the energy spectra of the mean circulation obtained by computing the energy spectrum from the average Hough expansion coefficients, $\overline{\chi_\nu}$,

$$\overline{\chi_\nu} = \frac{1}{N}\sum_{t=1}^{N} \chi_\nu(t) . \qquad (4.106)$$

Under the ergodicity assumption, it is easy to show that for each ν the difference between the energy of the time mean component $\overline{\chi_\nu}$, $I(\overline{\chi_\nu}) = 1/2\, gD_m \overline{\chi_\nu}\left[\overline{\chi_\nu}\right]^*$, and the time mean energy $\overline{I_k}$ corresponds to the transient circulation energy in mode ν, T_ν

$$T_\nu = \frac{1}{N}\sum_{t=1}^{N}\frac{1}{2}gD_m\left(\chi_\nu(t)-\overline{\chi_\nu}\right)\left(\chi_\nu(t)-\overline{\chi_\nu}\right)^*, \tag{4.107}$$

i.e. $\overline{I_\nu} = I(\overline{\chi_\nu}) + T_\nu$.

The transient circulation energy corresponds to the variance, which for a single mode ν is denoted $V_\nu = V_n^k(m)$,

$$V_\nu = \frac{1}{N-1}\sum_{t=1}^{N} gD_m\left(\chi_\nu(t)-\overline{\chi_\nu}\right)\left(\chi_\nu(t)-\overline{\chi_\nu}\right)^*, \tag{4.108}$$

and variability is its square root. Unit of V_ν is m^2/s^2. Notice that the difference to Eq. (4.107) is the factor 1/2 and the normalization by $(N-1)$ instead of N; in other words, if we use the biased variance, the following relation applies:

$$\overline{I_\nu} - I(\overline{\chi_\nu}) = \frac{1}{2}V_\nu\,. \tag{4.109}$$

For every vertical mode m, the globally integrated squared variability defined by Eq. (4.108) is equivalent to the integral in model space (after the vertical projection) of the variance S_m which at the point $(\lambda_i, \varphi_j, m)$ is defined as

$$S_m(\lambda_i,\varphi_j) = Var(u_m) + Var(v_m) + \frac{g}{D_m}Var(h_m)\,. \tag{4.110}$$

The three elements of the summation (4.110), $Var(u_m)$, $Var(v_m)$ and $Var(h_m)$, denote variances of the wind components and modified geopotential height after the vertical projection; for example, interannual variance in zonal wind, $Var(u_m)$, at the location $(\lambda_i, \varphi_j, m)$ is given by

$$Var(u_m) = \frac{1}{N-1}\sum_{t=1}^{N}\left(u_m(t)-\overline{u}_m\right)^2 = \frac{1}{N-1}\sum_{t=1}^{N}\Delta u_m^2\,, \tag{4.111}$$

where $\overline{u}_m$ is the mean zonal wind at the location and N is the number of data over time period T. The summation of variances on the regular Gaussian grid is performed using the Gaussian weights $w(\lambda_i, \varphi_j)$.

In order to show the equivalence between

$$\sum_m\sum_k\sum_n V_n^k(m) \quad\text{and}\quad \sum_m\sum_i\sum_j w(\lambda_i,\varphi_j)S_m(\lambda_i,\varphi_j), \tag{4.112}$$

we write the horizontal decomposition into Hough harmonics for a difference between the model state at time t and its time-mean over period T, after the vertical projection, $\mathbf{\Delta X}_m$, in terms of $\Delta\chi_n^k(m;t) = \Delta\chi_\nu(t) = \chi_\nu(t) - \overline{\chi_\nu}$:

$$\mathbf{\Delta X}_m = \mathbf{X}_m - \overline{\mathbf{X}}_m = \sum_{n=1}^{R} \sum_{k=-K}^{K} \Delta\chi_n^k(m)\,\mathbf{H}_n^k\,. \tag{4.113}$$

We take the inner product defined in Chapter 1 of (4.113) with itself and multiply from the left by gD_m and apply the vertical expansion and vertical orthogonality relationships. The left -hand side of (4.113) becomes

$$gD_m\langle \mathbf{\Delta X}_m, \mathbf{\Delta X}_m\rangle = \frac{1}{2\pi}\int_0^{2\pi}\int_{-1}^{1}\left(\Delta u_m^2 + \Delta v_m^2 + \frac{g}{D_m}\Delta h_m^2\right)d\mu\, d\lambda\,. \tag{4.114}$$

The right-hand side of (4.113), through the application of the orthogonality of the Hough harmonics, becomes

$$\sum_{n=1}^{R} \sum_{k=-K}^{K} gD_m\,\Delta\chi_\nu\,[\Delta\chi_\nu]^*\,. \tag{4.115}$$

Now we sum both results over N samples, divide by $N-1$ and use definitions of modal and model space variances given by Eq. (4.108) and Eq. (4.111), respectively. Using the Gaussian weights to calculate the right-hand side of (4.114) we obtain

$$\sum_k \sum_n V_n^k(m) \approx \frac{1}{2\pi}\sum_i \sum_j w(\lambda_i, \varphi_j) S_m(\lambda_i, \varphi_j), \tag{4.116}$$

which after summation over the vertical modes corresponds to (4.112) as we stated.

4.7.2 Spectrum of Model Bias

For a single mode ν, we denote the complex expansion coefficients for the climate model and verifying analysis χ_ν^c and χ_ν^a, respectively. Their time-averaged difference is computed as

$$\overline{\Delta\chi_\nu} = \frac{1}{N}\sum_{t=1}^{N}\left[\chi_\nu^c(t) - \chi_\nu^a(t)\right] = \overline{\chi_\nu^c} - \overline{\chi_\nu^a}\quad . \tag{4.117}$$

If differences between the model and analysis were random with zero mean, their time average should be close to zero. A systematic difference, however, is referred to as a bias. Equation (4.117) thus formally defines model bias in every zonal wavenumber, meridional and vertical mode. For any ν or a range of indices, $\overline{\Delta\chi_\nu}$ can be filtered from modal space back to physical space in the same way as for the full fields. As the

NMF decomposition is complete, filtering of all modes should ideally correspond to the averaged difference in physical space for the wind components, geopotential and surface pressure fields. The validity of this statement can be shown in a similar way as done for variance.

By multiplying the amplitudes of bias coefficients $\overline{\Delta\chi_\nu}$ by a factor gD_m we obtain the spectrum of bias variance. In order to see this, we define specific modal bias variance B_ν for a single mode ν as

$$B_\nu = gD_m\overline{\Delta\chi_\nu}\left[\overline{\Delta\chi_\nu}\right]^* \quad . \tag{4.118}$$

The variance of bias at the point $(\lambda_i, \varphi_j, m)$ is defined as

$$Q(\lambda_i, \varphi_j, m) = \overline{\Delta u_m}^2 + \overline{\Delta v_m}^2 + \frac{g}{D_m}\overline{\Delta h_m}^2 , \tag{4.119}$$

with bias for individual variables u_m, v_m and h_m at point $(\lambda_i, \varphi_j, m)$ defined as (for the zonal wind)

$$\overline{\Delta u_m} = \frac{1}{N}\sum_{t=1}^{N}\left[u_m^c(t) - u_m^a(t)\right] = \overline{u_m^c} - \overline{u_m^a} \quad , \tag{4.120}$$

and similarly for v_m and h_m.

We can show that the summation of Eq. (4.118) over all (k, n) is equivalent to the horizontal integral of the squares of the biases in wind and geopotential height fields after the vertical projection, as defined by Eq. (4.119), that is calculated as a double summation using the Gaussian weights $w(\lambda_i, \varphi_j)$:

$$\sum_k\sum_n B_n^k(m) \approx \sum_i\sum_j w(\lambda_i, \varphi_j)\, Q(\lambda_i, \varphi_j, m) \,. \tag{4.121}$$

The proof can be found in Žagar et al. (2020).

By evaluating spatio-temporal variability in climate models in comparison with reanalyses in modal space, we can discuss properties of model biases for the Rossby and IG regimes across scales. Let us see this. First, we denote the average energy spectra for the model simulation and reanalysis data $\overline{I_k^c}$ and $\overline{I_k^a}$, respectively. Using Eq. (4.109) to evaluate $\overline{I_\nu^c} - \overline{I_\nu^a}$ and expressing $I(\overline{\chi_\nu^c})$ as $I(\overline{\chi_\nu^c}) = I(\overline{\Delta\chi_\nu} + \overline{\chi_\nu^a})$, we arrive at the following expression coupling the bias with variance and energy distribution in the model and verifying analyses:

$$\overline{I_\nu^c} - \overline{I_\nu^a} = \frac{1}{2}\left[V_\nu^c - V_\nu^a\right] + \frac{1}{2}B_\nu + P(\overline{\Delta\chi_\nu}, \overline{\chi_\nu^a}), \tag{4.122}$$

with the covariance term $P(\overline{\Delta\chi_\nu}, \overline{\chi_\nu^a})$ defined as

$$P(\overline{\Delta\chi_\nu}, \overline{\chi_\nu^a}) = \frac{1}{2} g D_m \left(\overline{\Delta\chi_\nu} \left[\overline{\chi_\nu^a} \right]^* + \left[\overline{\Delta\chi_\nu} \left[\overline{\chi_\nu^a} \right]^* \right]^* \right) =$$

$$g D_m \left(\left[\overline{\Delta\chi_\nu} \right]_r \left[\overline{\chi_\nu^a} \right]_r + \left[\overline{\Delta\chi_\nu} \right]_i \left[\overline{\chi_\nu^a} \right]_i \right), \quad (4.123)$$

where the subscripts r and i denote the real and imaginary parts, respectively. The term $P(\overline{\Delta\chi_\nu}, \overline{\chi_\nu^a})$ describes the covariance between the mean state of the verifying reanalysis and bias in the same mode. Using (4.109) for the model and reanalysis, Eq. (4.122) can be rewritten as

$$I^c(\overline{\chi_\nu}) - I^a(\overline{\chi_\nu}) = \frac{1}{2} B_\nu + P(\overline{\Delta\chi_\nu}, \overline{\chi_\nu^a}) . \quad (4.124)$$

Equation (4.122) states that deficiency in simulated mean energy in mode ν can be expressed as a sum of three terms: deficiency in simulated variance, the bias variance and the covariance between the mean state of verifying reanalysis and the bias in the same mode ν. Equation (4.124) states that the misrepresentation of the climatological spatial variance (energy of the mean state) in the model with respect to the reanalysis is described by the 2-term bias expression that accounts for differences between the model and reanalysis in the variance and in the mean total energy.

4.7.3 Modal Filtering of Bias

Žagar et al. (2020) compared interannual variability in ERA-20C reanalysis data and numerical simulations performed by SPEEDY model (Molteni, 2003), a primitive-equation GCM with 8 terrain-following vertical levels (near pressures 950, 835, 685, 510, 340, 200, 95 and 25 hPa) and horizontal resolution T30. The model was compared with the ERA-20C reanalyses at the same horizontal and vertical grid. In one setup, the model was used with the prescribed SST from ERA-20C (SPEEDY run) whereas another simulation applied a slab ocean model which updates SST in each forecast step (SPEEDY-SlO run). A relatively good representation of the average precipitation, winds and temperature in SPEEDY became worse in the coupled model (SPEEDY-SlO), as a consequence of unrealistically poor SST produced by the slab-ocean model.

Scale validation revealed that the model underestimates interannual variance at all scales but especially at large scales, and that the variance underestimation in SPEEDY-SlO oscillates among the seasons compared to the SST-forced run. The comparison of the first two terms on the right-hand side of (4.122), missing variance in the model and bias variance is shown in Fig. 4.20 which visualizes the bias variance in all k using the logarithmic scale against the relative missing variance which is computed as a difference between the ERA-20C and simulated variance normalized by ERA-20C (or a difference between SPEEDY and SPEEDY-SIO normalized by SPEEDY). Therefore, values smaller than 0 in Fig. 4.20 imply the variance overestimation whereas values between 0 and 1 denoted increasingly missing variance.

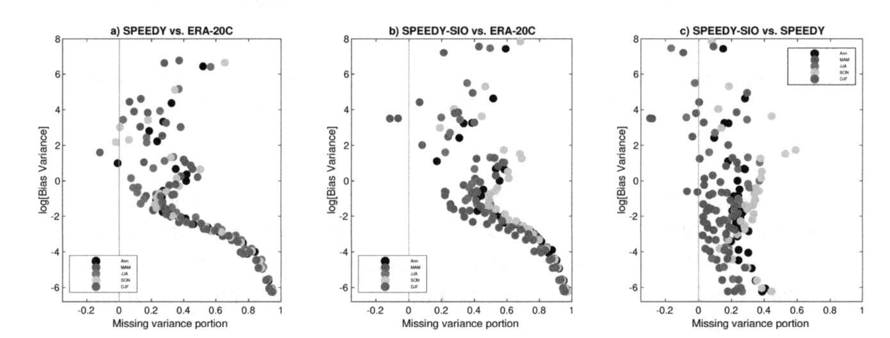

Fig. 4.20 Scatterplot of the logarithm of the bias variance, $\log B_k$ versus the normalized missing model variance, computed as $(V_k^a - V_k^c)/V_k^a$, in annual and seasonal means. (a) SPEEDY and (b) SPEEDY-SIO evaluated against ERA-20C and (c) SPEEDY-SIO evaluated against SPEEDY. Different dots correspond to different zonal wavenumbers k from $k = 0$ (top dots with the largest bias variance) to $k = 30$ (bottom dots with the smallest bias variance). From Žagar et al. (2020).

Biases in $k = 0$ are represented by the dots at the top of the three panels in Fig. 4.20. Dots towards the bottom of each panel correspond to successively larger k suggesting an increasing lack of relative variability in the model as the spatial scale reduces (Fig. 4.20a-b). Biases introduced by the slab-ocean model increase spread in missing portions of variance at the synoptic scales among the seasons that is seen in the area with $\log(B) \approx -2$ (Fig. 4.20b vs. Fig. 4.20a). On the other hand, Fig. 4.20c suggests that the missing portion of variance is greatly reduced in the perfect model scenario although large biases remain at planetary scales due to a very poor SST. The large reduction in missing variance at subsynoptic scales is the most noticeable feature in Fig. 4.20c in support to high-resolution climate modelling providing variability information regardless of the amplitude of large-scale biases.

However, the modal decomposition alone can not tell us what exactly to change in the model. It can guide the physical space analysis and sensitivity studies that should lead to improved models. We can use NMF to filter bias in physical space for any $k > 0$ or a combination of (k, n, m) indices is obtained by setting $\overline{\Delta\chi_n^k(m)} = 0$ for all other k, n and m indices and solving the inverse projection. Such modal filtering relies on the idea that normal-mode functions, although not the normal modes of the model, represent physically meaningful spatial patterns, especially in comparison to

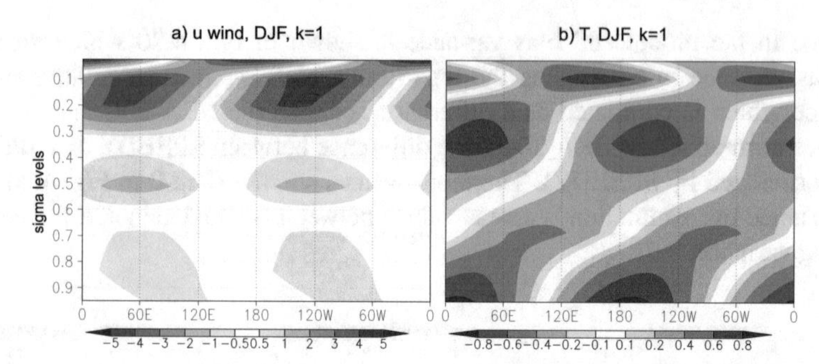

Fig. 4.21 Longitudinal structure of the bias in $k = 1$ along latitude circle 16.7°N in DJF in (a) zonal wind (in m/s) and (b) temperature (in Kelvins). From Žagar et al. (2020).

statistical analysis such as EOFs. In the next two chapters of this monograph we shall refer to a number of studies that successfully employed this assumption to investigate different aspects of dynamics. An example of such filtering is shown in Fig. 4.21 for the vertical structure of bias along a latitude circle in the subtropics. It shows the baroclinic vertical structure of the zonal wind bias and the coupling between biases in remote regions, the surface temperature bias in the Atlantic and western Pacific with the upper troposphere biases in temperature in the opposite hemisphere.

References

Ahlquist, J. (1982). Normal-mode global Rossby waves: Theory and observations. *J. Atmos. Sci.*, 39(1):193–202.

Bartello, P. (1995). Geostrophic adjustment and inverse cascades in rotating stratified turbulence. *J. Atmos. Sci.*, 52:4410–4427.

Bechtold, P. M., Koehler, M., Jung, T., Doblas-Reyes, F., Leutbecher, M., Rodwell, M. J., Vitart, F., and Balsamo, G. (2008). Advances in simulating atmospheric variability with the ECMWF model: from synoptic to decadal time-scales. *Q. J. R. Meteorol. Soc.*, 134:1337–1351.

Chapman, S. and Lindzen, R. S. (1970). Atmospheric tides: Thermal and gravitational. *Gordon and Breach/Sci. Pub.*, page 200 pp.

Charney, J. G. (1947). The dynamics of long waves in a baroclinic westely current. *J. Meteorol.*, 4:135–162.

Chen, T.-C. and Wiin-Neilsen, A. (1978). On nonlinear cascades of atmospheric energy and enstrophy in a two-dimensional spectral index. *Tellus*, 30:313–322.

Dee, D. P. and Coauthors (2011). The ERA-Interim reanalysis: configuration and performance of the data assimilation system. *Q. J. R. Meteorol. Soc.*, 137:553–597–1841.

Dewan, E. (1979). Stratospheric wave spectra resembling turbulence. *Science*, 204:832–835.

Dickinson, R. E. (1968). On the exact and approximate liner theory of vertically propagating planetary Rossby waves forced at a spherical lower boundary. *Mon. Wea. Rev.*, 96:405–415.

Eliasen, E. and Machenhauer, B. (1965). A study of the fluctuation of atmospheric planetary flow patterns represented by spherical harmonics. *Tellus*, 17:220–238.

Errico, R. M. (1989). Theory and application of nonlinear normal mode initialization. NCAR Technical Note, NCAR/TN-344+IA. 145 pp.

Fritts, D. C. and Alexander, J. M. (2003). Gravity wave dynamics and effects in the middle atmosphere. *Reviews of Geophysics*, 41(1):doi:10.1029/2001RG000106.

Fritts, D. C. and Coauthors (2016). The deep propagating gravity wave experiment (DEEPWAVE): An airborne and ground-based exploration of gravity wave propagation and effects from their sources throughout the lower and middle atmosphere. *Bull. Amer. Meteor. Soc.*, 97:425–453.

Garcia, R. R. and Geisler, J. E. (1981). Stochastic forcing of small amplitude oscillation in the stratosphere. *J. Atmos. Sci.*, 38:2187–2197.

Gubenko, V. N., Pavelyev, A. G., and Andreev, V. E. (2008). Determination of the intrinsic frequency and other wave parameters from a single vertical temperature or density profile measurement. *J. Geophys. Res.*, 113(D8):dos:10.1029/2007JD008920.

Hamilton, K. (1991). Climatological statistics of stratospheric inertia-gravity waves deduced from historical rocketsonde wind and temperature data. *J. Geophys. Res*, 96(20):831–20.

Hayasaki, M. and Tanaka, H. L. (1999). A study of drastic warming in the troposhere: A case study for the winter of 1989 in Alaska. *Tenki*, 46:123–135.

Hayashi, Y. (1980). Estimation of nonlinear energy transfer spectra by the cross-spectral method. *J. Atms. Sci.*, 37:2299–307.

Hayashi, Y. (1982). Space-time spectral analysis and its applications to atmospheric waves. *J. Meteor. Soc. Japan*, 60:156–171.

Hirota, I. (1971). Excitation of planetary Rossby waves in the winter stratosphere by periodic forcing. *J. Meteor. Soc. Japan*, 49:439–449.

Holmström, I. (1963). On a method for parametric representation of the state of the atmosphere. *Tellus*, 15:127–149.

Hough, S. S. (1898). On the application of harmonic analysis to the dynamical theory of the tides - Part II. On the general integration of Laplace's dynamical equations. *Phil. Trans. Roy. Soc. London*, A191:139–185.

Kao, S. K. (1968). Governing equations and spectra for atmospheric motion and transports in frequency wavenumber space. *J. Atmos. Sci.*, 25.

Kasahara, A. (1976). Normal modes of ultralong waves in the atmosphere. *Mon. Wea. Rev.*, 104:669–690.

Kasahara, A. (1977). Numerical integration of the global barotropic primitive equations with hough harmonic expansions. *J. Atmos. Sci.*, 34:687–701.

Kasahara, A. (1978). Further studies on a spectral model of the global barotropic primitive equations with hough harmonic expansions. *J. Atmos. Sci.*, 35:2043–2051.

Kasahara, A. (1980). Effect of zonal flows on the free oscillations of a barotropic atmosphere. *J. Atmos. Sci.*, 37:917–929. Corrigendum, J. Atmos. Sci., 38 (1981), 2284–2285.

Kasahara, A. (1984). The linear response of a stratified global atmosphere to a tropical thermal forcing. *J. Atmos. Sci.*, 41:2217–2237.

Kasahara, A. and Puri, K. (1981). Spectral representation of three-dimensional global data by expansion in normal mode functions. *Mon. Wea. Rev.*, 109:37–51.

Kitamura, Y. and Matsuda, Y. (2006). The k^{-3} and $k^{-5/3}$ energy spectra in stratified urbulence. *Geophys. Res. Lett.*, 111:L05809, doi:10.1029/2005GL024996.

Kitamura, Y. and Matsuda, Y. (2010). Energy cascade processes in rotating stratified turbulence with application to the atmospheric mesoscale. *Geophys. Res. Lett.*, 115:L11104, doi:10.1029/2009JD012368.3013–3030.

Kung, E. (1988). Spectral energetics of the general circulation and time spectra of transient waves during the FGGE year. . *J. Clim.*, 1:5–19.
Kung, E. C. (1972). A scheme for kinematic estimate of large-scale vertical motion with an upper-air network. *Quart. J. Roy. Meteor. Soc.*, 98:402–411.
Kung, E. C. and Tanaka, H. (1983). Energetics analysis of the global circulation during the special observation periods of fgge. *J. Atmos. Sci.*, 40:2575–2592.
Kung, E. C. and Tanaka, H. (1984). Spectral characteristics and meridional variations of energy transformations during the first and second special observation periods of fgge. *J. Atmos. Sci.*, 41:1836–1849.
Lamb, H. (1932). Hydrodynamics. Dover Pub.
Leith, C. E. (1971). Atmospheric predictability and two-dimensional turbulence. *J. Atmos. Sci.*, 28:145–161.
Lindborg, E. (2006). The energy cascade in a strongly stratified fluid. *J. Fluid Mech.*, 550:207–242.
Lindborg, E. and Cho, J. (2001). Horizontal velocity structure functions in the upper troposphere and lower stratosphere: 2. theoretical considerations. *J. Geophys. Res.*, 106:10233–10241.
Lindzen, R. S., Straus, D. M., and Katz, B. (1984). An observational study of large-scale atmospheric Rossby waves during FGGE. *J. Atmos. Sci.*, 41:1320–1335.
Longuet-Higgins, M. S. (1968). The eigenfunctions of Laplace's tidal equations over a sphere. *Phil. Trans. Roy. Soc. London, Series A. Mathematical and Physical Sciences*, 262:511–607.
Lorenz, E. N. (1955). Available potential energy and the maintenance of the general circulation. *Tellus*, 7:157–167.
Madden, R. (1978). Further evidence of traveling planetary waves. *J. Atmos. Sci.*, 35:1605–1618.
Matsuno, T. (1966). Quasi-geostrophic motions in the equatorial area. *J. Meteor. Soc. Japan*, 44:25–42.
Merilees, P. E. (1966). On the linear balance equation in terms of spherical harmonics. *Tellus*, 20:200–202.
Miura, H., Satoh, M., Nasuno, T., Noda, A. and Oouchi, K. (2007). Madden-Julian oscillation event realistically simulated using a global cloud-resolving model. *Science*, 318:1763–1765.
Molteni, F. (2003). Atmospheric simulations using a GCM with simplified physical parametrizations. I: model climatology and variability in multi-decadal experiments. *Clim. Dyn.*, 20:175–195.
Nastrom, G. D. and Gage, K. S. (1985). A climatology of aircraft wavenumber spectra observed by commercial aircraft. *J. Atmos. Sci.*, 42:950–960.
Nasuno, T., Tomita, H., Iga, S., Miura, H. and Satoh, M. (2007). Multi-scale organization of convection simulated with explicit cloud processes on an aquaplanet. *J. Atmos. Sci.*, 64:1902–1921.
O'Brien, J. J. (1970). Alternative solutions to the classical vertical velocity problem. *J. Appl. Meteor.*, 9:197–203.
Onogi, K., Tsutsui, J., Koide, H., and Coauthors (2007). The JRA-25 Reanalysis. *J. Meteorol. Soc. Japan*, 85:369–434.

Salby, M. L., Garcia, R. R., O'Sullivan, D. and Tribbia, J. (1990). Global transport calculations with an equivalent barotropic system. *J. Atmos. Sci.*, 47:188–214.

Salby, M. (1981). Rossby normal modes in nonuniform background configurations, part ii: Equinox and solstice conditions. *J. Atmos. Sci.*, 38:1827–1840.

Saltzman, B. (1957). Equations governing the energetics of the larger scales of atmospheric turbulence in the domain of wavenumber. *J. of Meteorology*, 14:513–523.

Sasaki, Y. K. and Chang, L. P. (1985). Numerical solution of the vertical structure equation in the normal mode method. *Mon. Wea. Rev.*, 113:782–793.

Sato, K. (1994). A statistical study of the structure, saturation, and sources of inertia-gravity waves in the lower stratosphere observed with the mu radar. *J. Atmos. Terr. Phys.*, 56:755–774.

Satoh, M., Matsuno, T., Tomita, T., Miura, H., Nasuno, T. and Iga, S. I. (2008). Nonhydrostatic icosahedral atmospheric model (nicam) for global cloud resolving simulations. *J. of Computational Physics*, 227:3486–3514.

Shigehisa, Y. (1983). Normal modes of the shallow water equations for zonal wavenumber zero. *J. Meteor. Soc. Japan*, 61:479–493.

Silva Dias, P. L. and Schubert, W. H. (1979). The dynamics of equatorial mass-flow adjustment. Atmos. Sci.

Swarztrauber, P. N. and Kasahara, A. (1985). The vector harmonic analysis of Laplace tidal equations. *SIAM J. Stat. Comput.*, 6:464–491.

Takahashi, Y. O., Hamilton, K. and Ohfuchi, W. (2006). Explicit global simulation of the mesoscale spectrum of atmospheric motions. *J. Geophys. Lett.*, 33:L12812.

Tanaka, H. (1985). Global energetics analysis by expansion into three-dimensional normal-mode functions during the FGGE winter. *J. Meteor. Soc. Japan*, 63:180–200.

Tanaka, H. L. and Kung, E. (1988). Normal-mode expansion of the general circulation during the FGGE year. *J. Atmos. Sci.*, 45:3723–3736.

Tanaka, H. L. (1984). On the amplification and vertical propagation of zonal wavenumber 1 for january 1979. *Gross Wetter (in Japanese)*, 22:17–25.

Tanaka, H. L. and Kasahara, A. (1992). On thenormal modes of Laplace's tidal equations for zonal wavenumber zero. *Tellus*, 44A:18–32.

Tanaka, H. L. and Milkovich, M. F. (1990). A heat budget analysis of the polar troposphere in and around Alaska during the abnormal winter of 1988/89. *Mon. Wea. Rev.*, 118:1628–1639.

Tanaka, H. L. and Sun, S. (1990). A study of baroclinic energy source for large-scale atmospheric normal modes. *J. Atmos. Sci.*, 47:2674–2695.

Tanaka, H. L. and Terasaki, K. (2005). Energy spectrum and energy flow of the Arctic oscillation in the phase speed domain. *SOLA*, 1:65–68.

Tanaka, H. L. and Yatagai, A. (2000). Comparative study of vertical motions in the global atmosphere eveluatedby various kinematical schemes. *J. Meteor. Soc. Japan*, 78:289–298.

Tanaka, H., Kung, E. C. and Baker, W. E. (1986). Energetics analysis of the observed and simulated general circulation using three-dimensional normal mode expansion. *Tellus*, 38A:412–428.

Terasaki, K. and Tanaka, H. L. (2007). An analysis of the 3-D atmospheric energy spectra and interactions using analytical vertical structure functions and two reanalyses. *J. Meteor. Soc. Japan*, 85:785–796.

Terasaki, K., Tanaka, H. L., and Žagar, N. (2011). Energy spectra of Rossby and gravity waves. *SOLA*, 11:45–48.

Terasaki, K., Tanaka, H. L., and Satoh, M. (2009). Characteristics of the kinetic energy spectrum of nicam model atmosphere. *SOLA*, 5:180–183.

Tomita, H., Miura, H., Iga, S. I., Nasuno, T. and Satoh, M. (2005). A global cloud-resolving simulation: Preliminary results from an aqua planet experiment. *Geophys. Res. Lett.*, 32:L08805.

Tribbia, J. (1979). Non-linear initialization on an equatorial beta plane. *Mon. Wea. Rev.*, 107:704–713.

Tsuda, T., Nishida, M., Rocken, C., and Ware, R. (2000). A global morphology of gravity wave activity in the stratosphere revealed by the GPS occultation data (GPS/MET). *J. Geophys. Res.*, 105:doi:10.1029/1999JD901005.

Uppala, S., Kallberg, P., Simmons, A., Andrae, U., da Costa Bechtold, V., Fiorino, M., Gibson, J., Haseler, J., Hernandez, A., Kelly, G., Li, X., Onogi, K., Saarinen, S., Sokka, N., Allan, R., Andersson, E., Arpe, K., Balmaseda, M., Beljaars, A., van de Berg, L., Bidlot, J., Bormann, N., Caires, S., Chevallier, F., Dethof, A., Dragosavac, M., Fisher, M., Fuentes, M., Hagemann, S., Holm, E., Hoskins, B., Isaksen, L., Janssen, P., Jenne, R., McNally, A., Mahfouf, J.-F., Morcrette, J.-J., Rayner, N., Saunders, R., Simon, P., Sterl, A., Trenberth, K., Untch, A., Vasiljevic, D., Viterbo, P., and Woollen, J. (2005). The ERA-40 re-analysis. *Q. J. R. Meteorol. Soc.*, 131:2961–3012.

Van Zandt, T. E. (1982). A universal spectrum of buoyancy waves in the atmosphere. *Geophys. Res. Lett.*, 9:575–578.

Žagar, N., Boyd, J., Kasahara, A., Tribbia, J., Kallen, E., Tanaka, H., and Yano, J.-I. (2016). Normal modes of atmospheric variability in observations, numerical weather prediction, and climate models. *Bulletin of the American Meteorological Society*, 97:ES125–ES128.

Žagar, N., Jelić, D., Blaauw, M., and Bechtold, P. (2017). Energy spectra and inertia-gravity waves in global analyses. *J. Atmos. Sci.*, 74:2447–2466.

Žagar, N., Kasahara, A., Terasaki, K., Tribbia, J., and Tanaka, H. (2015). Normal-mode function representation of global 3D datasets: open-access software for the atmospheric research community. *Geosci. Model Dev.*, 8:1169–1195.

Žagar, N., Kosovelj, K., Manzini, E., Horvat, M., and Castanheira, J. (2020). An assessment of scale-dependent variability and bias in global prediction models. *Clim. Dyn.*, pages 1–20.

Waite, M. L. and Snyder, C. (2009). The mesoscale kinetic energy spectrum of a baroclinic life cycle. *J. Atmos. Sci.*, 66:883–901.

Waite, M. L. and Snyder, C. (2013). Mesoscale energy spectra in moist baroclinic waves. *J. Atmos. Sci.*, 70:1242–1256.

Wiin-Nielsen, A. (1967). On the annual variation and spectral distribution of atmospheric energy. *Tellus*, 19:540–559.

William, E. and Richard, C. (1977). Elementary differential equations and boundary value problems. Third Edition, Wiley.

Chapter 5

Generalization of Baroclinic Instability and Rossby Wave Saturation Theory

Hiroshi L. Tanaka

Abstract This chapter describes a problem of barotropic and baroclinic instability in realistic zonal and non-zonal basic state on a sphere, using three-dimensional (3D) spectral primitive equations derived by the 3D NMFs. The most unstable mode in a realistic zonal basic state, called Charney mode, appears in mid-latitudes associated with the baroclinicity of the subtropical jet. In a zonally-varying basic state, we find that the unstable modes are modified by the regionality of the local baroclinicity of the basic state. Given the zonally varying barotropic basic state, we find that the barotropically most unstable standing mode appears to be the Arctic Oscillation (AO) mode. The eigensolution of the linear baroclinic model (LBM) in this section is regarded as a generalized extension of the 3D normal mode at the motionless atmosphere to those of an arbitrary climate basic state. Some applications of the nonlinear primitive equation models are presented for the nonlinear life-cycle experiment of baroclinic waves. The saturation of the growing Rossby waves produces the characteristic energy spectrum obeying the squared phase speed of Rossby waves. We call it as the Rossby wave saturation theory which explains the observed energy spectrum of the geostrophic turbulence in the phase speed domain.

5.1 Barotropic and Baroclinic Instability in Realistic Basic State

5.1.1 Introduction

The problem of large amplification of planetary waves attracts increasingly more attention in association with low-frequency variabilities in the atmosphere. Ener-

N. Žagar and J. Tribbia (eds.), *Modal View of Atmospheric Variability*,
Mathematics of Planet Earth 8, https://doi.org/10.1007/978-3-030-60963-4_5

getics analyses of the observed circulation field often assert an enhanced baroclinic conversion in the amplified planetary waves, implying that the amplification involves a process of baroclinic instability (e.g. Schilling, 1986). However, the linear stability analyses of planetary waves with zonal basic states (e.g. Simmons and Hoskins, 1976; Hartmann, 1979) indicate major discrepancies with observation. The expected growth rate is insufficient to explain the amplification, the unstable planetary waves propagate eastward, and the wave structure of the most unstable mode is confined to the troposphere, even though less unstable mode may penetrate into the deep atmosphere.

Comprehensive case studies of blocking episodes during the First GARP (Global Atmosphere Research Program) Global Experiment (FGGE) indicate that synoptic baroclinic waves feed large amount of energy into the planetary waves by means of an up-scale, nonlinear energy cascade (Hansen and Chen, 1982; Kung and Baker, 1986; Holopainen and Fortelius, 1987). Long-term statistics of blocking formations also suggest an important role of high-frequency transient eddies in reinforcing the vorticity field of blocking waves (Colucci, 1986; Mullen, 1987). Although energetics analyses and the enstrophy budget may be helpful in understanding the cause of blocking formations, the characteristic structure and behavior of blockings are not fully explained by the energy redistribution due to the wave-wave interaction.

Unstable eigenmodes in a zonally varying basic state were numerically investigated by Simmons et al. (1983), using a barotropic model. They demonstrated that various tropical forcings tend to excite a unique unstable normal mode in the model atmosphere. Moreover, with his two-layer, quasi-geostrophic model, Frederiksen (1982) showed that the zonal asymmetry of the basic state reorganizes the synoptic baroclinic waves to yield the Atlantic and Pacific storm tracks. Presumably his high-frequency synoptic disturbances are associated with Charney type instability in a zonal basic state. He also found blocking-like unstable modes with a dipole structure among a number of unstable solutions. He proposed that the dipole unstable mode explains the onset of the blocking formation. However, the physical explanation of the dipole modes seems less clear than the high-frequency synoptic disturbances. Although Frederiksen and Bell (1987) have extended his model to a five-layer tropospheric model, the structure and behavior of their eigenmodes in planetary waves are too complicated to relate to the well-known unstable Charney and Green type modes in a zonal basic state. It is desirable to compare both the unstable modes in a zonal basic state and those in a zonally varying basic state. It is also desirable to use primitive equations because the quasi-geostrophic equations involve quasi-nondivergent and quasi-geostrophic assumptions, which may affect eigensolutions for planetary waves.

The objective of this section is to investigate the low-frequency, unstable planetary waves in realistic global basic states of January 1979 during the FGGE as described by Tanaka and Kung (1989) (hereafter TK1989). The eigenfrequencies, modal structures, and energetics of low-frequency unstable solutions are contrasted for zonal and zonally varying basic states in order to examine the effect of zonal asymmetry of the basic state on the low-frequency, unstable planetary waves. For that purpose, we have solved linearized, three-dimensional spectral primitive equations with a basis of three-dimensional normal mode functions (3-D NMFs) of motionless atmosphere. The use of the normal mode expansion is advantageous for stability analysis in that the matrix size for the eigenvalue problem can be effectively reduced by retaining only the rotational mode basis, and excluding the gravity mode basis. Thus, it is possible to analyze atmospheric eigenmodes with primitive equations, not only for zonal basic states but also for zonally varying basic states.

First, the governing equations in terms of the three-dimensional, spectral primitive equations are reduced to the eigenvalue problem as described by Tanaka and Seki (2013) (hereafter TS2013). The results for the zonal basic states are presented to identify the the unstable modes with Charney (1947) and Green (1960) modes in previous research. We then examine how the zonal asymmetry of the basic state modulates the unstable planetary waves. Finally, we discuss the relation between the results of low-frequency unstable modes and the observed large-scale Pacific blocking and the amplification of wavenumber 1 during the winter season.

5.1.2 Governing Equations

A system of primitive equations with a spherical coordinate of longitude λ, latitude θ, pressure p, and time t may be reduced to three prognostic equations of horizontal motions and thermodynamics for three dependent variables of $\mathbf{U}=(u, v, \phi)^T$. Here, u and v are the zonal and meridional components of the horizontal velocity V. The variable ϕ is a departure of the local isobaric geopotential from the reference state geopotential ϕ_0, which is related through the hydrostatic equation to the reference state temperature T_0, and the superscript T denotes a transpose. Refer to the governing equations in Chapter 4.1 derived for energetics analysis. Here, we describe the equations in more detail for theoretical derivations. Using a matrix notation, these primitive equations may be written as

$$\mathbf{M}\frac{\partial \mathbf{U}}{\partial t} + \mathbf{L}\,\mathbf{U} = \mathbf{N} + \mathbf{F}, \tag{5.1}$$

where

$$\mathbf{U} = (u, v, \phi)^T, \tag{5.2}$$

$$\mathbf{M} = diag(1,\ 1,\ -\frac{\partial}{\partial p}\frac{p^2}{R\gamma}\frac{\partial}{\partial p}), \tag{5.3}$$

$$\mathbf{L} = \begin{pmatrix} 0 & -2\Omega\, sin\theta & \frac{1}{a\, cos\theta}\frac{\partial}{\partial\lambda} \\ 2\Omega\, sin\theta & 0 & \frac{1}{a}\frac{\partial}{\partial\theta} \\ \frac{1}{a\, cos\theta}\frac{\partial}{\partial\lambda} & \frac{1}{a\, cos\theta}\frac{\partial()cos\,\theta}{\partial\theta} & 0 \end{pmatrix}, \tag{5.4}$$

$$\mathbf{N} = \begin{pmatrix} -\mathbf{V}\cdot\nabla u - \omega\frac{\partial u}{\partial p} + \frac{tan\theta}{a}uv \\ -\mathbf{V}\cdot\nabla v - \omega\frac{\partial v}{\partial p} - \frac{tan\theta}{a}uu \\ \frac{\partial}{\partial p}(\frac{p^2}{R\gamma}\mathbf{V}\cdot\nabla\frac{\partial\phi}{\partial p} + \omega p(\frac{p}{R\gamma}\frac{\partial\phi}{\partial p})) \end{pmatrix}, \tag{5.5}$$

$$\mathbf{F} = (F_u,\ F_v,\ \frac{\partial}{\partial p}(\frac{pQ}{C_p\gamma}))^T. \tag{5.6}$$

The left-hand side of (5.1) represents linear terms with matrix operators **M** and **L** and the dependent variable vector **U**. The matrix **M** is referred to as a mass matrix which is nonsingular and positive definite under a proper boundary conditions. The right-hand side represents a nonlinear term vector **N** and a diabatic term vector **F**, which includes the zonal F_u and meridional F_v components of frictional forces and a diabatic heating rate Q. The symbol a represents the earth's radius, Ω the angular speed of the earth's rotation, R the specific gas constant, C_p the specific heat at a constant pressure, γ the static stability parameter, ∇ the horizontal del-operator, and $\omega = dp/dt$ is diagnostically related to divergence.

The boundary conditions of the system are vanishing kinematical wind (u, v, w) at $p = p_s$ and the sum of kinetic energy K and available potential energy A integrated over the whole vertical column of the atmosphere is finite:

$$(u, v, w) = 0 \quad at \quad p = p_s,$$
$$\int_0^{p_s} K + A\,dp < \infty, \tag{5.7}$$

where $w = dz/dt$ is kinematical vertical velocity. These boundary conditions are imposed for the computation of the normal mode expansion basis so that the atmospheric variables become the subspace of the system. Refer to Tanaka (1998) for the detail of the model description and variables.

In order to obtain a system of spectral primitive equations, we expand the vectors **U** and **F** in 3-D normal mode functions in a resting atmosphere, $\mathbf{\Pi}_{knm}(\lambda, \theta, p)$:

$$\mathbf{U}(\lambda, \theta, p, t) = \sum_{knm} w_{knm}(t)\mathbf{X}_m\mathbf{\Pi}_{knm}(\lambda, \theta, p), \tag{5.8}$$

$$\mathbf{F}(\lambda, \theta, p, t) = \sum_{knm} f_{knm}(t)\mathbf{Y}_m\mathbf{\Pi}_{knm}(\lambda, \theta, p). \tag{5.9}$$

Here, the dimensionless expansion coefficients $w_{knm}(t)$ and $f_{knm}(t)$ are the functions of time alone. The subscripts represent zonal wavenumbers k, meridional indices n, and vertical indices m. They are truncated at K, N, and M, respectively. The vertical indices m=0 and $m \neq 0$ represent barotropic (external) and baroclinic (internal) modes.

The scaling matrices should be defined for each vertical index as:

$$\mathbf{X}_m = diag(\sqrt{gh_m}, \sqrt{gh_m}, gh_m), \tag{5.10}$$

$$\mathbf{Y}_m = diag(2\Omega\sqrt{gh_m}, 2\Omega\sqrt{gh_m}, 2\Omega), \tag{5.11}$$

where *diag* represents diagonal matrix and the entries $\sqrt{gh_m}$ is a phase speed of gravity waves in shallow water, associated with the equivalent height h_m for the vertical mode m.

The expansion basis of the 3-D normal mode functions $\mathbf{\Pi}_{knm}(\lambda, \theta, p)$ is obtained as eigensolutions of a homogeneous partial differential equation, putting zero for the right-hand side of (5.1). By the method of separation of variables, we obtain a set of vertical structure equation and horizontal structure equation (Laplace's tidal equation), associated with the linear operators **M** and **L**, respectively:

$$-\frac{d}{dp}\left(\frac{p^2}{R\gamma}\frac{dG_m}{dp}\right) = \frac{1}{gh_m}G_m, \tag{5.12}$$

$$(Y_m^{-1}\mathbf{L}X_m)\,\mathbf{H}_{knm} = i\sigma_{knm}\mathbf{H}_{knm}, \tag{5.13}$$

where the equivalent height h_m appears as the separation constant, representing the vertical scale of a vertical normal mode due to the self-adjyoint property of the system. The dimensionless eigenfrequency σ_{knm} is referred to as Laplace's tidal frequency, which represents the scale of a Hough mode in the primitive equations.

The 3-D normal mode functions are given by a tensor product of vertical structure functions (vertical normal modes, G_m) and Hough harmonics (horizontal normal modes, $\mathbf{H}_{knm}$) as $\mathbf{\Pi}_{knm} = G_m\,\mathbf{H}_{knm}$. It is known that they form a complete set and satisfy an orthonormality condition under an inner product $<,>$ defined as:

$$\begin{aligned} <\mathbf{\Pi}_{knm}, \mathbf{\Pi}_{k'n'm'}> &= \tfrac{1}{2\pi p_s}\int_0^{p_s}\int_{-\pi/2}^{\pi/2}\int_0^{2\pi}\mathbf{\Pi}^*_{knm}\cdot\mathbf{\Pi}_{k'n'm'}cos\theta d\lambda d\theta dp \\ &= \delta_{kk'}\delta_{nn'}\delta_{mm'}, \end{aligned} \tag{5.14}$$

where the asterisk denotes the complex conjugate, the symbols δ_{ij} is the Kronecker delta, and the surface pressure p_s is treated as a constant near the earth's surface.

Using the orthonormality condition (5.14) of the 3-D normal mode functions, the expansion coefficients of the state variables w_{knm} and external forcing f_{knm} in (5.8) and (5.9) may be computed by the set of inverse Fourier transforms:

$$\begin{aligned} w_{knm} &=< \mathbf{U}(\lambda, \theta, p, t),\ \mathbf{X}_m^{-1}\, \mathbf{\Pi}_{knm}(\lambda, \theta, p) >, \\ f_{knm} &=< \mathbf{F}(\lambda, \theta, p, t),\ \mathbf{Y}_m^{-1}\, \mathbf{\Pi}_{knm}(\lambda, \theta, p) > . \end{aligned} \tag{5.15}$$

Applied to the same inner product for (5.1), the weak form of the primitive equation becomes

$$< \mathbf{M}\frac{\partial \mathbf{U}}{\partial t} + \mathbf{L}\,\mathbf{U} - \mathbf{N} - \mathbf{F},\ \mathbf{Y}_m^{-1}\mathbf{\Pi}_{knm} >= 0. \tag{5.16}$$

Substituting (5.8) and (5.9) into (5.16), rearranging the time-dependent variables, and evaluating the remaining terms, we obtain a system of 3-D spectral primitive equations in terms of the spectral expansion coefficients:

$$\frac{dw_i}{d\tau} + i\sigma_i w_i = -i\sum_{jk} r_{ijk} w_j w_k + f_i, \quad i = 1, 2, 3, \ldots \tag{5.17}$$

where τ is a dimensionless time scaled by $(2\Omega)^{-1}$ and r_{ijk} is the interaction coefficients for nonlinear wave-wave interactions. The triple subscripts are shortened for simplicity as $w_{knm} = w_i$. There should be no confusion in the use of i for a subscript even though it is used for the imaginary unit.

In order to derive (5.17) from (5.16), we first show the following relation for the linear terms.

$$< \mathbf{M}\frac{\partial \mathbf{U}}{\partial t} + \mathbf{L}\,\mathbf{U},\ \mathbf{Y}_m^{-1}\mathbf{\Pi}_{knm} >= \frac{dw_i}{d\tau} + i\sigma_i w_i, \quad i = 1, 2, 3, \ldots \tag{5.18}$$

The vertical differential operator $\mathbf{M}$ may be replaced by its eigenvalue based on the relation of the vertical structure equation (5.12):

$$\mathbf{M}\,\mathbf{\Pi}_i = diag(1, 1, \frac{1}{gh_i})\,\mathbf{\Pi}_i. \tag{5.19}$$

By substituting (5.8) in (5.18), using the relation in (5.13) and (5.19), we obtain

$$\begin{aligned} \sum_j < 2\Omega \mathbf{Y}_j^{-1}\mathbf{M}\mathbf{X}_j\mathbf{\Pi}_j, \mathbf{\Pi}_i > \tfrac{dw_j}{d\tau} + < \mathbf{Y}_j^{-1}\mathbf{L}\mathbf{X}_j\mathbf{\Pi}_j, \mathbf{\Pi}_i > w_j \\ = \tfrac{dw_i}{d\tau} + i\sigma_i w_i, \end{aligned} \tag{5.20}$$

which completes the proof of (5.18).

The proof for the external forcing **F** in (5.16) to be transformed to f_i in (5.17) is straightforward by the relation (5.15).

Finally, we derive the specific form of the nonlinear interaction coefficients r_{ijk} in (5.17). As noted before, the 3-D normal mode function is given by the tensor products of G_m and $\mathbf{H}_{knm}$ as $\mathbf{\Pi}_{knm} = \mathbf{H}_{knm} G_m$, in which the Hough harmonics are given by the tensor products of the meridional normal modes (Hough vector functions) and longitudinal normal modes (complex-valued trigonometric functions): $\mathbf{H}_{knm} = (U_{knm}, -iV_{knm}, Z_{knm})^T e^{ik\lambda}$. The computational method of the Hough vector functions $(U_{knm}, -iV_{knm}, Z_{knm})^T$ are detailed by Swarztrauber and Kasahara (1985), and that of the vertical normal mode G_m by Kasahara (1984). We assume that those basis functions are already available.

By taking the inner products of the nonlinear term **N** and the 3-D normal mode functions, we can prove the following relation for the nonlinear interaction coefficients:

$$< \mathbf{N}, \mathbf{Y}_m^{-1} \mathbf{\Pi}_{knm} >= -i \sum_{jk} r_{ijk} w_j w_k, \quad i = 1, 2, 3, \ldots \tag{5.21}$$

The running indices i, j, k represent combinations of the 3-D wavenumbers. We need to distinguish them respectively as $k_i n_i m_i$, $k_j n_j m_j$, and $k_k n_k m_k$. Likewise, the equivalent heights and vertical structure functions are also distinguished similar way with the subscripts of i, j, k. Substituting (5.5) and (5.11) into (5.21), the inner product to be computed becomes:

$$< \mathbf{N}, \mathbf{Y}_i^{-1} \, \mathbf{\Pi}_i >= \frac{1}{2\pi p_s} \int_0^{p_s} \int_{-\pi/2}^{\pi/2} \int_0^{2\pi} \begin{pmatrix} \frac{1}{2\Omega\sqrt{gh_i}} & U_i \, G_i e^{-ik_i\lambda} \\ \frac{1}{2\Omega\sqrt{gh_i}} & (iV_i) \, G_i e^{-ik_i\lambda} \\ \frac{1}{2\Omega} & Z_i \, G_i e^{-ik_i\lambda} \end{pmatrix}^T \begin{pmatrix} -\mathbf{V} \cdot \nabla u - \omega \frac{\partial u}{\partial p} + \frac{\tan\theta}{a} uv \\ -\mathbf{V} \cdot \nabla v - \omega \frac{\partial v}{\partial p} - \frac{\tan\theta}{a} uu \\ \frac{\partial}{\partial p} [\mathbf{V} \cdot \nabla (\frac{p^2}{R\gamma} \frac{\partial \phi}{\partial p}) + \omega p \frac{\partial}{\partial p} (\frac{p}{R\gamma} \frac{\partial \phi}{\partial p})] \end{pmatrix} \cos\theta d\lambda d\theta dp \,. \tag{5.22}$$

It is recognized that the nonlinear terms are at most the second order nonlinearity of the state variables. Using (5.8) we substitute the following expansion of the state variables in the nonlinear terms of (5.22):

$$\begin{pmatrix} u \\ v \\ \phi \end{pmatrix} = \sum_i w_i \begin{pmatrix} \sqrt{gh_i} & U_i \\ \sqrt{gh_i} & (-iV_i) \\ gh_i & Z_i \end{pmatrix} G_i \mathrm{e}^{ik_i\lambda}. \tag{5.23}$$

The vertical p-velocity ω may be expanded as the next form based on the continuity equation:

$$\omega = \sum_i w_i \, 2\Omega \int_0^p G_i dp(-i\sigma_i Z_i)\, e^{ik_i\lambda}. \tag{5.24}$$

The vertical integral in (5.24) can be replaced by the first order derivative derived from the integral of (5.12) as:

$$\int_0^p G_i dp = -\frac{gh_i}{R\gamma} p^2 \frac{dG_i}{dp}. \tag{5.25}$$

Moreover, the second order vertical derivative in (5.22) can be replaced by the first order derivative derived from (5.12) as:

$$-p\frac{d}{dp}\frac{p}{R\gamma}\frac{dG_i}{dp} = \frac{p}{R\gamma}\frac{dG_i}{dp} + \frac{G_i}{gh_i}. \tag{5.26}$$

With those preparations, the final form of the computation for the nonlinear interaction coefficients is summarized as the volume integral of the triple products of the normal mode functions:

$$\begin{aligned} <\mathbf{N}, \mathbf{Y}_i^{-1}\mathbf{\Pi}_i> &= -i\textstyle\sum_j\sum_k w_j w_k \frac{1}{2\pi p_s}\int_0^{p_s}\int_{-\pi/2}^{\pi/2}\int_0^{2\pi} \\ &\begin{pmatrix} U_i \\ V_i \\ Z_i \end{pmatrix}^T \begin{pmatrix} P_1(\frac{k_k\mathbf{U}_k}{\cos\theta} + \tan\theta V_k) & -P_1\frac{d\mathbf{U}_k}{d\theta} & P_2 U_k \\ P_1(\frac{k_k\mathbf{V}_k}{\cos\theta} + \tan\theta U_k) & -P_1\frac{d\mathbf{V}_k}{d\theta} & P_2 V_k \\ P_3\frac{k_k\mathbf{Z}_k}{\cos\theta} & -P_3\frac{d\mathbf{Z}_k}{d\theta} & -P_4 Z_k \end{pmatrix} \begin{pmatrix} U_j \\ V_j \\ \sigma_j Z_j \end{pmatrix} \\ &e^{i(-k_i+k_j+k_k)\lambda}\cos\theta d\lambda d\theta dp. \end{aligned} \tag{5.27}$$

Here, the triple products of the vertical structure functions are combined with the scaling parameters as:

$$\begin{aligned} P_1 &= \frac{\sqrt{gh_j}\sqrt{gh_k}}{2\Omega a\sqrt{gh_i}} G_i G_j G_k, \\ P_2 &= \frac{\sqrt{gh_k gh_j}}{\sqrt{gh_i}R\gamma} p^2 G_i \frac{dG_j}{dp}\frac{dG_k}{dp}, \\ P_3 &= \frac{\sqrt{gh_j}}{2\Omega a} G_i G_j G_k - \frac{\sqrt{gh_j gh_k}}{2\Omega a R\gamma} p^2 G_i \frac{dG_j}{dp}\frac{dG_k}{dp}, \\ P_4 &= G_i G_j G_k + \frac{gh_k}{R\gamma} p G_i G_j \frac{dG_k}{dp} + \frac{gh_j}{R\gamma} p G_i \frac{dG_j}{dp} G_k \\ &+ (\frac{gh_k}{R\gamma} - 1)\frac{gh_j}{R\gamma} p^2 G_i \frac{dG_j}{dp}\frac{dG_k}{dp}, \end{aligned} \tag{5.28}$$

which completes the description of the real-valued nonlinear interaction coefficients r_{ijk}, represented by the volume integral in (5.27).

As shown in (5.27), the nonlinear interactions are non-zero only when the zonal wavenumbers satisfy the relation $k_i = k_j + k_k$. In (5.27), there are many first derivatives of the normal modes which are obtainable analytically when these are evaluated in terms of a series expansion with the Associated Legendre functions. Hence, the computations for r_{ijk} are all analytical except for the volume integrals by means of the Gausian quadrature which is exact under the specified truncations of the Legendre polynomials. In this study, we applied the spectral expansion method by Kasahara and Puri (1981) for the computation of the vertical normal modes. It is worth noting that the spectral primitive equation (5.17) is as accurate as the original one in (5.1) with approximately 1‰ in error for the dynamics part.

5.1.3 Development of the Linear Baroclinic Model

Linear Baroclinic Model (LBM) developed in this study is based on a spectral primitive equation expanded by the three-dimensional normal mode functions Tanaka and Kung (1989). We linearize (5.17) by introducing a perturbation method to develop the LBM. Independent variables w_i and external forcing f_i are divided into the time-independent basic state $(\bar{w}_i, \bar{f}_i)$ and small perturbations superimposed on the basic state $(w_i, f_i$: the same symbols as the original variables are used for convenience), and substituted into (5.17). After disregarding nonlinear perturbation terms, we obtain a linear spectral primitive equation model.

$$\frac{dw_i}{d\tau} + i\sigma_i w_i = -i \sum_{j=1}^{L} \left(\sum_{k=1}^{L} \left(r_{ijk} + r_{ikj} \right) \bar{w}_k \right) w_j + f_i, \tag{5.29}$$

where the modal index k is used for the basic state, and i and j for the perturbations which interact with the basic state. Linear equation (5.29) is the governing equation of the LBM. In (5.27), the interaction coefficients r_{ikj} need to satisfy the triad combination for the zonal wavenumber $k_i = k_j + k_k$. Thus, if the basic state $\bar{w}_k$ is zonally symmetric $(k_k = 0)$, r_{ikj} is required to satisfy $k_i = k_j$. Due to the limitation of calculation resources, Tanaka and Kung (1989) solved the LBM only for zonal wavenumber 1 system, and Tanaka and Tokinaga (2002) and Seki et al. (2011) employed only the zonally symmetric basic state and conducted the linear stability analysis in such a highly idealized case. But, the real atmosphere is zonally asymmetric associated with the regionality of the westerly or the effect of land-sea distribution for the atmosphere. In this study, the LBM allows k_k to have all the zonal wavenumbers within its truncation so that we can conduct a linear stability analysis with a zonally-asymmetric basic state.

Since the negative zonal wavenumbers represent the complex conjugates of the positive zonal wavenumbers, we can rearrange (5.29) in a matrix form for $k_i \geq 0$ and $L_1 = (K + 1)(N + 1)(M + 1)$:

$$\frac{d\mathbf{w}}{d\tau} + i\mathbf{D}\mathbf{w} = -i\mathbf{B}\mathbf{w} - i\mathbf{C}\mathbf{w}^* + \mathbf{f}, \tag{5.30}$$

where

$$\mathbf{w} = (w_1, \cdots, w_i, \cdots, w_{L_1})^T, \quad \text{for} \quad k_i \geq 0 \tag{5.31}$$

$$\mathbf{f} = (f_1, \cdots, f_i, \cdots, f_{L_1})^T, \quad \text{for} \quad k_i \geq 0 \tag{5.32}$$

$$\mathbf{D} = diag(\sigma_1, \cdots, \sigma_i, \cdots, \sigma_{L_1}), \quad \text{for} \quad k_i \geq 0 \tag{5.33}$$

$$\mathbf{B} = \sum_{k=1}^{L_1} \left(r_{ijk} + r_{ikj}\right) \bar{w}_k, \quad \text{for} \quad k_j \geq 0 \tag{5.34}$$

$$\mathbf{C} = \sum_{k=1}^{L_1} \left(r_{ijk} + r_{ikj}\right) \bar{w}_k, \quad \text{for} \quad k_j < 0. \tag{5.35}$$

5.1.3.1 Eigenvalue Problem for Non-zonal Basic State

In order to perform the linear stability analysis for zonally-asymmetric basic state, a complex wave-type solution is substituted into (5.29). Because an inviscid and adiabatic eddy will be examined, we disregard **f** for perturbations. The solution may be assumed in the form:

$$\begin{bmatrix} \mathbf{w_R} \\ \mathbf{w_I} \end{bmatrix}(\tau) = \begin{bmatrix} \xi \\ \zeta \end{bmatrix} e^{\nu\tau}, \tag{5.36}$$

where $\mathbf{w_R}$ is a real part and $\mathbf{w_I}$ is a complex part of $\mathbf{w}$. ξ and ζ are structure vectors of the solution and ν is the eigenvalue of a frequency. Since both matrices **B** and **C** in (5.30) become complex full matrices, they are also split into real and imaginary parts. Then the eigenvalue problem becomes:

$$\nu \begin{bmatrix} \xi \\ \zeta \end{bmatrix} = \begin{bmatrix} \mathbf{B_I} + \mathbf{C_I} & \mathbf{B_R} - \mathbf{C_R} + \mathbf{D} \\ -\mathbf{B_R} - \mathbf{C_R} - \mathbf{D} & \mathbf{B_I} - \mathbf{C_I} \end{bmatrix} \begin{bmatrix} \xi \\ \zeta \end{bmatrix}. \tag{5.37}$$

The complex matrices **B** and **C** are determined by the basic state $\bar{w}_k$, and **D** is a constant matric containing the Laplace's tidal frequency. Therefore, the eigenvalues (ν) and eigenvectors (ξ, ζ) may be solved using a standard real-valued matrix solver. The real part of the eigen solution is expressed by

$$\begin{bmatrix} \mathbf{w_R} \\ \mathbf{w_I} \end{bmatrix}(\tau) = 2e^{\nu_R\tau} \left(\begin{bmatrix} \xi_\mathbf{R} \\ \zeta_\mathbf{R} \end{bmatrix} \cos \nu_I \tau - \begin{bmatrix} \xi_\mathbf{I} \\ \zeta_\mathbf{I} \end{bmatrix} \sin \nu_I \tau \right). \tag{5.38}$$

The real part of the eigenvalue ν_R represents a growth rate, and the complex part ν_I represents a frequency. An eigenmode with a positive growth rate is called an unstable mode. In this study, an eigenmode with the largest growth rate ν_R is regarded as the fastest growing baroclinic instability mode. The eigenvalue problem (5.37) was first solved by Tanaka and Kung (1989) only for the system of zonal wavenumber 1. The present study has extended the system to full interactions with all zonal wavenumbers.

As is clear from the solution, the eigenmode has a life-cycle in its structure with the period determined by ν_I. As a special case when the eigenvalue is real, the structure is geographically trapped with no life-cycle except for the positive and negative sign of the eigenvector. The mode is called a standing mode, which appears to be most interesting for the study of the Arctic Oscillation as discussed in this study.

5.1.3.2 Eigenvalue Problem for Zonal Basic State

For a basic state at rest ($\overline{w}_k$=0), both **B** and C vanish, thus the equation (5.30) satisfies Laplace's classical tidal theory for inviscid adiabatic motion.

For a zonal basic state ($\overline{w}_k \neq 0$ if k''=0), the matrix **B** becomes a real block diagonal ($b_{ij} \neq 0$ if k=k'), and **C** vanishes. Consequently, (5.30) can be solved for each (k,k') block:

$$\frac{d}{d\tau}\mathbf{W}_k = -i\mathbf{D}_k\mathbf{W}_k - i\mathbf{B}_k\mathbf{W}_k, \quad k = 1, 2, \ldots, K, \tag{5.39}$$

where the subscripts represent wavenumbers of (k,k') blocks associated with perturbation vectors $\mathbf{W}_k$. Because (5.39) is linear, we can assume the solution of $\mathbf{W}_k$ to be a function of dimensionless time τ:

$$\mathbf{W}_k(\tau) = \xi e^{-i\nu\tau}. \tag{5.40}$$

The initial value problem (5.39) is then reduced to an eigenvalue problem for a real matrix with eigenvactors ξ and eigenvalues ν as:

$$\nu\xi = (\mathbf{D}_k + \mathbf{B}_k)\xi. \tag{5.41}$$

We can solve this matrix eigenvalue problem (EVP) by using a standard eignvalue solver. The EVP (5.41) for the zonal basic state is a special case of (5.37) for the general climate basic state.

5.1.4 Applications

Tanaka (1985) and Tanaka and Kung (1989) (TK1989) applied the derived theory to the FGGE data and used twelve vertical structure functions (m=0-11) that were constructed numerically after Kasahara (1984) using Legendre function expansion method with 24 Gaussian levels. The vertical index m=0 is called an external (barotropic) modes, and m=1-11 are called internal (baroclinic) modes which have m nodes in the vertical. The numerical internal modes are the discrete approximation of the continuous spectrum. Since the higher order internal modes have a problem of aliasing in the stratosphere and above Sasaki and Chang (1985); Staniforth et al. (1985), only the first seven vertical structure functions (m=0-6) were applied. The Hough vector functions were computed using semi-normalized associated Legendre polynomials with 120 Gaussian latitudes so that the integral is exact up to the triple product of the basis functions. The meridional indices n contain two distinct modes: gravity-inertial modes and rotational modes of the Rossby-Haurwitz type. One advantage of Hough mode expansion is an efficient reduction of the matrix size of the instability problem by retaining the rotational modes alone, excluding the gravity mode basis. As is shown in the Appendix A of TK1989, excluding the gravity mode basis does not affect the unstable solution of planetary waves. Moreover, the unstable solutions have almost converged with meridional truncation N=18 for the rotational modes.

Unstable eigenmodes are examined for two different zonal basic states: a zonal wind profile of a 30°jet described by Simmons and Hoskins (1976) and the observed monthly-mean field of January 1979. The 30°jet profile has a separable structure in the vertical and the meridional, and is deliberately chosen to be barotropically stable Simmons and Hoskins (1976).

5.1.4.1 Growth Rates and Phase Speeds

TK1989 presented the growth rate and phase speeds of the first three unstable modes for zonal wavenumbers $k = 1 - 10$. This is a result of antisymmetric solutions with the 30°jet basic state. A pronounced Charney-type baroclinic instability (labeled M_c) appeared with the e-folding time of about 2.0 (day) at $k = 7 - 8$ with the phase speed of about 9 (°/day). Contrasted with the result by **?**, they found that the stable layer in the stratosphere reduces the growth rate for $k > 6$, and the maximum growth rates shift toward smaller wavenumbers. According to Ioannou and Lindzen (1986) analytical solutions, the unstable modes in the figure, M_c, M_2, and M_1 can be identified as Charney modes with different meridional structures. The phase speed of M_1 was very slow in the planetary waves, whereas that of M_2 was fast (about

9°/day). The distinct phase speeds of these unstable modes were also detected by Gall (1976) and Zhang and Sasamori (1985).

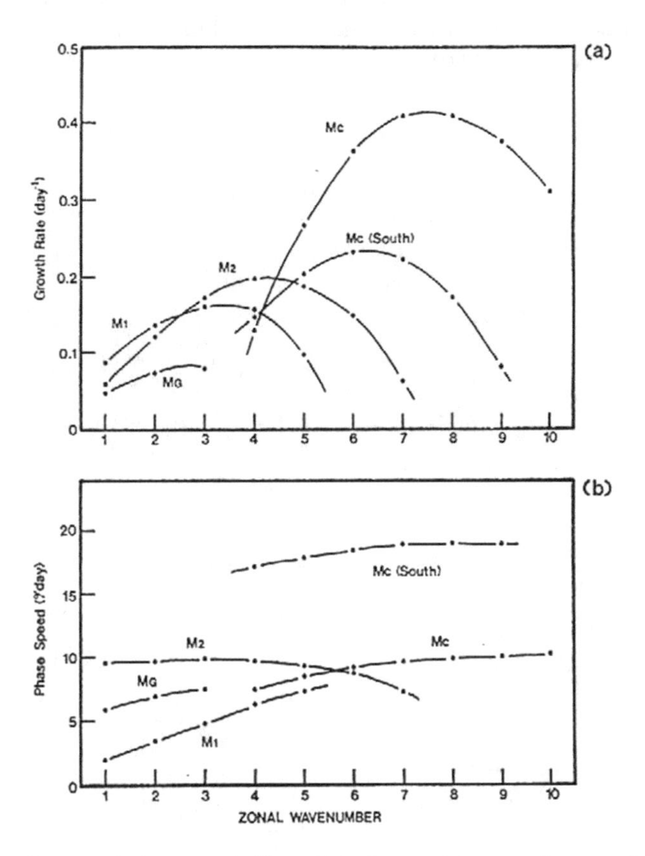

Fig. 5.1 Growth rate and phase speed of baroclinically unstable waves for a zonal basic state of January 1979. M_c is the shallow Charney mode, and M_1 is Polar mode, M_2 the dipole Charney modes. M_G is regarded as Green mode. M_c(south) is that appearing in Southern Hemisphere. From Tanaka (1993), Fig. 4.

Figure 5.1 illustrates the solutions for the January basic state. For the basic state having a global extension, the unstable modes are partitioned in the northern and southern hemispheric modes according to the analysis of their structure. The results in Fig. 5.1 are of the northern hemispheric modes, and a similar result has been obtained by a symmetric extension of the northern basic state toward the Southern Hemisphere. Contrasted with the results of TK1989 we find here the growth rate of M_c decrease in the planetary waves, and the maximum growth rates of M_1 and M_2 shift toward the planetary wave range. As a result, M_1 and M_2 become the dominant unstable modes at k=1-2, and k=3-4. Further experiments showed that the up-scale shift of M_1 and M_2 are caused mainly by the small meridional scale of the basic state in the mid- to lower-troposphere resulting from the tropical and polar easterlies. The results are consistent with those documented by Zhang and Sasamori (1985). The shallow Charney mode M_c seems to have changed to the Green mode M_G in the planetary waves (Green (1960)).

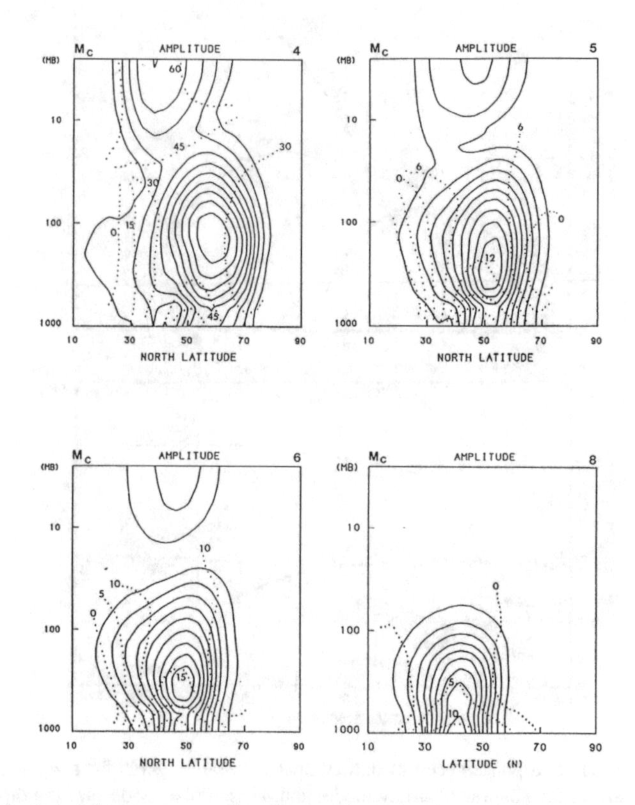

Fig. 5.2 Latitude-height structures of geopotential amplitudes (solid line in arbitrary unit) and phases (dashed line denoting longitudes of ridges) of unstable Charney modes M_c for zonal wavenumbers 4 to 8 (upper right). From Tanaka (1993), Fig. 5.

The latitude-height structures of the geopotential field for M_1, M_2, M_G, and M_c for the January basic state were illustrated in Tanaka (1993). The structure of M_c at k=6 in 5.2 shows an amplitude maximum at the upper troposphere in mid-latitudes, including a westward phase tilt with height. These results resemble those obtained by Hartmann (1979) in which a typical winter basic state is used. The structure of wavenumber 8 shows an amplitude maximum near the surface in mid-latitude, including a westward phase tilt with height. The tilt implies a northward heat transport, which would release zonal available potential energy. The structure, and thus the eddy heat and momentum transports, is consistent with Simmons and

Hoskins' results. At planetary waves, the growth rates in Fig. 5.1 are competed, and the solution is hardly obtainable with the time integration method as discussed by Simmons and Hoskins. That, however, is not the case for the present eigenvalue problem. The structure of M_2 shows a westward phase tilt with height, but indicates a characteristic northwest-southeast phase tilt in the horozontal plane. This phase structure distinguishes M_2 from M_c. The monopole structure of amplitude becomes dipole in the higher zonal waves as seen in Simmons and Hoskins' result. For the structure of M_2 at k=3, the amplitude maximum shifts northward near 60°N above the tropopause level, indicating a northwest-southeast phase tilt. The phase structure indicates a westward phase tilt with height at mid-latitude, but the structure is nearly barotropic and out of phase at high latitudes. The structure of M_1 at k=1 shows an amplitude maximum at 100 mb in high latitudes and another peak in the upper stratosphere, indicating a monotonic westward phase tilt with height. This structure also closely resembles the results obtained by Hartmann (1979) and Zhang and Sasamori (1985), even though the former identifies M_1 as the deep Charney mode whereas the latter identifies it as the Green modes. Tanaka (1993) identified M_1 and M_2 as deep Charney modes. In addition, M_1 was characterized by its large external component of m=0, which is typical of external Charney modes, contrasted with the dominant internal component of m=1 for Green mode. The high latitudes are necessary for deep Charney modes as suggested by Hartmann.

In summary, the NMF approach to baroclinicity identified at least four distinct unstable modes:

1. Shallow Charney mode M_c has its e-folding time of about 2 days, and the maximum growth rate is seen at the synoptic-scale zonal wavenumbers (k=7-8). This mode is shallow in its structure in a sense that it is trapped within the troposphere and will not penetrate into the deep atmophsere.

2. Dipole-Charney mode M_2 has one node in the meridional structure compared with the shallow Charney mode which has no node. This mode dominates at the zonal wavenumbers k=3-4.

3. Polar-Charney mode M_1 has largest meridional and vertical structure. The vertical structure indicates that the largest energy is contained at the external component, which is the common characteristics of Charney modes.

4. Green mode M_G indicates the largest amplitude within the middle atmosphere. The vertical energy spectrum indicates its energy peak at the first internal component.

These results suggest that not only the fastest growing modes but also the second and third fastest unstable modes are important for many atmospheric events. The shallow Chaney mode is important in exciting synoptic disturbances as the major energy source for atmospheric eddies. The dipole Chaney mode may be related to the formation of blocking anticyclone; the Polar Charney mode to the steady planetary waves; and the Green mode to the stratospheric sudden warming. Further study is expected to compare the unstable modes in the planetary waves with observed weather phenomena.

5.2 Generalization of the 3D-NMF in Realistic Climate Basic State

5.2.1 Background and Motivation

The analysis of the baroclinic instability for a zonally-varying basic state was pursued by Frederiksen (1982) in the quasi-geostrophic framework. He showed that the zonal asymmetry of the basic state reorganizes the synoptic scale baroclinic waves to yield the Pacific and Atlantic storm tracks. He also found a blocking-like unstable mode with a meridional dipole structure among a number of unstable solutions. These unstable modes indicate a life-cycle with changing 3D structure combined with the exponential growth as an unstable solution. Frederiksen also found a unique standing unstable mode with a geographically fixed 3D structure. However, the physical interpretation of the dipole modes and the standing mode was less clear compared with that of synoptic-scale unstable Charney modes.

The standing unstable mode was further investigated by Tanaka and Matsueda (2005) (hereafter TM2005), using a barotropic spectral model linearized about a non-zonal climate basic state. With an inclusion of the scale-dependent diffusion to the linear dynamical system, they found that the standing unstable mode shows a similar structure to the Arctic Oscillation (AO) proposed by Thompson and Wallace (1998). It is a barotropically unstable standing mode, but the growth rate becomes about zero by introducing a surface friction in the form of Rayleigh friction. Since the eignsolution has a zero eigenvalue with zero frequency and zero growth rate, the governing matrix of the linear system becomes singular, and the AO mode appears to be resonant to the arbitrary quasi-steady forcing.

The AO is an atmospheric annular pattern dominating in the Northern Hemisphere especially during winter. The AO is defined as a leading EOF (empirical orthogonal function) mode of the wintertime see-level pressure (SLP). The score time-series is referred to as the AO Index (AOI) which indicates the strength of the AO. When the

AOI is positive, a low-pressure anomaly appears in the poler region north of 60° N and a high-pressure belt appears in the mid-latitudes surrounding the low-pressure anomaly. Thus the polar jet near 60° N is intensified for the positive AOI having a barotropic structure.

One of the factors that intensifies the AO is an interaction with the baroclinic eddies by the zonal-wave interactions. Limpasuvan and Hartmann (2000) and Lorenz and Hartmann (2003) showed that the zonal-wave interactions intensify the zonal wind which is the relevant characteristics of the AO. Tanaka and Tokinaga (2002) and Seki et al. (2011) theoretically examined the interaction between the AO and the baroclinic instability by analyzing the eddy momentum transport for various zonal basic states with positive and negative AOI. Tanaka and Tokinaga (2002) conducted a linear instability analysis for the atmospheric basic state with the extremely positive and negative AOI. They found that the basic state with the positive AOI tends to excite the baroclinic modes which transport more eddy momentum poleward to intensify the polar jet. Such a baroclinic instability appearing in planetary scale was called as a polar mode (M_1). On the other hand, the ordinary baroclinic instability appearing in a synoptic scale is known as a Charney mode (M_C) showing a single amplitude maximum in mid-latitude. The Charney mode with a meridional dipole structure in geopotential is called as a dipole Charney mode (M_2). The positive feedback between the AO and baroclinically unstable modes is further investigated by Seki et al. (2011) for the wide range of the AOI. As the AOI becomes large positive, the structure of M_2 is modified to that of M_1, and the ridge axes of the M_C change to tilt from northeast to southwest. The eddy momentum is then transported more poleward than the ordinary situation to intensify the polar jet. These results explain that the baroclinically unstable modes interact with the AO in a sense of positive feedback by the modification of the features of the eddy momentum transport.

The previous studies by Tanaka and Tokinaga (2002) and Seki et al. (2011) adopted the zonally symmetric basic state for the linear stability analysis due to the limitation of the computational resources. Thus the experiment does not include the wave-wave interactions, and the atmospheric variable does not feel the local effects of the large mountain ranges or the large-scale land-sea distribution. Eichelberger and Hartmann (2007) shows that the leading mode of variability describes latitudinal shifting of the eddy-driven jet when the jets are well separated. Chang and Fu (2002); Chang and Fu (2003) find that the AO is well correlated with the storm tracks especially in the Atlantic sector where the westerly tends to become the double-jet shape. Although there are many studies on the variability in the zonal-eddy interactions associated with the locality of the atmospheric state, the spatial relationship between the AO and the baroclinic instability has not been studied theoretically for the non-zonal basic state.

The purpose of this section is to develop a general Linear Baroclinic Model (LBM) from the 3D spectral primitive equation model as described by Tanaka and Seki (2013) (hereafter TS2013). The nonlinear equations are linearized about various non-zonal basic states such as positive and negative AOI. The local change in baroclinic-barotropic instability is theoretically investigated for various phases of the positive and negative AOI. The LBM developed in this study enables us to conduct a linear stability analysis with the zonally-varying basic state, so that we can investigate the regionality of the positive feedback and the spatial structure of the baroclinically unstable modes.

5.2.2 Data

Tanaka and Seki (2013) used the data from NCEP/NCAR reanalysis of four-times-daily zonal wind, meridional wind and geopotential height on constant pressure levels since 1950 Kalnay et al. (1996). Figure 5.3a illustrates the 30 year mean zonal mean zonal wind for December to February (DJF) during 1971 to 2000 used as a basic state for the linear stability analysis. The structure is reconstructed by the truncated basic state of $\bar{w}_k$.

For the 3D spectral model, the wavenumbers are truncated at K=20, N=20, M=6, with a boundary condition of an equatorial wall, solving for the Northern Hemisphere. The vertical structure functions are calculated for 24 vertical layers by mean of the spectral method Kasahara and Puri (1981), and only the lower order vertical modes up to M are used in this study because the higher order modes have a problem of aliasing.

For the Arctic Oscillation Index (AOI), we used the NCEP/NCAR reanalysis since 1950, computed with the barotropic component of the expansion coefficient Tanaka (2003). To represent the basic state modified by the AO, we added (or subtracted) the anomalies of dependent variables regressed onto the normalized AOI to (or from) the climatology. For example, adding three times anomaly (Fig. 5.3b) to the climatology (Fig. 5.3a), we can construct the basic state for the AOI $+3\sigma$ (σ: standard deviation). In this way, zonal means of the the zonally-varying basic states for the AOI from -3σ to $+3\sigma$ are represented (Fig. 5.3c, d). Because of its zonally-varying character, the basic states are different at each longitude. For instance, the distributions of the zonal wind tend to have a double-jet shape in the Atlantic sector and a single-jet shape in the Pacific sector in the climatology. The structures regressed on the AOI also vary for each longitude. The eigenmodes are affected by these localities embedded in the basic states as will be confirmed in the later section.

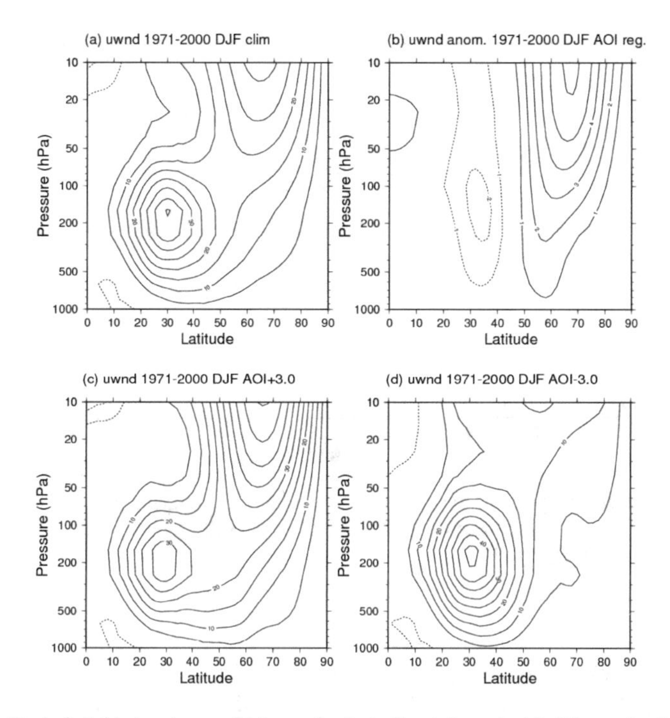

Fig. 5.3 Latitude-height structures of (a) zonal wind climatology (m/s), (b) zonal wind anomaly regressed onto the AOI (m/s), (c) the basic state of zonal wind (m/s) for the AOI +3.0σ, and (d) as in (c) but for the AOI -3.0σ. The climate data are averaged for DJF from 1971 to 2000 in the Northern Hemisphere. Solid (dotted) line is positive (negative) value. From Tanaka and Seki (2013), Fig. 1.

5.2.3 LBM for a Zonal Basic State

The linear stability analysis is performed first for the zonal basic state. The basic state for January rather than DJF is used here in order to compare with the results by Seki et al. (2011). The baroclinic instability for a zonal basic state is compared for the solutions of the general form (5.37) and the zonal form (5.41) in order to confirm the methodology of the LBM in the general form. Figure 5.4 plots the scatter diagram of the complex-valued eigenvalues, i.e., spectrum, obtained as the solution for the LBM (5.37) for the zonal basic state as in Fig. 5.1a. The abscissa denotes the frequency ν_I and the ordinate denotes the growth rate ν_R in dimensionless values. According to the result, we find a dominant unstable Charney mode M_C as connected by a solid line at the lower frequency range at 0.1 which corresponds to the period $1/(2\nu_I)$ = 5 day. For the zonal wavenumber 8, the period corresponds to the phase speed 360°/(8× 5 day)= 9°/day. The peak growth rate of 0.04 corresponds to $4\pi\nu_R$ = 0.5 /day in a dimensional value. The growth rate and frequency coincide with the values in previous studies.

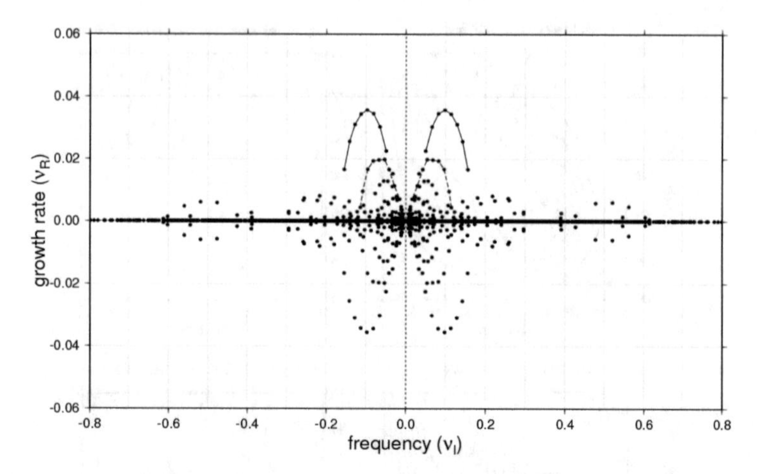

Fig. 5.4 The scatter diagram of the growth rate (ν_R) and the frequency (ν_I) given the complex eigenvalues for the zonally symmetric Janaury basic state. Solid (dotted) line is M_C (M_2). From Tanaka and Seki (2013), Fig. 3.

Since the system matrix governing the instability (5.37) is real, the complex conjugate of the solution is also the solution of the system. Therefore, the scatter diagram becomes symmetric about the frequency in positive and negative domain. For the zonal basic state, the system can be solved as (5.62). In this case, the complex conjugate of the unstable solution is also the solution, which corresponds to the decaying mode. Therefore the scatter diagram becomes symmetric also for the positive and negative growth rates. The second unstable modes are connected by dotted lines, and they are identified as the dipole Charney modes M_2, There are much higher order Charney modes at the lower frequency range. There are some weak unstable modes at frequencies 0.2 (period 2.5 day) and 0.5 (period 1 day). It may be important to note that there is no standing mode at ν_I=0. All solutions are travelling modes, and the direction is expressed by the eigenvector. When the basic state is at rest, all modes become neutral with zero growth rate because there is no energy in the basic state, and the spectrum aligns along the abscissa as a normal mode for resting atmosphere.

Figure 5.5 illustrates the horizontal structures of the geopotential for the Charney mode M_C at zonal wavenumber $k = 6$ and that for the dipole-Charney mode M_2 at $k = 2$, reconstructed by the barotropic component of the solutions. Such a height distribution is called barotropic height. Although the structure indicates a westward vertical tilt as seen by Fig.5.2, the barotropic structure explains the characteristics of the eddy momentum transport. The Charney mode M_C of k=6 shows the amplitude maximum at the mid latitude about 45°N, and the trough and ridge axes tilt opposite direction in lower and higher latitudes so that the eddy momentum flux converges at about 45°N. Note that the polar jet is decelerated in this case. The baroclinic

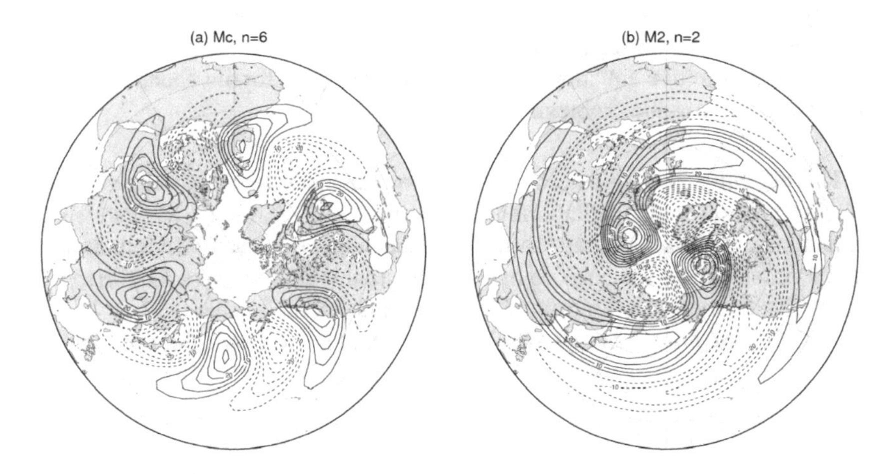

Fig. 5.5 Barotropic height (in arbitrary unit) of (a) M_C at zonal wavenumber 6, and (b) M_2 at zonal wavenumber 2 in the Northern Hemisphere. Solid (dotted) line is positive (negative) value. From Tanaka and Seki (2013), Fig. 4.

wave moves eastward with the phase speed about 10°/day maintaining the same 3D structure. In contrast, the dipole Charney mode M_2 of k=2 shows two amplitude maxima at 45°N and at 70°N, and the trough and ridge axes tilt from northeast to southwest so that the eddy momentum flux converges at 70°N. The polar jet is accelerated in this case. Seki et al. (2011) examined the detail of the meridional dipole structure, and found that the dipole Charney mode M_2 is modified to the polar mode M_1 for large AOI. In the following subsections, we use the fact that the structure of M_2 has a similar property with M_1 in the eddy momentum flux. The baroclinic wave moves eastward with a constant phase speed maintaining the same 3D structure. Hence, it is confirmed that the linear stability solution by LBM in (5.37) reproduces the same baroclinic instability for the zonal basic state by (5.41).

5.2.3.1 LBM for a Zonally-Varying Basic State

The linear stability analysis using the 3D spectral primitive equation model is now extended to a zonally-varying basic state compiled as the 30 year mean January climate of the observed atmosphere. The solutions are recognized as a generalization of the 3D normal modes from the resting atmosphere to a general 3D basic state on the sphere. The basic state has three meteorological variables of zonal wind, meridional wind, and geopotential deviation from the global mean on the isobaric surface. In general, when the wind and mass fields are not in geostrophic balance, gravity waves are generated rapidly associated with the geostrophic adjustment. Because of this imbalance problem, a linear stability problem has been difficult to solve in the primitive equation, and only the quasi-geostrophic equation has been solved as an eigenvalue problem.

The difficulty is overcome by the 3D normal mode method, because we can effectively eliminate the unimportant gravity waves by means of a truncation of the basis functions. If desired, we can include Kelvin modes or other important gravity modes depending on the purpose of the study. In this study, we are interested in the Arctic Oscillation as an atmospheric unstable normal mode on a realistic 3D basic state. Therefore, gravity modes are eliminated from the beginning of the analysis.

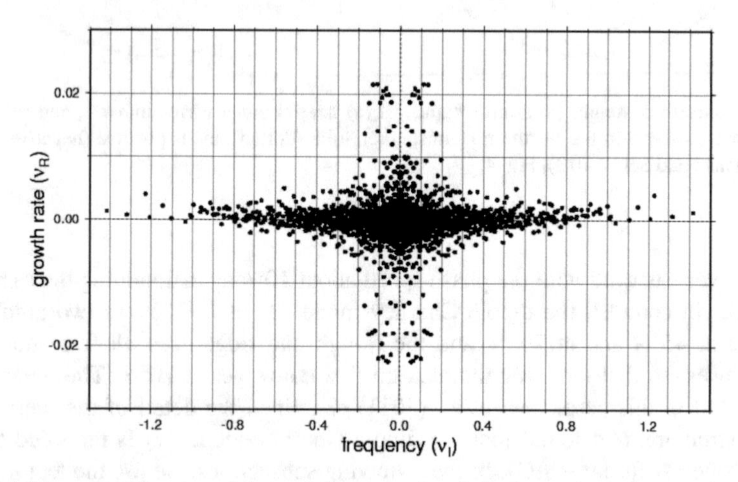

Fig. 5.6 The scatter diagram of the growth rate (ν_R) and the frequency (ν_I) given the complex eigenvalues for non-zonal January basic state. From Tanaka and Seki (2013), Fig. 5.

Figure 5.6 plots the scatter diagram of the complex-valued eigenvalues obtained as the solution for the LBM (5.37) for the zonally-varying basic state of the 30-year mean January data. The abscissa denotes the frequency ν_I, and the ordinate denotes the growth rate ν_R in dimensionless values. According to the result, we can identify a dominant unstable Charney mode M_C at the lower frequency range about 0.1 which corresponds to the period 5 day. Since the system matrix is not decoupled in zonal wavenumbers, one unstable mode has all contributions from different zonal wavenumbers. Thus we cannot argue the property of each zonal wavenumber, nor we can calculate the phase speed of the unstable mode. Instead, all transient modes have a life-cycle described by (5.38). Note that there are some standing modes with real eigenvalues, although the growth rate is small.

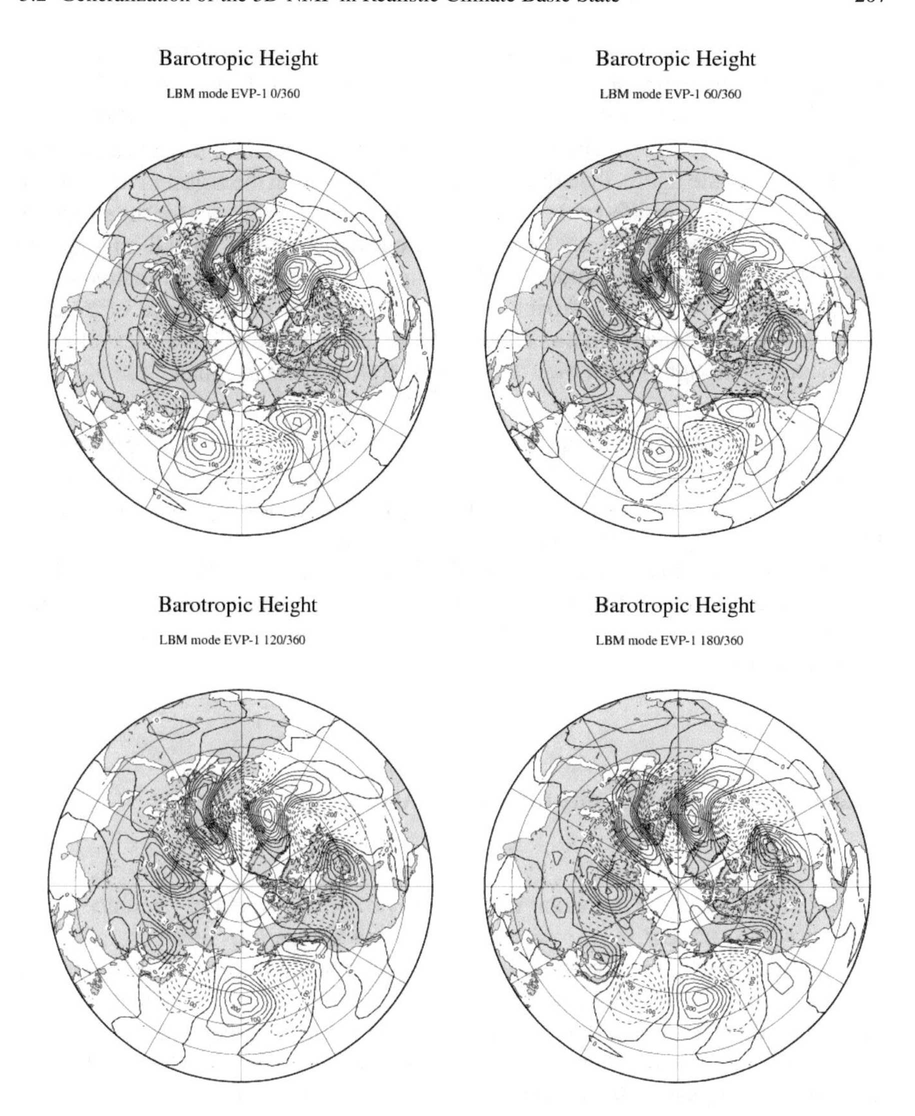

Fig. 5.7 Life-cycle of the barotropic height for the leading unstable mode in the non-zonal basic state. The phase of the life-cycle is expressed by degree for (a) 0°, (b) 60°, (c) 120°, (d) 180°out of 360°. From Tanaka and Seki (2013), Fig. 6.

Figure 5.7 illustrates the life-cycle of the most unstable mode with the growth rate 0.27 /day and the period 4.5 day. The phase of the life-cycle is denoted by the angle of 0° 60° 120° and the 180° in Fig. 5.7a to Fig. 5.7d. Note that Fig. 5.7d is just the opposite sign of Fig. 5.7a. The reader can track the movement of troughs and ridges for one cycle in the figure, then the next cycle will start from the adjacent troughs and ridges. By tracking the life-cycle of troughs and ridges, a reader can see the structural change not only for the one sector of the life-cycle, but also for the all longitudinal excursion around the mid-latitudes by tracking six life-cycles. The baroclinic waves are most amplified over the Atlantic Ocean showing the wedge shape of troughs, converging eddy momentum flux about 60°N to intensify the polar jet. A southward excursion of a trough is seen over the Far East from Siberia to Japan.

TS2013 furthermore presented the most unstable standing mode appearing in the LBM for the non-zonal basic state. The growth rate was only 0.04 /day and the period is infinity. The modal structure is geographically locked by the existence of the complex conjugate term **C**. There is a large negative anomaly in polar region with four troughs extending toward 45°E, 150°E, 120°W, and 30°W. The structure of the standing mode can have the opposite sign of the solution due to the property of the eigenvector. This standing mode is connected to the standing mode in the barotropic basic state in the next.

5.2.3.2 Baroclinic Instability for Various AO Indices

The modification of the dominant unstable mode is investigated in reference to the AOI varying from -3σ to +3σ shown by Fig. 5.8 for the DJF mean climate. The present study is unique compared with our former study in that the basic state is non-zonal, and all zonal waves of the state variables interact with all zonal waves of the basic state. We selected the two fastest growing modes as the dominant baroclinic instability for each basic state, which are the low-frequency-side unstable mode (M_L, and the high-frequency-side unstable mode (M_H. While the AOI varies from -3σ to +3σ, the M_L and M_H exist continuously. Their growth rates tend to increase about 0.005 as the AOI becomes large positive.

Figure 5.8 illustrates the barotropic height of the M_L and M_H for the AOI from -3σ to +3σ. For the modified climatological basic states (Fig. 5.3c, d), the M_L and M_H have the wedge shape like M_C structures which prevail in the mid-latitudes. Comparing the Atlantic sector and Pacific sector, the Atlantic baroclinic modes are more prominent and shift poleward than the Pacific baroclinic modes due to the local difference of the basic state (Fig. 5.3c, d). In this study, the Atlantic sector extends from 300°E to 0°E, and the Pacific sector spans from 150°E to 240°E. In particular, the Atlantic jet axis locates about 50° N (Fig. 2a of TK2013), however the

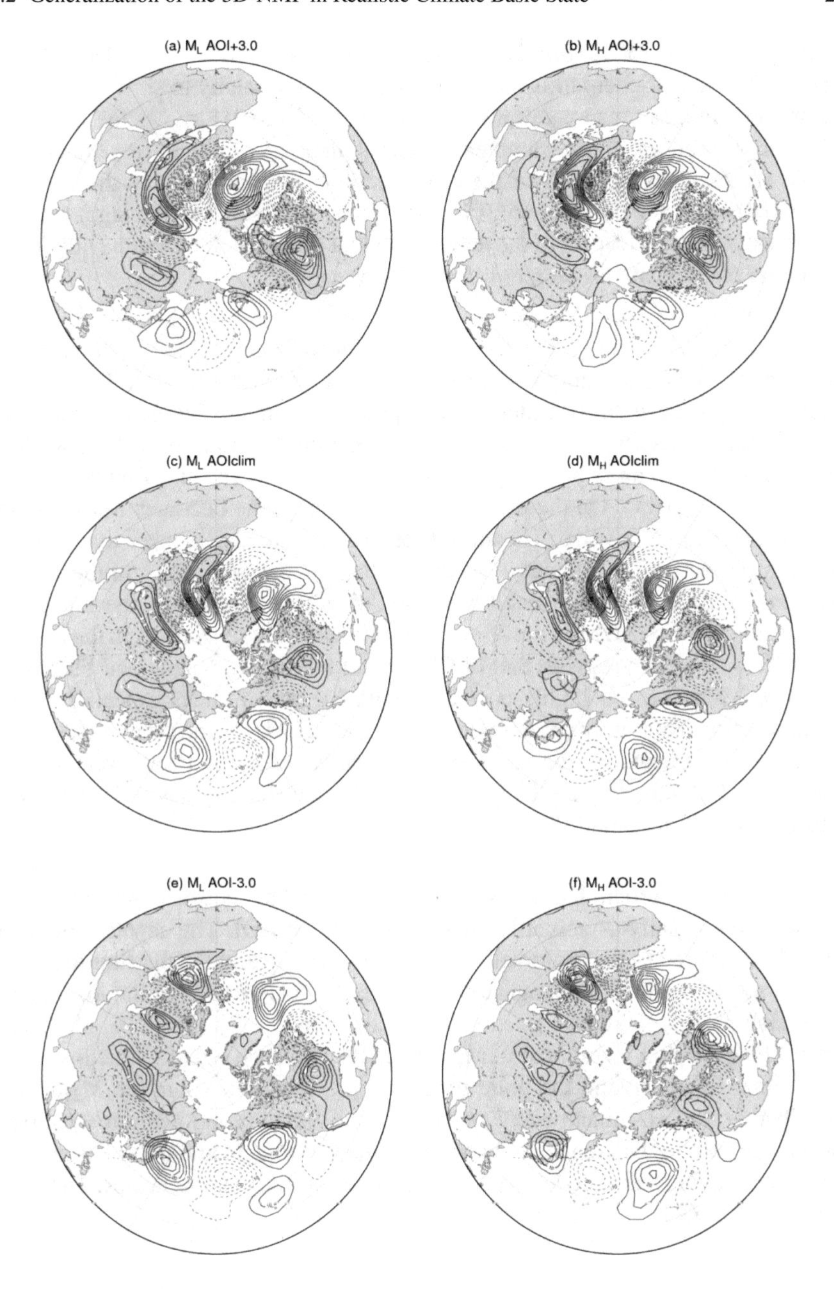

Fig. 5.8 Barotropic height (in arbitrary unit) of (a) M_L in AOI +3.0σ, (b) as in (a) but for M_H, (c) M_L in climatology, (d) as in (c) but for M_H, (e) M_L in AOI -3.0σ, (f) as in (e) but for M_H in the Northern Hemisphere. Solid (dotted) line is positive (negative) value. From Tanaka and Seki (2013), Fig. 15.

Pacific jet is about 35° N (Fig. 2b of TK2013). Therefore, the Atlantic baroclinicity is strengthened in higher latitudes than the Pacific baroclinicity to shift the center of the unstable modes much poleward. Figure 16a, b of TS2013 shows the longitude-height cross sections of the geopotential height anomaly of M_L in the Atlantic and Pacific sectors. The latitudes of the cross sections is located where the center of the M_L exists. In the climatology (Fig.16c, d), the Atlantic trough (ridge) axis tilts more westward with respect to the height than that of the Pacific trough (ridge). The amplitude maxima are also larger in the Atlantic sector.

When the AOI is large positive, the Atlantic M_C (Fig. 5.8a, b) modifies its trough axes tilted from northeast to southwest at the mid-latitudes. The Atlantic amplitude maxima are also shifted to more poleward than the Pacific maxima. Thus the baroclinic instability in the Atlantic sector is modified to transport the westerly eddy momentum to more poler region, as suggested by Tanaka and Tokinaga (2002) and Seki et al. (2011). On the other hands, the Pacific M_C becomes weaker as the AOI becomes large positive. The amplitude maxima are also weaker than those in the Atlantic sector. Throughout a life cycle, the M_L and M_H maintain the specific structure of the baroclinic instability at the Atlantic sector so that the interaction between the positive AOI and the baroclinic instability mainly appears in the Atlantic sector. This result agrees with that of Chang and Fu (2002); Chang and Fu (2003), who reveal that the AO highly correlates with the storm tracks especially in the Atlantic sector. This study confirms the result with LBM. On the other hands, when the AOI is large negative, the M_C appears in the Atlantic and Pacific sectors (Fig. 5.8e, f, and Fig. 16e, f of TS2013). Each baroclinic instability has a similar wedge shaped structure in both of the sectors.

Therefore, the baroclinically unstable modes excited by the zonally-varying basic state have the zonally-asymmetric interaction with the AOI. The asymmetric interaction is attributed to the asymmetric basic state, specifically the distribution of the zonal wind. With the climatological basic state, the distributions of the M_C depend on the spatial structure of the jet streams which differs locally (Fig. 5.3a, b). Tanaka and Tokinaga (2002) theoretically investigated the response of baroclinic instability to the polar jet and found the polar mode M_1 prevails rather than the M_2 or M_C. Eichelberger and Hartmann (2007) revealed that the eddy momentum selectively converges to strengthen the zonal wind when the subtropical and polar jets are well separated showing a double-jet structure. The present study supports these facts with a linearized model.

5.2.3.3 Barotropic Instability for a Zonally-Varying Basic State

The linear stability analysis for the non-zonal basic state represented by (5.30) of the LBM may be solved as a barotropic model, using only the barotropic component $m = 0$ for the state variable $w_i = w_{knm}$. The matrix size to be solved for the barotropic model is reduced substantially with the new matrix dimension of $L_0 = (K+1)(N+1)$. The property of the solution is identical to the baroclinic model (5.39), and the solution becomes a standing mode when the eigenvalue is real.

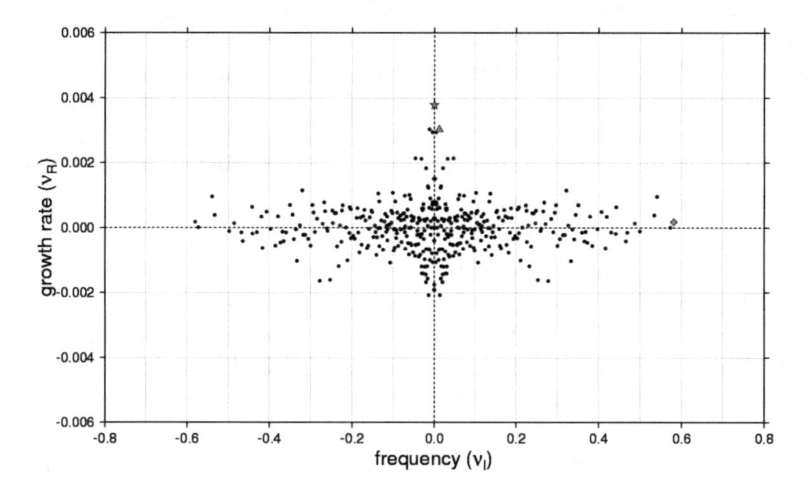

Fig. 5.9 The scatter diagram of the growth rate (ν_R) and the frequency (ν_I) given the complex eigenvalues for the barotropic Janaury basic state without damping. The star indicates the AO mode and the triangle indicates the second growing mode. From Tanaka and Seki (2013), Fig. 9.

Figure 5.9 plots the scatter diagram of the complex-valued eigenvalues obtained as the solution for the barotropic model with the zonally-varying basic state of the 30-year mean January data. It is worth noting that the most unstable mode is a standing mode at ν_I =0 (see star symbol in Fig. 5.9). The growth rate is only 0.004 (=0.05 /day) which is supported by the barotropic instability from the non-zonal component of the jet stream.

The most unstable barotropic mode M_{AO} was extensively investigated by Tanaka and Matsueda (2005) by means of the singular value decomposition (SVD) technique as well as the eigenvalue problem (EVP). They introduced a linear damping $d_i w_i$ represented by a hyper diffusion for the viscosity and by Rayleigh friction for the surface friction:

$$d_i w_i = -k_D c_i^{-4} w_i - \nu_s w_i \tag{5.42}$$

A diffusion coefficient k_D is set as $k_D(2\Omega a^8) = 2.7 \times 10^{40}\, m^8 s^{-1}$ and a linear damping coefficient ν_s is set as 1.5×10^{-3}. Here, c_i is the phase speed of the Hough mode representing the 3D scale dependency.

The result of the scatter diagram of the spectrum was presented by TS2013. All solutions are neutral or damping modes. The Arctic Oscillation mode M_{AO} appeared as the standing mode ν_I =0 and the growth rate ν_R becomes zero by the inclusion of the linear damping. It is found that the AO mode M_{AO} in Fig. 5.9 is modified to the barotropic height. This result coincides with the result of EVP and SVD by Tanaka and Matsueda (2005), and the observational analysis by Thompson and Wallace (1998). The scale dependent diffusion modifies the structure of the standing barotropic mode to the structure of the Arctic Oscillation, and the surface friction shifts the growth rate to zero maintaining the same structure of the eigenvector. The origin of the AO mode M_{AO} is the most unstable barotropic instability mode for the non-zonal barotropic basic state.

Since the linear dynamical system becomes singular by the appearance of the zero eigenvalue mode, the mode is called a singular eigenmode of the AO. According to the SVD analysis of the steady solution in response to the steady forcing, the AO mode with the zero eigenvalue becomes resonant for arbitrary quasi-steady forcing to cause an excitation of the AO structure. All climate models exhibit a common structure of the AO as the most dominant internal variability Ohashi and Tanaka (2009). The result supports that the AO is a dynamical eigenmode of the atmosphere which can be resonantly excited by any quasi-steady forcing.

In summary, we confirmed that the LBM was applicable to the linear barotropic and baroclinic instability problem, and the results are consistent with previous studies. We found the spatial characteristics of the positive feedback between the positive/negative AOI and the modification in the baroclinically unstable modes. The positive feedback dominates at the Atlantic sector for positive AOI due to the enhanced double jet structure there. On the other hand, when the AOI is negative, M_C converges the eddy momentum flux at the mid-latitudes to shift the subtropical jet poleward in the Atlantic and Pacific sectors due the intensified baroclinic instability. For a zonal climate basic state, we confirmed that the traditional Charney mode M_C and dipole-Charney mode M_2 appear as the most dominant unstable modes in the synoptic to planetary scales. For a zonally-varying basic state, we found that these unstable modes are modified by the regionality of the local baroclinicity of the basic state. Moreover, the zonal asymmetry of the basic state allows the existence of a standing mode with zero frequency, although the growth rate is smaller than the dominant baroclinic instability.

When a zonally-varying barotropic basic state is given to the LBM, we found the most dominant unstable mode is the standing mode excited by the barotropic instability of the non-zonal basic state. The structure of the standing mode is geographically fixed showing a negative anomaly over the Arctic which is surrounded by a positive anomaly in mid-latitudes. There are three troughs and ridges in mid-latitudes trapped by the zonal asymmetry of the basic state. Although the standing mode has a fixed structure, the eigenvector allows the structure with opposite signs as an alternative solution. The standing mode appears to be similar to the Arctic Oscillation with its structure and behavior. We demonstrated in this study that the barotropically most unstable standing mode becomes the Arctic Oscillation mode as proposed by Tanaka and Matsueda (2005) by the inclusion of diffusion and surface friction into the linear dynamical system.

5.3 3D Spectral Prognostic Modelling with the Basis of 3D-NMF

5.3.1 Background

In this section, we apply the three-dimensional (3D) spectral primitive equation model for the life-cycle experiment of the nonlinear baroclinic waves as an initial value problem. A numerical simulation of a life-cycle of nonlinear baroclinic waves was first conducted by Simmons and Hoskins (1978). According to their simulation, an initial perturbation superimposed on a zonal field grows exponentially by means of the baroclinic conversion, drawing zonal available potential energy to the perturbations. The basic process controlling the eddy growth is baroclinic instability Charney (1947). The amplified baroclinic disturbances start to transfer the energy back to the zonal kinetic energy by means of the barotropic conversion. The nonlinear baroclinic disturbances appear to accelerate the zonal jet at the end of their life-cycle as an expense of the zonal available potential energy. The energy flow analyzed by the simulated eddy activities is consistent with the understanding of the energy flow in the observed general circulation Lorenz (1955); Oort (1964); Saltzman (1970); Kung (1988).

There should be no contradiction between the relaxed meridional temperature gradient and the accelerated zonal jet with reference to the thermal wind relation. The acceleration of zonal wind is supposed to occur at the lower troposphere and deceleration at the tropopause level so that the vertical wind shear diminishes. The role of the baroclinic disturbances on the structure change in the zonal motion is demonstrated by Hoskins (1983). An initial zonal mean flow is modified at day 15 by a single life-cycle of the wavenumber 6 disturbances so that the structure of the jet becomes more barotropic. The isentropic surface at day 15 indicates a relaxed meridional gradient in the mid-latitude, which implies a reduced zonal available potentialenergy. The core of the jet has moved north and the zonal kinetic energy has increased as will be confirmed later.

In this regard, more detailed analysis would be meaningful in order to understand the role of the baroclinic disturbances with a specific attention to the changes in the vertical structures of zonal and eddy fields.

In the present section, a life-cycle experiment is carried out for Simmons and Hoskins' 45° jet with initial perturbations of zonal wavenumber $k = 6$ by integrating a 3-D spectral primitive equation model as described by Tanaka (1995) (hereafter T1995). Here, the vertical spectral expansion is applied for the discretization of governing primitive equations. The spectral model in the vertical direction is straightforward for the analysis of the energy redistribution in the vertical spectral domain. We analyze the energy transfer within the vertical spectral domain associated with the life-cycle of the baroclinic disturbances. The energy evolution and corresponding energy transformations are presented in the framework of a baroclinic-barotropic decomposition of atmospheric energy.

5.3.2 Model Description

The governing spectral primitive equations are described in the previous subsections. The resulting spectral primitive equations of (5.17) become a system of ordinary differential equations for Fourier expansion coefficients of variables:

$$\frac{dw_i}{d\tau} + i\sigma_i w_i = -i \sum_{j=1}^{L} \sum_{k=1}^{L} r_{ijk} w_j w_k + f_i, \quad i = 1, 2, ..., L. \tag{5.43}$$

Here, the system is truncated by the total wavenumber L. In this section, we consider diffusion, DF, as a single physical process for f_i. The scale dependency of diffusion is parameterized using the 3-D scale index σ_i based on the wave dispersion relating the wave scale and wave frequency. We approximate biharmonic-type diffusion for the Rossby (rotational) wave dispersion (for wavenumber $k \neq 0$) by:

$$(DF)_i = -K_D \left(\frac{k}{\sigma_i}\right)^2 w_i, \tag{5.44}$$

where K_D is a diffusion coefficient. Haurwitz waves on a sphere have phase speeds represented by the total wavenumber of the spherical surface harmonics $\hat{k}$ (see

Swarztrauber and Kasahara (1985)):

$$c = \frac{-1}{\hat{k}(\hat{k}+1)} \simeq \frac{\sigma_i}{k}. \tag{5.45}$$

Since the diffusion is often approximated with $\hat{k}(\hat{k}+1)$, the present form of diffusion in (5.44) tends to be the biharmonic-type diffusion for higher order Rossby modes. For the zonal component, the meridional index n_R is substituted for $\hat{k}$.

We define n element of total energy, E_i, for each basis function is in a dimensional form by

$$E_i = \frac{1}{2} p_s h_m |w_i|^2. \tag{5.46}$$

In order to explore the origin of the energy supply for the unstable baroclinic waves, an energy flow box diagram describing energy interactions between barotropic and baroclinic components is constructed. By differentiating (5.46) with respect to time and substituting (5.17), we obtain for eddy:

$$\begin{aligned} \frac{dE_i}{dt} &= 2\Omega p_s h_m \textstyle\sum_j Re(ib^*_{ij} w^*_j w_i)_{m''=0} \\ &+ 2\Omega p_s h_m \textstyle\sum_j Re(ib^*_{ij} w^*_j w_i)_{m''\neq 0} \\ &= C(B_{m''=0}, E_i) + C(B_{m''\neq 0}, E_i). \end{aligned} \tag{5.47}$$

Note that the linear term in the left-hand side of (5.43) does not contribute to the energy balance equation. The first term of the right-hand side of (5.47) stands for energy transformations from the barotropic component of the zonal field $B_{m''=0}$ into E_i, and the second term represents those from the baroclinic components of the zonal field $B_{m''\neq 0}$ into E_i (refer to Tanaka and Kung (1989)). By adding all indices separately for m=0 and $m \neq 0$, (5.47) becomes:

$$\frac{dE_{m=0}}{dt} = C(B_{m''=0}, E_{m=0}) + C(B_{m''\neq 0}, E_{m=0}), \tag{5.48}$$

$$\frac{dE_{m\neq 0}}{dt} = C(B_{m''=0}, E_{m\neq 0}) + C(B_{m''\neq 0}, E_{m\neq 0}). \tag{5.49}$$

The system of the nonlinear equations (5.43) is truncated to include only the Rossby modes for m=0-6, k=0 and 6, and n_R=0-20. Note that the truncation is imposed in the frequency domain as well as in the wavenumber domain by excluding high-frequency gravity modes.

The initial condition is a northern zonal field of the 45° jet which is assumed to be symmetric about the equator. The most unstable linear mode is obtained by solving a matrix eigenvalue problem of the primitive equations linearized for this zonal field. Then, a small amplitude unstable normal mode is superimposed on the initial zonal field. The time integration is based on a combination of leap-frog and a periodic use of Euler-backward scheme. By virtue of the closure with the low-frequency subspace of the atmospheric modes, our model requires no implicit scheme and no artificial smoothing.

5.3.3 Results of Numerical Simulations

The set of model equation (5.43) is integrated in time with respect to the initial condition discussed above. The meridional height cross sections of geopotential amplitude for zonal wavenumber 6 are illustrated in Fig. 5.10. At day 2 in Fig. 5.10a, the maximum of the geopotential amplitude is seen at the mid troposphere over 50°N. The arrows in the figures describe the EP flux associated with the disturbances (e.g., Edmon and McIntyre (1980); Hoskins (1983)). The EP flux is upward indicating large heat transport at the lower level. The structure is fundamentally the same as the most unstable linear mode (see Tanaka and Kung (1989)). At day 4 in Fig. 5.10b, the wave attains almost at its mature stage. The EP flux directs upward indicating acceleration of the zonal wind near the surface. The EP flux direction shifts for south at the upper troposphere decelerating the southern part of the zonal jet. At day 6 in Fig. 5.10c, the wave has passed its mature stage. The EP flux directs southward by this time showing dominant barotropic conversion.

The results shown here agree well with the previous analysis by Hoskins (1983). The EP flux diverges near the surface and converges near the tropopause level, indicating the acceleration and deceleration of the zonal flow, respectively. As the result, the zonal jet becomes more barotropic.

The resulting energy flow box-diagrams are presented in Fig. 5.11 for the most unstable Charney mode at $k = 6$. Upper boxes ($m'' \neq 0, m \neq 0$) denote the baroclinic components and lower boxes ($m'' = 0, m = 0$) the barotropic components for k=0 and 6, respectively. In the figure the energy interactions in (5.48) and (5.49) are rearranged further so that the energy is transferred along the lines denoted in the energy flow box diagram based on the fact that there is no barotropic-baroclinic interaction for the zonal geostrophic (rotational) modes.

It is evident from the result that large proportions of energy are transformed from zonal baroclinic energy to eddy baroclinic energy then to eddy barotropic energy for the growing modes. The result evaluated with FGGE observations is also illustrated

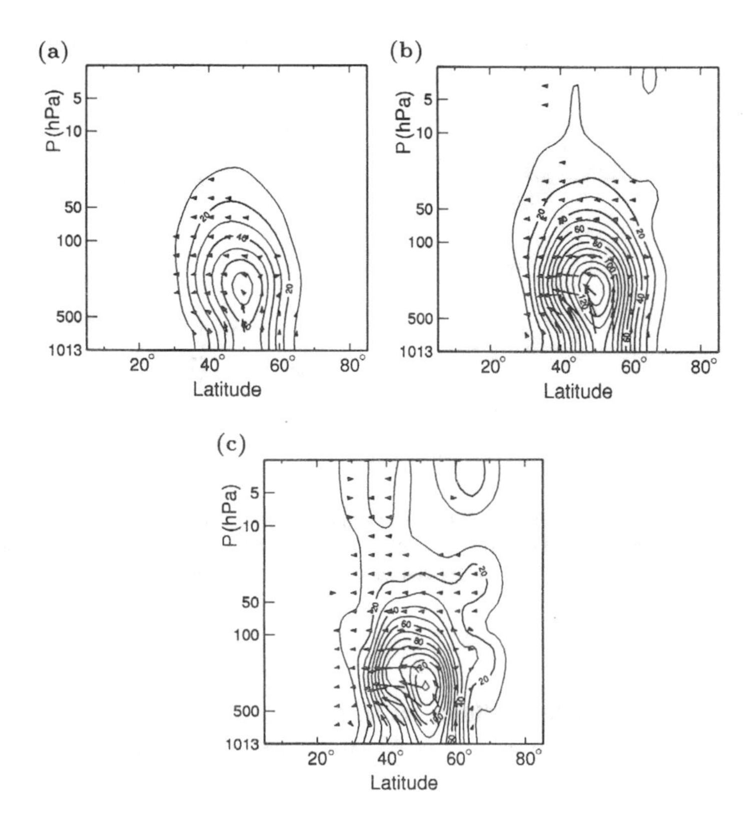

Fig. 5.10 Latitude-height cross sections of geopotential amplitude for wavenumber k=6 and the associated EP flux during the life-cycle of the baroclinic wave. Shown in (a), (b), and (c) are for days 2, 4, and 6, respectively. From Tanaka (1995), Fig. 2.

in Fig. 5.11 for comparison. The observed atmospheric energy flow clearly represents the characteristics of energy flow due to the baroclinically unstable modes.

Figure 5.12 shows how the initial perturbations of $k = 6$ grow exponentially drawing zonal baroclinic energy. This early stage of the evolution is reasonably explained by linear baroclinic instability of the 45° jet. Both of baroclinic energy and barotropic energy of $k = 6$ increase simultaneously because the unstable mode maintains its consistent structure to grow. The energy flow is characterized as from zonal baroclinic energy via eddy baroclinic energy to eddy barotropic energy. These energy transformations are also synchronized because they are proportional to the eddy energy levels in the linear framework.

The energy flow is characterized as from zonal baroclinic energy via eddy baroclinic energy to eddy barotropic energy. The barotropic conversion follows when

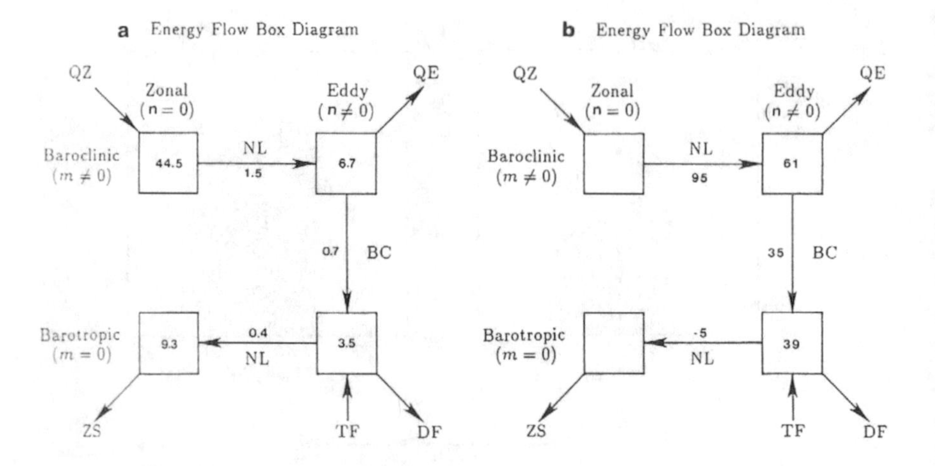

Fig. 5.11 Energy flow box diagram within zonal and eddy energies decomposed in barotropic and baroclinic components. (a) FGGE observation, (b) unstable Charney mode for $k = 6$. Units are $10^5 \mathrm{Jm}^{-2}$ for energy and Wm^{-2} for energy conversions. Percentile contributions are substituted for (b). From Tanaka (1995), Fig. 3.

the waves reach the finite amplitude, transferring the accumulated eddy barotropic energy toward zonal barotropic energy. As a result, zonal barotropic energy increases when the synoptic waves decay.

The important process in baroclinic instability is the zonal-wave interaction due to the eddy heat flux and the simultaneous baroclinic conversion at each zonal wavenumber. This baroclinic conversion is fundamentally a linear process. In contrast, the upscale zonal-wave interaction of the barotropic conversion is essentially a nonlinear process. It is found from Fig. 5.12 that the important baroclinic-barotropic interactions are coupled with linear baroclinic instability rather than the nonlinear barotropic conversion.

Figure 5.12 illustrates also the time evolutions of energy and energy conversions for several life-cycles. During this simulation, the largest meridional components of the zonal baroclinic field ($m = 4$) are treated as steady to maintain the meridional temperature gradient due to the differential heating. The diffusion coefficient is slightly increased in order to balance with the increased energy supply. As a result, we find that the baroclinic disturbances repeat the life-cycle for several times, drawing the energy from the zonal baroclinic components and feeding the zonal barotropic jet. We can clearly observe the time lag (about 2 days) between the relaxation of the meridional temperature gradient and the acceleration of the zonal jet due to the life-cycle of the nonlinear baroclinic disturbances. The jet is accelerated so that the structure becomes more barotropic in the vertical. This means that the zonal wind is

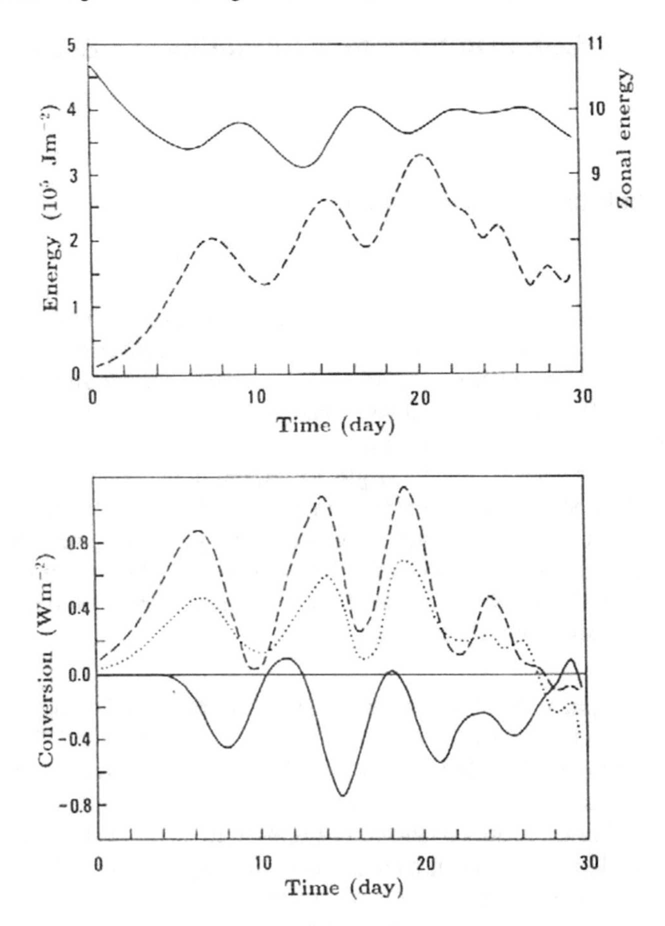

Fig. 5.12 (a) Time evolutions of eddy energy (dashed line) and zonal barotropic energy (solid line with its scale at right ordinate) for $k = 6$. Units are 10^5Jm^{-2}. (b) Time evolutions of energy conversions from zonal baroclinic to eddy baroclinic energies (dashed line), from eddy baroclinic to eddy barotropic energies (dotted line), and from eddy barotropic to zonal barotropic energies (solid line). Units are Wm^{-2}. From Tanaka (1995), Fig. 5.

accelerated at the lower troposphere, where the frictional dissipation is most efficient in a real atmosphere.

5.4 Parametrization of Baroclinic Instability in a Barotropic Model

5.4.1 Background and Motivation

Rossby wave breaking (or planetary wave breaking) has long been studied extensively, especially in middle atmosphere in relation to stratospheric sudden warming. A wave-mean flow interaction associated with the stratospheric sudden warming is a realization of Rossby wave breaking, where amplified planetary waves ultimately break down to deposit the easterly momentum into the mean flow. The dynamical stability of robust polar vortex draws great attention in relation to a study of ozone hole in the Southern Hemisphere (e.g. Jucks and McIntyre, 1987; Jucks, 1989). Intrusion of a tongue of low potential vorticity (PV) into the Arctic results in an enhanced material mixing in and outside the polar vortex which prevent the extreme low temperature in the Northern Hemisphere. Such a material mixing is another realization of Rossby wave breaking.

In the troposphere the Rossby wave breaking has been studied in the context of nonlinear life cycle of baroclinic waves Simmons and Hoskins (1976); Thorncroft and McIntyre (1993); Whitaker and Snyder (1993); Govindasamy and Garner (1997); Balasubramanian and Garner (1997); Hartmann and Zuercher (1998). The detailed studies of the life cycle are conducted in response to latent heat release, spherical geometry, and barotropic shear, among others. Rossby wave breaking draws more attention in conjunction with the onset of blocking in the troposphere. According to a model simulation of blocking by Tanaka (1998), a breaking Rossby wave leads to the onset of blocking, and the blocking itself causes subsequent break down of travelling Rossby waves. In that model, travelling Rossby waves grow exponentially by means of parameterized baroclinic instability, so the waves must break down somewhere. A Rossby wave, which grows critical in amplitude, breaks down at a topographically induced stationary ridge and is captured by the stationary ridge. Then, overturning of high and low PV centers takes place there to create a blocking. An example of blocking formation triggered by a breaking Rossby wave is illustrated in Fig. 5.13. In the figure, contours of shallow water PV are plotted with latitude in the ordinate and longitude in the abscissa in descending order, respectively, in order to mimic the progression and breaking of waves in analogy of shallow water system at the shore. The high PV in the polar region is hatched to illustrate the breaking waves at the surf zone. The elongation of trough and ridge axes from northwest to southeast and the anti-clockwise overturning of the vortex pair are the major characteristics of Rossby wave breaking induced by baroclinic instability.

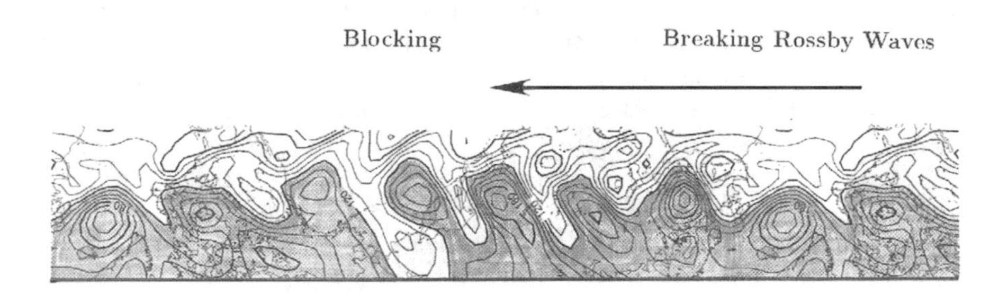

Fig. 5.13 A blocking formation triggered by breaking Rossby waves, redrafted from the distribution of potential vorticity (PV) in the barotropic model atmosphere. Ordinate and abscissa are the latitudes (70~20°N) and longitudes (360~0°E), respectively, in descending order. The contours of PV are in the units 10^{-10} m^{-1} s^{-1}. The high PV in polar region exceeding the value of 120 is hatched. From Tanaka and Watarai (1999), Fig. 1.

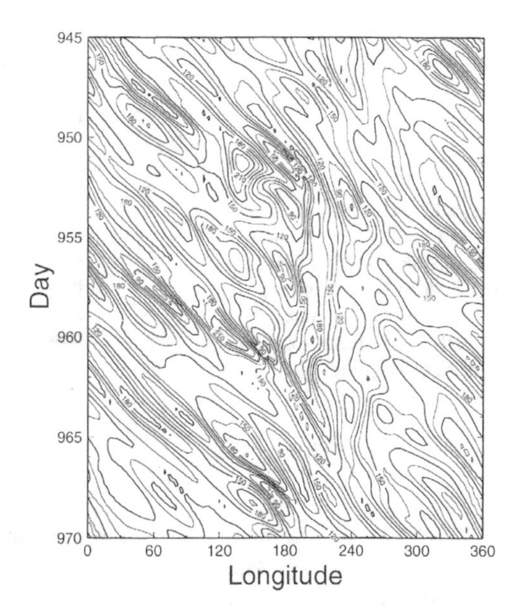

Fig. 5.14 Longitude-time section of potential vorticity along 58°N for a life-cycle of a blocking in the barotropic model atmosphere. The units are 10^{-10} m^{-1}s^{-1}. From Tanaka (1998), Fig. 17.

Figure 5.14 illustrates longitude-time section of potential vorticity along 58°N for a life-cycle of a blocking in the barotropic model atmosphere from Tanaka (1998). In the model atmosphere, topographic forcing produces stationary ridges along the west coast of the major continents. A progressively travelling Rossby wave was captured by the topographic ridge and it breaks down there to create a blocking. Subsequent Rossby waves are then blocked by the decelerated waves exhibit meridional stretch in PV field upstream of the blocking. It eventually breaks down to deposit the fresh low PV into the main body of blocking high and high PV into the cut-off low south of it to maintain the blocking system Shutts (1983); Tanaka (1998). In this blocking theory, the Rossby wave breaking appears to play the key role both for the onset and maintenance of blocking. Therefore, it is important to understand the mechanism and criterion of the Rossby wave breaking in more detail.

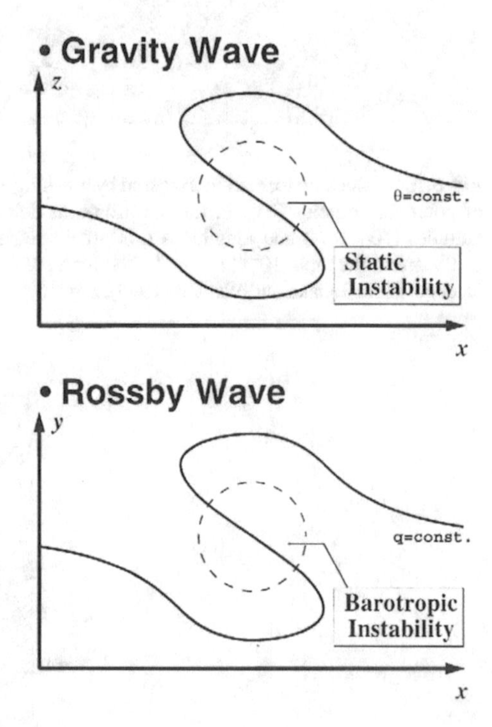

Fig. 5.15 Schematic illustrations of gravity wave breaking in the vertical section (upper) and Rossby wave breaking in the meridional section (lower). From Tanaka and Watarai (1999), Fig. 3.

Rossby wave breaking was discussed as an analogy of gravity wave breaking by many authors (e.g. McIntyre and Palmer, 1983, 1984, 1985; Leovy et al., 1985; Robinson, 1988; Garcia, 1991). The nonlinear behavior of overturning waves near the critical layer was analytically investigated by Warn and Warn (1978). Figure 5.15 schematically compares gravity wave breaking in the vertical section and Rossby wave breaking in the meridional section. When the isentropic surface overturns wrapping up the contours, negative vertical gradient of potential temperature appears as indicated in the figure. The convective instability is the principal mechanism for gravity wave breaking which eliminate the instability through the convective mixing Fritts (1984). Once the gravity waves become convectively unstable, they will be dissipated at a rate just sufficient to prevent further amplitude growth. This situation is referred to as wave saturation (e.g. Lindzen, 1981). The wave breaking and wave saturation seem to be indistinguishable for gravity wave.

For the Rossby wave, the dynamical analogy is illustrated here with potential vorticity contours in the meridional plane. When the high and low PV centers roll up to exhibit surf zone structure, negative meridional gradient of potential vorticity appears as indicated in the figure, this being the necessary condition for barotropic-baroclinic instability of the flow (e.g. Charney and Stern, 1962; Haynes, 1989). According to Garcia (1991), the breaking criterion for Rossby wave was defined as the wave amplitude for which the magnitude of the perturbation PV exceeds the background PV. This is a reasonable criterion at which negative meridional gradient of PV appears in the domain.

Although the analogy of Rossby wave breaking is perfectly clear, the theoretical basis of Rossby wave breaking is not well established. For instance, laboratory and field observations indicate that inertial gravity wave breaking leads to the generation of three-dimensional turbulence, whereas the Rossby wave breaking may be treated within the framework of two-dimensional turbulence (e.g. McIntyre and Palmer, 1985). The former would produce a number of small eddies through energy cascade by the gravity wave breaking. The latter, on the other hand, would produce even larger coherent vortices through the inverse energy cascade by the Rossby wave breaking. Namely, synoptic-scale wave breaking may result in excitation of planetary waves. A splitting jet and zonalization are the realization of such a Rossby wave breaking. Two paradigms of cyclonic and anticyclonic evolutions of the vortex pair overturning lead to opposite direction of eddy momentum flux Thorncroft and McIntyre (1993); Whitaker and Snyder (1993); Akahori and Yoden (1996); Govindasamy and Garner (1997); Balasubramanian and Garner (1997); Hartmann and Zuercher (1998). Further study is thus desirable to understand the Rossby wave breaking to confirm if Garcia's analogy is applicable to the study of the blocking onset.

The purpose of this section is to examine the breaking Rossby waves in the barotropic atmosphere using a simple barotropic model which implements parameterization of baroclinic instability as described by Tanaka (1998) and Tanaka and Watarai (1999) (hereafter T1998 and TW1999, respectively). Within the linear framework, the parameterized unstable wave with a small amplitude grows exponentially by the baroclinic instability. The exponential growth must, however, terminate at a finite amplitude leading to a wave breaking at certain energy level. We examine the nonlinear evolution of the growing Rossby waves and the criterion of the wave breaking in the barotropic atmosphere. The energy flows associated with the Rossby wave breaking are extensively examined in the wavenumber domain.

5.4.2 Formulation of a Barotropic Model

5.4.2.1 Parametarization of Baroclinic Instability

We start from the fully nonlinear equation retrieving the wave-wave interactions and external forcing disregarded in (5.43).

$$\frac{dw_i}{d\tau} + i\sigma_i w_i = -i\sum_{j=1}^{L}(\sum_{k=1}^{L}(r_{ijk} + r_{ikj})\overline{w}_k)w_j - i\sum_{jk} r_{ijk}w_j w_k + f_i, \quad i = 1, 2, 3, ..., L. \tag{5.50}$$

The first term in the right hand side represents linear zonal-wave interactions as appeared in (5.29), and the second term represents the rest of nonlinear wave-wave interactions for perturbations. Note that the state variables here are finite-amplitude deviations from the time-independent zonal basic state.

Supposing that the eigenspace for the matrix ($\mathbf{D}_k$+$\mathbf{B}_k$) in (5.41) is full rank without any multiple roots, we have L linearly independent non-orthogonal eigenvectors ξ_l, l=1,2,3,...,L. The state variable w_i may then be expanded in the basis of the eigenvectors ξ_l for each zonal wavenumber:

$$w_i(\tau) = \sum_l a_l(\tau)\xi_{li}, \tag{5.51}$$

where the amplitude coefficients $a_l(\tau)$ are supposed to be determined by solving a linear system. Since $w_i(\tau)$ is a function of time, so is the amplitude coefficient $a_l(\tau)$. Substituting (5.51) into the first term of the right hand side of (5.50), the linear operators associated with the zonal-wave interactions may be reduced to their eigenvalues of ν_n in reference to (5.41) for each zonal wavenumber:

$$\frac{dw_i}{d\tau} + i\sigma_i w_i = -i\sum_l(\nu_l - \sigma_i)a_l\xi_{li} - i\sum_{jk} r_{ijk}w_j w_k + f_i, \quad i = 1, 2, 3, ..., L. \tag{5.52}$$

Note that the complicated zonal-wave interactions are represented by the summation of eigenmodes. The real part of eigenvalues are modified by σ_i.

When the model equation (5.52) is integrated from infinitesimal white noise of w_i superimposed on the basic state, the most unstable mode $a_1\xi_{1i}$ would soon dominate the other modes, growing exponentially with the growth rate determined by ν_1, as predicted by the linear theory. The barotropic components grow in proportion to baroclinic components maintaining the normal mode structure ξ_{1i}. It is in this process of zonal-wave interactions where the baroclinic energy in the basic state is converted to barotropic energy in eddies. When the waves reach finite amplitudes, the nonlinear wave-wave interactions play the role to saturate the exponential growth. For this reason, the most unstable mode ξ_{1i} is anticipated to explain the largest fraction of the zonal-wave interactions.

We now attempt in this section to construct a barotropic spectral model, using only the barotropic components (*i.e.*, m=0) of w_i for (5.52). When a barotropic subset is written for (5.52), linear terms retain the same form, but the nonlinear wave-wave interactions are divided into interactions among barotropic components and those between barotropic and baroclinic components.

We assume first that the important energy supply to the barotropic component of the atmosphere is accomplished by the zonal-wave interactions represented by $a_l\xi_{li}$. The barotropic-baroclinic interactions are dominated by baroclinic instability associated with the most unstable mode $a_1\xi_{1i}$. All contributions from higher order eigenmodes are assumed to be less important due to the slow growth rates. With these assumptions, we attempt to close the system using only the barotropic components of w_i and ξ_{1i} for each zonal wavenumber as follows:

$$\frac{dw_i}{d\tau} + i\sigma_i w_i = -i \sum_{jk} r_{ijk} w_j w_k - i\nu_1 a_1 \xi_{1i} + f_i, \quad i = 1, 2, 3, ..., L. \quad (5.53)$$

The first term of the right hand size represents the nonlinear wave-wave interactions among barotropic components of w_i. The second term in the right hand side of (5.53) represents the wavemaker introduced in this section to amplify barotropic eddies in synoptic scale. Only the growing parts are considered to supply energy for the barotropic atmosphere.

The amplitude coefficient $a_1(\tau)$ is determined by an orthogonal projection of the state variable w_i onto the most unstable mode ξ_{1i}. Because the meaning of the amplitude coefficient a_1 is a fraction of the state variables w_i represented by ξ_{1i}, we performed the orthogonal projection within the barotropic atmosphere as follows:

$$w_i(\tau) = a_1(\tau)\xi_{1i} + \varepsilon_i(\tau), \quad (5.54)$$

where the vectors are reduced to include only the barotropic component, ξ_{1i} is normalized to have a unit norm $\sum_i \xi^*_{1i}\xi_{1i}$ =1, and $\varepsilon_i(\tau)$ is the orthogonal complement of the projection, i.e., $\sum_i \xi^*_{1i}\varepsilon_i$=0. The amplitude coefficient $a_1(\tau)$ is thus evaluated every time step by a vector inner product for complex numbers:

$$a_1(\tau) = \sum_i \xi^*_{1i} w_i. \quad (5.55)$$

With the amplitude coefficient so obtained, the resulting wavemaker based on baroclinic instability is given as:

$$(BC)_i = -i\nu_1 a_1(\tau)\xi_{1i}. \tag{5.56}$$

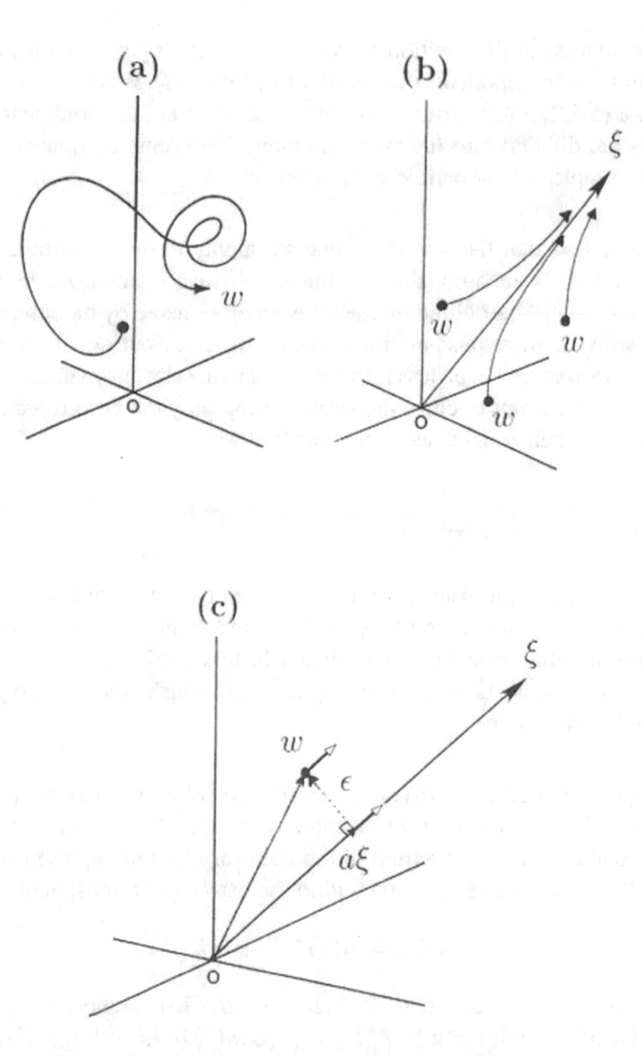

Fig. 5.16 Schematic illustration of a time evolution of the solution trajectory W in a phase space (a) for the nonlinear system (4); (b) for the unstable direction (5.56); and (c) for the orthogonal projection of W onto the unstable direction ξ (5.54). From Tanaka and Watarai (1999), Fig. 4.

Figure 5.16a schematically illustrates the evolution of the trajectory W for the nonlinear system (5.53) in multi-dimensional phase space. The trajectory moves chaotically around the origin. When the system is linearized with respect to a zonal basic state as in (5.50) and small perturbations are superimposed on it, the trajectories of the small perturbations would behave as in Fig. 5.16b. Here, the trajectories starting at random initial points approach eventually to the most unstable direction ξ_1 and grows along the vector ξ_1 at the speed of the given growth rate. The orthogonal projection is a reasonable representation of the exponential growth of the linear baroclinic instability as long as the nonlinear term is negligible. A structure expected from baroclinic instability emerges exponentially from the infinitesimal white noise.

The linear theory no longer holds when the nonlinear term becomes comparable to the forcing term by the baroclinic instability. As will be shown in the results, the exponential growth of the small perturbation eventually saturates when the nonlinear wave-wave interactions begin to play the role. The energy supply due to the instability appears to balance with nonlinear scattering toward the different scales.

The parameterized baroclinic instability $(BC)_i$ described here was compared with the initial exponential growth of baroclinic instability using a fully nonlinear baroclinic model (Tanaka (1991)). These two exponential growths agree quite well as long as the linearity is satisfied. However, the parameterized growth overestimates the true growth to some extent at the finite amplitudes. It was found that the orthogonal projection (5.54) tends to overestimate the amplitude factor, $a(\tau)$, for such a finite amplitude . Empirically, we need to reduce the growth rate when the eddy energy exceeds the basic state energy at which the perturbation assumption no longer holds. Such a reduction facter f was given by $f = (1 - \frac{E_k}{\beta E_0})$. Here, E_k is the sum of the total energy for a zonal wavenumber k, E_0 the total energy of the zonal basic state, and βE_0 is a threshold at which the growth rate vanishes. In this section, $\beta = 0.2$ is used for the value of $E_0 = 56.7 \times 10^5 \mathrm{Jm}^{-2}$. Since the magnitude of the observed E_k is approximately of the order $1 \times 10^5 \mathrm{Jm}^{-2}$, the growth rate is reduced about 10 % by this reduction constant. The justification of this wavemaker is assessed by observing the structure and behavior of the eddies produced by this method.

5.4.2.2 External Forcing

We consider only the following four physical processes as an external forcing:

$$f_i = (TF)_i + (DF)_i + (DZ)_i + (DE)_i, \tag{5.57}$$

where $(TF)_i$ represents the topographic forcing, $(DF)_i$ the biharmonic diffusion, $(DZ)_i$ the zonal surface stress, and $(DE)_i$ the Ekman pumping for eddies. Apart from the energy source of the topographic forcing, the sole energy source of the model is $(BC)_i$ induced by the baroclinic instability. The rest of the three physical processes are the energy sinks of the model.

5.4.2.3 Topographic Forcing

A kinematic uplift of an air column by the surface topography H originates with a horizontal component of the normal stress of the topography. It has been parameterized by a forced upward motion $w = dz/dt$, which is induced by the barotropic flow $\mathbf{V}_0$. We parameterized the topographic effect with the barotropic component w_0 as

$$w_0 = \int_0^1 w\, d\left(\frac{p}{p_s}\right) = K_T\, \mathbf{V}_0 \cdot \nabla H, \tag{5.58}$$

where K_T is a topographic constant depending on the profile of w. Charney and Eliassen (1949) original work uses K_T=0.4. The spectral representation of (5.58) becomes the topographic forcing $(TF)_i$ in (5.57):

$$(TF)_i = \left\langle \left(0, 0, \frac{w_0}{h_0}\right)^T, \mathbf{Y}_0^{-1}\, \mathbf{\Pi}_{kn0} \right\rangle. \tag{5.59}$$

There is some arbitrariness in the choice of K_T; *e.g.*, Chen and Trenberth (1988) use 850 hPa, and Valdes and Hoskins (1989) use the true surface wind. We use K_T=0.2.

5.4.2.4 Diffusion

For large-scale atmospheric motions, an eddy momentum transfer by the Reynolds stress dominates the viscous stress. The Reynolds stress is initially and in reality a nonlinear contribution in (5.53) for unresolvable (high-frequency) motions. In this section, we attempted to parameterize the scale dependency of diffusion using the 3-D scale index σ_i based on the wave dispersion relating the wave scale and wave frequency. As was described in the baroclinic model of (5.44), we approximate biharmonic-type diffusion for the Rossby (rotational) wave dispersion (for wavenumber $k \neq 0$) by:

$$(DF)_i = -K_D \left(\frac{k}{\sigma_i}\right)^2 w_i, \tag{5.60}$$

where K_D is a diffusion coefficient and $K_D(2\Omega a^4) = 2.0 \times 10^{16} m^4 s^{-1}$. Haurwitz waves on a sphere have phase speeds represented by the total wavenumber of the spherical surface harmonics $\hat{l}$ (see Swarztrauber and Kasahara (1985)):

$$c = \frac{-1}{\hat{k}(\hat{k}+1)} \simeq \frac{\sigma_i}{k}. \tag{5.61}$$

The rotational Hough modes are known to converge to Haurwitz waves in the limit as an equivalent height h_m tends to infinity. For the barotropic component, m=0, a realistic temperature profile gives $h_0 = 9623.9$ m. Since the diffusion is often approximated with $\hat{k}(\hat{k}+1)$, the present form of diffusion in (5.60) approaches the biharmonic-type diffusion for higher order rotational modes. For the zonal component, the meridional index n_R is substituted for $\hat{k}$. Note that the same form of diffusion is automatically imposed on geopotential field as well as wind field.

5.4.2.5 Zonal Surface Stress

As the waves grow, the nonlinear zonal-wave interaction begins to accelerate the zonal flow. A northward shift of the subtropical jet occurs due to the northward eddy momentum transport induced by the baroclinic waves. For the barotropic flow, the important physics that must be included to balance the northward shift of the jet is the zonal surface stress Hoskins (1983). The westerly deceleration in the mid-latitudes and the easterly deceleration in the low latitudes by the surface wind $\overline{u}_s$ should balance with the meridional convergence of the northward eddy momentum flux $-\frac{\partial \overline{u'v'}}{\partial y}$. With this fact, we adopt the following parameterization for the zonal surface stress $(DZ)_i$:

$$(DZ)_i = -K_Z[w_i - (\overline{w}_i - \alpha \overline{s}_i)] \quad \text{for } k = 0, \tag{5.62}$$

where $\overline{w}_i$ is a prescribed reference state of w_i and $\overline{s}_i$ is the spectral representation of the surface wind $\overline{u}_s$ of the reference state. In this formula, zonal wind u tends to relax to $\overline{u}$ when $-\frac{\partial \overline{u'v'}}{\partial y}$ balances with the zonal surface stress. We used the monthly mean FGGE data for January 1979 for $\overline{w}_i$ and $\overline{s}_i$ with constants of α=1.5 and $K_Z(2\Omega) = 3.87 \times 10^{-6} s^{-1}$, which corresponds to the relaxation time of three days. Similar relaxation is seen for barotropic models used by Charney and DeVore (1979), Jucks (1989), and Salby et al. (1990).

5.4.2.6 Ekman Pumping

The Ekman pumping is considered for the surface friction of eddy motions. According to Charney and Eliassen (1949), the upward motion at the top of the Ekman layer w_E is proportional to the vorticity $\zeta = \mathbf{k} \cdot \nabla \times \mathbf{V}$ of the free atmosphere:

$$w_E = h_0 \frac{K_E}{f_0} \zeta \quad \text{for } k > 0, \tag{5.63}$$

where f_0 is a Coriolis parameter at a mid-latitude and K_E is an Ekman pumping constant. The vorticity can be evaluated by wind field. Therefore, the final form of the Ekman pumping $(DE)_i$ may be derived by the spectral representation of the upward motion w_E:

$$(DE)_i = \left\langle \left(0, 0, \frac{w_E}{h_0}\right)^T, \mathbf{Y}_0^{-1} \, \mathbf{\Pi}_{kn0} \right\rangle . \tag{5.64}$$

The Ekman pumping constant used in this section is K_E/f_0=0.005, which is rather small compared with the value 0.02 of Charney and Eliassen. We demonstrated that synoptic disturbances tend to die out with such a large damping constant, although the large value may be suitable for obtaining a steady state.

5.4.2.7 Energetics

In the model, the total energy (i.e. the sum of kinetic energy and available potential energy) is monitored as one of the fundamental variables representing the global state of the atmosphere. In the spectral domain, the total energy is simply the sum of the energy elements E_i, which is defined by the squared magnitude of the state variable w_i. By differentiating it with respect to time and substituting (5.53), we obtain the energy balance equation:

$$\frac{dE_i}{dt} = NL_i + BC_i + TF_i + DF_i + DZ_i + DE_i \, . \tag{5.65}$$

We use similar symbols as in (5.57), but without the parenthesis for real-valued energy variables of dynamical and physical processes. The nonlinear wave-wave interaction is designated by NL_i. Note that the linear term in the left-hand side of (5.53) does not contribute to the energy balance equation.

The present spectral barotropic model on a sphere is equivalent to a shallow water system on a sphere Kasahara (1977)) which may be described as follows:

$$\frac{\partial u}{\partial t} - 2\Omega \sin\theta v + \frac{1}{a\cos\theta}\frac{\partial \phi}{\partial \lambda} = -\mathbf{V}\cdot\nabla u + \frac{\tan\theta}{a}uv + F_x \tag{5.66}$$

$$\frac{\partial v}{\partial t} + 2\Omega \sin\theta u + \frac{1}{a}\frac{\partial \phi}{\partial \lambda} = -\mathbf{V}\cdot\nabla v - \frac{\tan\theta}{a}uu + F_y \tag{5.67}$$

$$\frac{\partial \phi}{\partial t} + gh_0\nabla\cdot\mathbf{V} = -\mathbf{V}\cdot\nabla\phi + F_z\,. \tag{5.68}$$

Note that the equivalent height h_0 corresponds to the mean depth of the fluid, which is assumed to be constant, and the continuity equation has a source term of F_z.

Since the present barotropic model has the same form with a shallow water equation model, the potential vorticity of the system may be defined as:

$$q = \frac{\mathbf{k}\cdot\nabla\times\mathbf{V} + 2\Omega\sin\theta}{h_0 + z}. \tag{5.69}$$

The potential vorticity q should be conserved when the external forcing and the additional barotropic-baroclinic interaction terms in (5.53) are absent.

5.4.3 Results for the Model with Wavenumber 6

5.4.3.1 Control Run

The model equation (5.53) is integrated in time, starting from an infinitesimal white noise superimposed on a zonal flow. In order to isolate the exponential amplification of unstable modes, the baroclinic instability is imposed in this section only at the zonal wavenumber 6. Figure 5.17 illustrates the time evolution of the total energy for the zonal wavenumber 6 and the zonal motions. The energy level of the initial white noise superimposed on a zonal flow is of the order of 10 J m^{-2}. The energy level decreases for the first 20 days to 2 J m^{-2} due to the frictional dissipation. The energy level then starts to increase exponentially for days from 25 to 50 as expected from the linear theory. We can infer that the solution trajectory W is in parallel with ξ as discussed in Fig. 5.16b. The exponential growth, however, must terminate when the wave amplitude becomes finite so that the nonlinear wave-wave interactions become comparable to the linear terms in (5.53). This may be the stage of wave saturation. The eddy energy is equilibrated at 2× 10^5 J m^{-2}. According to observation Tanaka (1985), the corresponding mean value for k=6 is 3× 10^4 J m^{-2}, i.e., about 10 times smaller than the saturation level here.

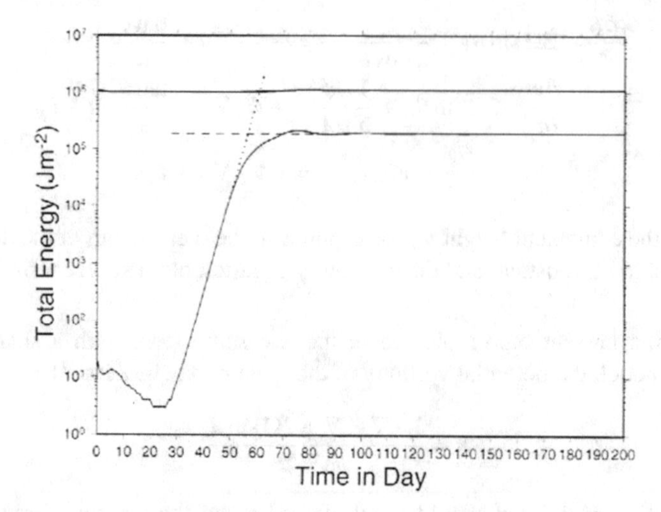

Fig. 5.17 Time series of total energy in the zonal (thick solid line) and eddy (thin solid line) componets during the first 200 days. The crossing of dotted and dashed line represents the saturation point. From Tanaka and Watarai (1999), Fig. 5.

By drawing two lines from the equilibrium level (dashed) and the exponential growth (dotted), we can determine the date of the wave saturation as day 55 in this example. Accordingly, the growth rate may be evaluated as 0.20 day $^{-1}$ from the fact of increase by 5 orders of magnitude during 30 days in energy level. Since the growth rate for k=6 is given as 0.34 day $^{-1}$ by the the linear stability analysis, we found that an appreciable fraction of growth rate has lost by the viscosity and frictional damping. Therefore, planetary waves of k=1 to 4 may be hardly amplified by the linear instability alone since the growth rate is too small. The zonal barotropic energy is approximately constant at 1×10^6 J m^{-2}. It is interesting to note that the eddy energy reaches its equilibrium when the energy level becomes 20% of the zonal barotropic energy.

Figure 5.18 illustrates hemispheric distributions of shallow water potential vorticity (PV) for days 48, 50, 52, 54, 100, 101, 102, and 200. A rotating wavenumber 6 emerges from infinitesimal noise in mid- to high-latitudes. When the wave amplitude is sufficiently small as for day 48, the structure is similar to the vertical mean of unstable Charney mode anticipated from baroclinic instability. Trough axis tilts from northwest to southeast at the northern flank of the Charney mode near 45°N, whereas it tilts from southwest to northeast at the southern flank of the Charney mode. The eddy momentum flux, thus, converges to mid-latitudes accelerating the westerly jet. When the amplitude becomes large enough as for day 54, high PV in the Arctic spreads away to lower latitudes maintaining the trough axis from northwest to southeast. At the same time, low PV in lower latitudes intrudes toward the Arctic. A positive and negative vorticity pair appears to rotate anti-clockwise at the each

Fig. 5.18 Hemispheric distributions of potential vorticity with the wave-6 model for model days on 48, 50, 52, 54, 100, 101, 102, and 200. The units are 10^{-10} $m^{-1}s^{-1}$ with the contour interval 15. From Tanaka and Watarai (1999), Fig. 6.

sector. This is the time when wave saturation takes place according to the result in Fig. 5.19. The wave is about to break down, indicating a characteristic "surf zone" shape as shown in Fig. 5.15. Evidently, the meridional gradient of potential vorticity indicates negative area at the surf zone latitudes. The existence of the negative PV gradient is the necessary condition for the pattern to be unstable (see Charney and Stern (1962); Garcia (1991)). Such an area appears on day 48 when the exponential growth of the unstable mode deviates from the theoretical straight line in Fig. 5.17.

However, it is interesting to note that the wave breaking has not occurred in this experiment as seen from PV distributions for days 100, 101, and 102. The surf zone shape of PV distributions attains a steady configuration, and it advects simply from west to east as is confirmed from the map of day 200. Although the meridional PV gradient is negative at some locations, the surf zone shape is stabilized in present model without developing into the wave breaking. Since the wave amplitude has saturated by definition as demonstrated in Fig. 5.17, the result of this section suggests that there is a case of equilibrium in the model atmosphere where the energy supply into the system balances with small frictional damping of harmonic waves. The regular movement of the wave-6 indicates that the waves are saturating but not breaking.

5.4.3.2 Sensitivity Experiments

In sensitivity runs, the regular travelling waves are intentionally destroyed by increasing the growth rate, Im(ν), until the waves break down. The first symptom of the Rossby wave breaking occurs when the growth rate is magnified by the factor 1.3. This time, the regularity in the phase speed is lost at day 50 and simultaneously, the wave amplitude is reduced substantially. The characteristics are consistent with the breaking Rossby wave discussed in Fig. 5.14 in relation to the onset of blocking in the model atmosphere. The travelling wave-6 once recovers the regular progression around day 60. Yet, the wave breaking takes place repeatedly after day 80 indicating larger amplitudes and irregular phase speed.

Figure 5.19 shows the hemispheric distributions of PV as in Fig. 5.18, but for the growth rate magnification factor 1.3 for days 49 to 52 when the Rossby wave undergoes breaking. For this experiment, the six cores of PV peaks in high latitudes exhibit different configurations, so it is realized that the sectorial symmetry of PV is now lost. Some well developed vortices undergo overturning of high and low PV distributions. Therefore, the wave breaking is characterized by the loss of sectorial symmetry and the overturning of the high and low PV distributions.

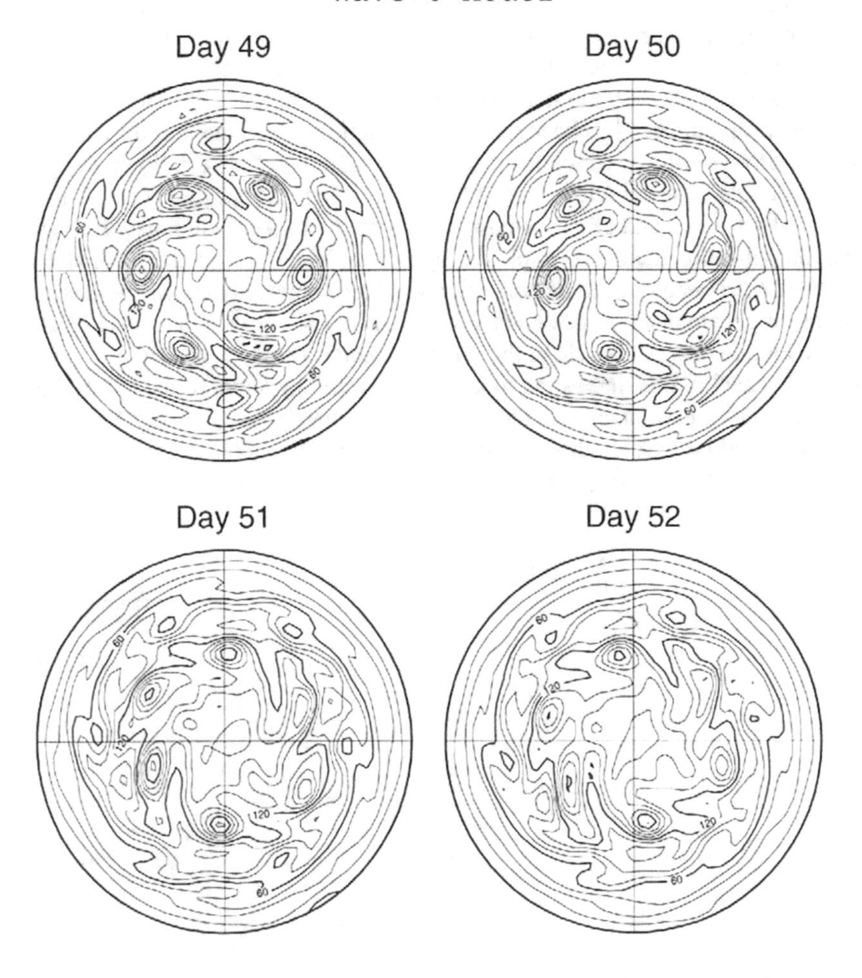

Fig. 5.19 As in Fig. 5.18, but for the experiment run with 1.3 time the growth rate for days 49 through 52 when the first wave-breaking occurs. From Tanaka and Watarai (1999), Fig. 9.

5.4.3.3 Energetics

Figure 5.20 presents the mean energy spectra over the zonal wavenumber domain for the wave-6 experiment during days 30 to 100 with various magnification factors for the growth rate. For the control run with the magnification factor 1.0, the energy spectrum indicates sharp spectral peaks at k=6, 12, and 18. Because of the amplified wave-6, its harmonics of k=12 and 18 are excited through the weak nonlinear interactions. As shown in Fig. 5.18 the control run maintains the steady configuration of the surf zone structure by the combination of those harmonic waves. The background noise energy in this case is of the order 10^2 J m^{-2}. When the growth rate is magnified by a factor 1.3, the three spectral peaks at k=6, 12, and 18 remain recognizable. However, the main difference is seen in the high noise level at the rest of wavenumbers. The background noise in this case is enhanced by the strong nonlinear wave-wave interactions reaching the order of 10^4 J m^{-2}. When the growth rate is doubled, the noise level becomes comparable to the spectral peaks. The spectral peaks at k=12 and 18 are almost filled with the background noise. A result for the magnification factor 5.0 is plotted by a thick line. The spectral peak at k=6 is recognizable, but its harmonics at k=12 and 18 are completely lost in this case. More energy is accumulated in planetary waves than in k=6, especially at k=1 by means of the inverse energy cascade of the 2-D turbulence. The flow pattern (not shown) is strongly chaotic. It is interesting to note that the energy levels of the spectral peaks at k=6 and its harmonics remain at the approximately same energy level despite the increased background noise level by the increased magnification factors for the growth rates. It is inferred from the result that the saturation level for Rossby wave is independent of the speed of its exponential growth.

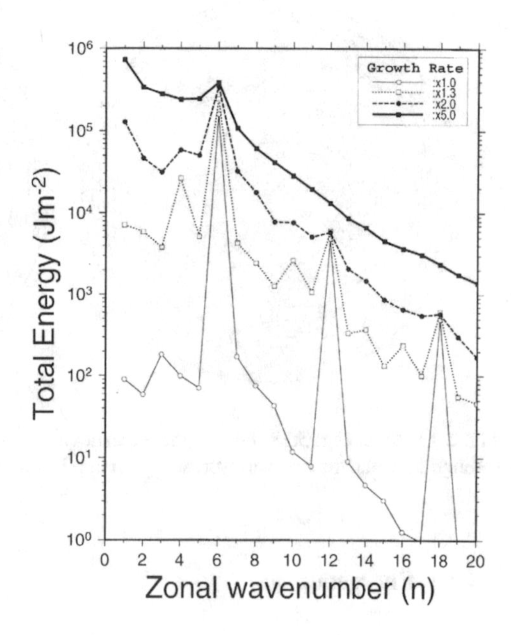

Fig. 5.20 Mean energy spectra over the zonal wavenumber domain for the wave-6 model during days 30 to 100 with various magnification for the growth rates. From Tanaka and Watarai (1999), Fig. 13.

Since the energy source is imposed only at k=6 for this model, its harmonics and also the background noise must be excited by the unstable wave through the weak and strong nonlinear wave-wave interactions, respectively. Details of energy flows within the zonal spectral domain is analyzed following the energetics scheme for the 3-D spectral model are presented in Tanaka (1991). The steady configuration of travelling Rossby waves presented in Fig. 5.18 is maintained by the energy balance, as seen in detailed energetics over zonal wavenumber in tables in Tanaka (1991).

5.4.4 Simulations with all Waves without Mountains

Now the model equation (5.53) is integrated for 1000 days with all baroclinic waves without mountains in order to demonstrate the performance of the wavemaker to excite synoptic eddies. Hereafter, this experiment will be referred to as no-mountain run. For the no-mountain run, the primary source of q is obviously the wavemaker described in (5.53). Small random noise seeded in the initial data are selectively amplified toward the most unstable structure of baroclinic instability. Since the maximum growth rate is seen at the zonal wavenumber 7-8 Tanaka and Kung (1989), the initial growth is dominated by the wavenumber 8 as expected. Travelling troughs and ridges of the wavenumber 8 start to grow at about day 20 to form synoptic waves. We refer to it as a Rossby wave for convenience even though a baroclinic instability wave would be more appropriate. The exponential growth of the Rossby waves reaches the saturation point at about day 30 due to the nonlinear wave-wave interactions of finite amplitude waves. Then, the well-developed Rossby waves start to break down. The role of the nonlinear wave-wave interactions during the wave breaking is to remove energy from the source region at the synoptic waves to the sink regions at both large and small scale motions.

In this no-mountain run, Rossby waves repeat exponential excitation due to the wavemaker and breakdown due to the nonlinear wave-wave interactions. As a result of the repeated Rossby wave breaking, the dominant scale of eddies shift toward planetary waves, and blocking-like anticyclones appear occasionally in high latitudes. Yet, the structure and behavior of the anticyclones are far from a typical blocking.

5.4.5 Simulations with all Waves and Mountains

Next, the model equation (5.53) is integrated with mountains for 1000 days. This experiment will be referred to as the mountain run. The energy level of planetary waves is sufficiently high and is comparable with observation. The energy levels for short waves are substantially lower than observation, indicating less down-scale energy cascade from the source region. The energy-level drop at both high- and low-frequency ranges quite consistent with the observations. Compared with the results from the no-mountain run, it is found that the spectral peak at the synoptic wave is produced by the wavemaker and that at planetary wave by topographic forcing.

The sole energy source into the model atmosphere is the wavemaker derived from baroclinic instability BC at the synoptic waves centered at the zonal wavenumber 7 for the no-mountain run. The topographic forcing TF becomes additional energy source for the mountain run. As seen in the no-mountain run, the energy supply at the synoptic waves is redistributed by the nonlinear wave-wave interaction NL, which is negative at the synoptic waves and is positive at both the short waves and planetary waves in addition to the zonal motions. The sum of NL over all waves becomes zero, as is required from the energy conservation law. Therefore, we can realize two branches of the energy flow in the wavenumber domain: one from synoptic waves to short waves and the other from synoptic waves to planetary waves and to zonal motions. The former may be regarded as the down-scale energy cascade and the latter the up-scale energy cascade which may be important for the blocking formation. By the down-scale cascade, an eddy breaks down to smaller eddies toward a turbulent flow. In contrast, small eddies merge to create a larger coherent vortex by the up-scale cascade. In other words, a coherent vortex emerges from turbulent flow, or an order is created from disorder by the up-scale cascade. Most of the up-scale energy cascade goes to the zonal motions, which are ultimately dissipated by the zonal surface stress. The circumpolar vortex by zonal jet may be therefore regarded as one of the largest coherent vortices created from geostrophic turbulence. The present barotropic model with a wavemaker is suitable to represent the up-scale cascade in a pure form.

When mountains are introduced, the topographic forcing amplifies planetary waves, especially the wavenumber 2. Then, the up-scale energy cascade by NL toward the planetary waves is combined with the energy supplied by the topographic forcing. Although the topographic forcing is approximately steady, the up-scale energy cascade is highly transient, causing occasional rapid amplification of planetary waves Tanaka (1991).

Figure 5.21a illustrates the Hovmöller diagram with mountain. When the wavemaker is introduced, progressive Rossby waves grow and pass the quasi-stationary troughs and ridges. When compared with the case of the no-mountain run, the forced quasi-stationary planetary waves and free progressive Rossby waves can interact with each other. However, the detail is complicated by the two types of waves. Figure 5.21b illustrates potential vorticity q for the same period. The progression of the Rossby waves is even more clear for q than the geopotential height. As seen from the no-mountain run, the mobile Rossby waves amplify exponentially, so they must eventually break down somewhere. When the topography is included, the eastward-moving Rossby waves tend to grow faster around the quasi-stationary troughs and

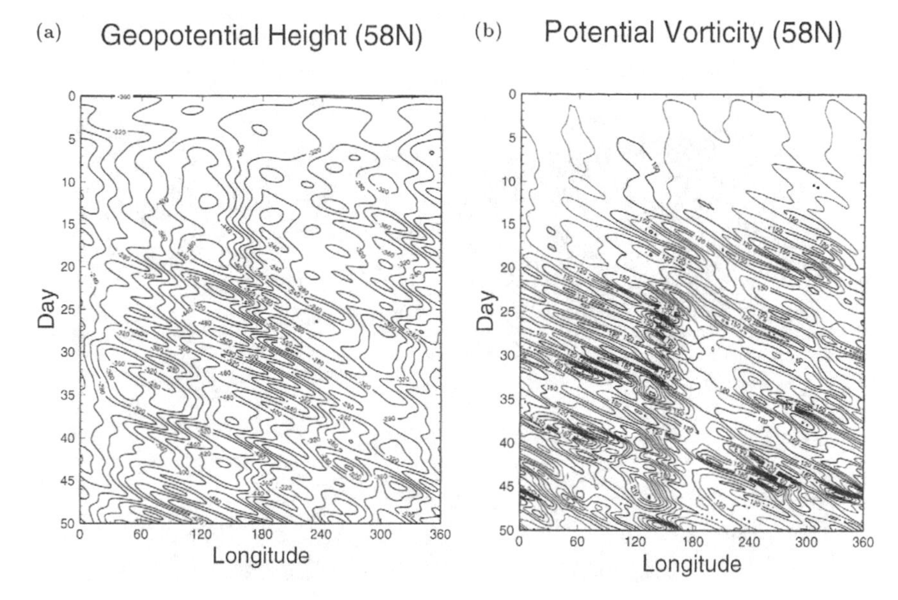

Fig. 5.21 Longitude-time section of (a) for geopotential height z (unit: m, interval: 40 m) and (b) for potential vorticity q (PV units, interval: 15 PV)along 58°N during the first 50 days for the mountain run with the wavemaker. From Tanaka (1998), Fig. 10.

then break down at the windward of the quasi-stationary ridge. Obviously, there are some geographical locations where the waves prefer to break down by the topographic effect.

5.4.6 Formation of the Blocking

Large-scale blocking occurs in the simulations with mountains once in a while when the eddy energy increases to a high level. T1988 found that about 10 marked blocking events occur during the 1000-day integration, each of which persists more than 10 days. Owing to the initial random noise superimposed on the zonally symmetric flow, the subsequent development of the atmosphere differs completely from a run to another, demonstrating the theory of chaos. We repeated similar long-term integrations with slightly different initial conditions to search for many blocking events in the model atmosphere. The large-scale blocking appears repeatedly in a sense of probability at the similar locations, exhibiting similar configuration and behavior.

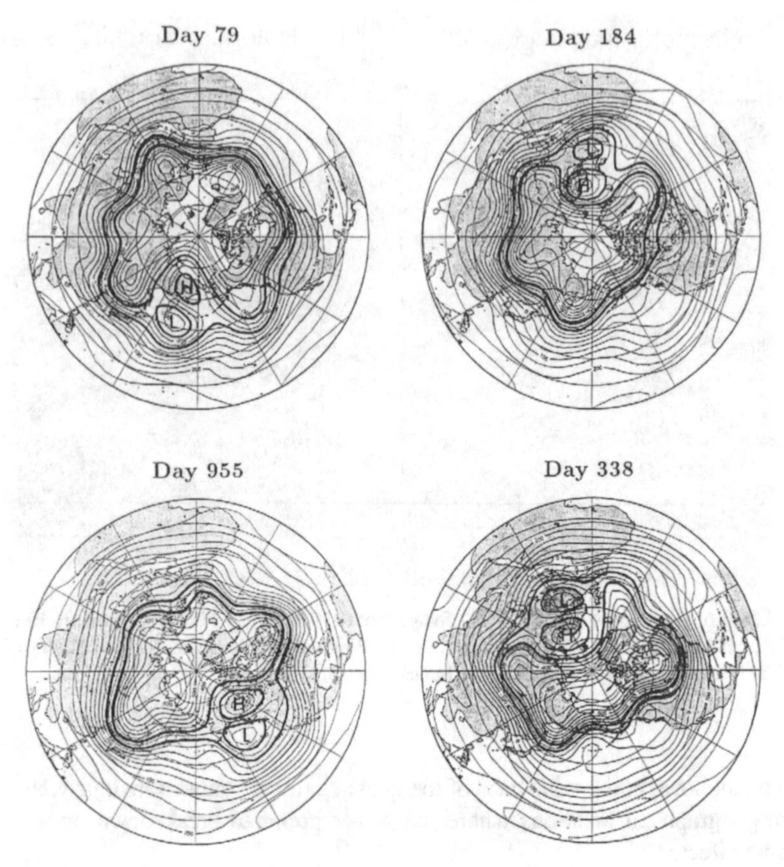

Fig. 5.22 Geopotential height (unit: m, interval: 50 m) for pronounced blockings appeared in the Pacific sector (days 79 and 955) and in the Atlantic sector (days 184 and 338). Splitting jetstream and the high-low vortex pair are marked by thick lines. From Tanaka (1998), Fig. 12.

Figure 5.22 illustrates some examples of pronounced dipole blocking that appeared in the Pacific sector (days 79 and 955) and Atlantic sector (days 184 and 338). The date is noted just for convenience since these are the collection from several different experiments. The jetstream splits in two branches over the blocking region. The well-known definition of blocking by Rex (1950) requires that 1) the basic westerly current must split into two branches, 2) each branch current must transport an appreciable mass, 3) the double-jet system must extend over at least 40° of longitude, 4) a sharp transition from zonal-type flow upstream to meridional-type downstream must be observed across the current split, and 5) the pattern must persist with recognizable continuity for at least 10 days. Evidently from the figures, conditions 1) through 4) are satisfied. We can confirm that those anticyclones persist more than 10 days. Hence, the characteristic features of blocking are well simulated by present barotropic model.

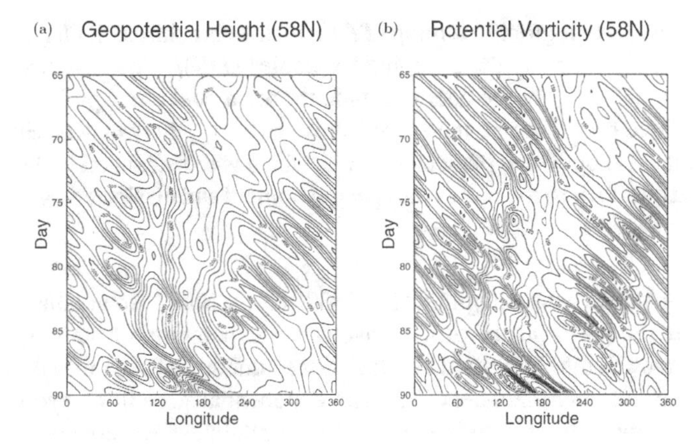

Fig. 5.23 Longitude-time section of (a) geopotential height z (unit: m, interval: 50 m) and (b) potential vorticity q (PV units, interval: 15 PV) along 58°N during days 65 to 90 for the mountain run with the wavemaker. A blocking is clearly identified at 180°E. From Tanaka (1998), Fig. 14.

Figure 5.23 illustrates the longitude-time section of geopotential height z and potential vorticity q along 58°N for a period from 65 to 90 days. The progressive troughs and ridges excited by the wavemaker approach the stationary ridge near 180°E which is induced by the topographic forcing. The blocking high is characterized as low values of q, and the trough upstream of the blocking high has a high value of q. A blocking high appears along the 180°E line for days 70 to 83. It is evident from the figure that the eastward move of free Rossby waves is blocked by the blocking high. The pattern of the Hovmüller diagram for q is not simply the superposition of free Rossby waves and stationary topographic ridge.

According to the result from the PV analysis, the travelling Rossby waves approaching at the western flank of the blocking undergo meridional stretch and break down, depositing a low potential vorticity at the north and high potential vorticity at the south. The results are consistent with the eddy straining hypothesis proposed by Shutts (1983).

5.4.7 Summary

In this section, we provided an overview of numerical experiments that investigated the baroclinic waves and atmospheric blocking using a simple barotropic model of Tanaka (1991) that featured a synoptic-scale wavemaker derived from baroclinic instability. The model considers only five physical processes: wavemaker as baroclinic instability $(BC)_i$, topographic forcing $(TF)_i$, biharmonic diffusion $(DF)_i$, zonal surface stress $(DZ)_i$ and Ekman pumping for eddies $(DE)_i$. The first two processes, *i.e.*, $(BC)_i$ and $(TF)_i$, are the energy source for the model atmosphere, whereas the remaining three processes are the energy sink.The energy flow may be characterized by the input at synoptic and planetary waves, which is redistributed in the wavenumber domain by $(NL)_i$ to the rest of the spectral domain and dissipated by the diffusion and surface friction.

First, the model was integrated for 1000 days without mountains in order to demonstrate the performance of the wavemaker as baroclinic instability. Since the maximum growth rate is located at zonal wavenumbers 7-8, the initial growth was dominated by the wavenumbers 7-8. The exponential growth of synoptic disturbances was soon saturated to balance with the nonlinear wave-wave interactions, which transfer the accumulated energy at the synoptic eddies toward the rest of the wavenumber domain. We demonstrated that the wavemaker in this section is useful to excite the synoptic eddies at the right location with the right structure.

Next, topography was introduced in the model and integrated for 1000 days for several times. The up-scale energy cascade from synoptic eddies to planetary waves is combined with the energy supply by the topographic forcing at the zonal wavenumber 2. The topographic forcing is approximately steady, whereas the up-scale energy cascade is highly transient, causing occasional rapid amplification of planetary waves. It is in this period when pronounced blockings occur in the model as the result of the increased up-scale energy cascade and Rossby wave breaking. Blocking occurs at two preferred locations over Alaska and North Europe, as is consistent with observations. The typical dipole configuration extends about 90° in zonal direction and remains at the same geographical location more than 10 days. The westerly jet is decelerated upstream of the block and split abruptly into two meridional branches. The inherent characteristics and behavior of blocking are well simulated by present model with an improved wavemaker.

The analysis of potential vorticity (PV) suggests that the onset of the blocking is triggered by Rossby wave breaking. As demonstrated by the no-mountain run, travelling Rossby waves are amplified exponentially by the wavemaker, accumulating eddy PV as they travel eastward. When topography is introduced, the travelling wave is amplified over the stationary trough and becomes critical to break down at the stationary ridge. The wave breaking proceeds at the windward of the stationary ridge in a form of an anti-clockwise overturning of PV field, which may be considered as the onset of blocking. In the real atmosphere, explosive cyclogenesis can be a trigger for the wave breaking as suggested by Colucci and Alberta (1996). Once a Rossby wave breaks down and is captured by the stationary ridge, subsequent progressive Rossby waves, carrying low PV, break down one after another at the western frank of the blocking, depositing fresh low PV into the main body of the blocking anti-cyclone. This process is consistent with the eddy straining mechanism proposed by

Shutts (1983). We suggest that the Rossby wave breaking is important for the onset of blocking, and the eddy straining mechanism is important for the maintenance of blocking.

Finally, detailed analysis is presented only for Pacific blockings in this section. We note here that Atlantic blocking in the model appears after a marked Rossby wave breaking as in the case of the Pacific blocking. However, the blocking is often maintained without clear evidence of the eddy straining mechanism at the western flank of the blocking. Dynamical instability of planetary waves may be an alternative explanation for the maintenance of Atlantic blocking Buizza and Molteni (1996). Nakamura (1994) and Nakamura and Anderson (1997) discussed the importance of the Rossby wavetrain across the Atlantic for Atlantic blocking. Compared with our former model in Tanaka (1991), we have shown in this section that the life-cycle of a blocking is improved in its geographical locations, configurations, and persistence.

5.5 Theory of Rossby Wave Saturation and Normal Mode Energy Spectrum

5.5.1 Background

Energy spectrum of the large-scale atmospheric circulation has been characterized by the -3 power law with respect to the horizontal wavenumber k over the synoptic to sub-synoptic scales Wiin-Nielsen (1967); Boer and Shepherd (1983); Nastrom et al. (1984). Using dimensional analysis, Kraichnan (1967) predicted a k^{-3} power law for 2D, isotropic and homogeneous turbulence in a downscale enstrophy cascading inertial subrange on the short-wave side of the scale of energy injection. It was shown by Tung and Orlando (2003) that not only enstrophy but also energy cascade down from the synoptic to meso scale. The downscale energy cascade is responsible for a $k^{-5/3}$ spectrum on the short-wave side where the energy cascade exceeds the enstrophy cascade.

Tung and Orlando (2003) demonstrated also that the energy injected at the synoptic scale cascades up to planetary waves and zonal motions where another dissipation exists. Contrasted to the k^{-3} law over the synoptic to sub-synoptic scales, there is no appropriate theory to describe the spectral characteristics at synoptic to planetary scales because of the existence of the energy source due to baroclinic instability. Welch and Tung (1998) argued that the theory of nonlinear baroclinic adjustment (Stone (1978)) is responsible to determine the spectrum over the energy source range.

They introduced a saturation criterion proposed by Garcia (1991) to determine the upper bound in meridional heat flux by the disturbances. According to the criterion, a Rossby wave breaks down when a local meridional gradient of potential vorticity is negative, i.e., $\partial q/\partial y < 0$, somewhere in the domain. The excessive energy is then transferred nonlinearly toward the so-called Rhines scale of cascade arrest in the large scale (Rhines (1975)). They suggested that the energy spectrum of the -3 power law holds by this Rossby wave breaking and saturation even for the energy source range. Their argument is, however, restricted only to the zonal wavenumbers for the synoptic waves, and planetary waves are beyond the scope of the theory.

The spectral characteristics of the synoptic to planetary waves are described by Tanaka (1985) and Tanaka and Kung (1988) by means of the 3D normal mode decomposition, including the vertical spectrum. The scheme is referred to as normal mode energetics. In their analysis, the scale of the 3D normal mode is represented by the eigen frequency of Laplace's tidal equation σ instead of the wavenumber k. The modal frequency is related to the scale by the wave dispersion relation. According to the result of the normal mode energetics analysis, the energy spectrum of the barotropic component of the atmosphere obeys characteristic slope of 2 to 3 power of the eigen frequency σ. Moreover, it is found by Tanaka and Kasahara (1992) that the energy spectrum is uniquely determined as a sole function of the phase speed of the Rossby mode c when the energy spectrum is plotted against c instead of σ.

The spectral peak over the scale parameter of c may be explained by the Rhines scale which separates the distinct slopes of turbulence and wave regimes as discussed by Tanaka (2003). The atmospheric energy is first converted from the baroclinic to the barotropic component at the synoptic scale when the baroclinicity is removed by the baroclinic instability. The accumulated barotropic energy is then cascades up to the larger scale obeying a specific power law governed by the 2D fluid mechanics. The upscale cascade is, however, arrested at the Rhines scale beyond which the linear term dominates due to the increased c. Although the spectral peak is clearly understood, there is no theory so far to explain the characteristic slope at the turbulence regime over the scale parameter of c. It is intriguing to demonstrate whether such a slope can be explained by the Rossby wave saturation as suggested by Welch and Tung (1998).

The purpose of this section is to understand the spectral slope observed in the large-scale atmospheric motions characteristic to the normal mode energetics, as described by Tanaka and Kanda (2004) (hereafter T2004). Attention is concentrated on the barotropic component of the atmosphere where most of the low-frequency variabilities, such as blocking, teleconnections, and the Arctic Oscillations, are contained. A theoretical derivation of the spectral peak and slope is attempted based on the criterion of the Rossby wave breaking and saturation proposed by Garcia (1991).

5.5.2 Governing Equations and Data

The governing equations used in this section are the 3D spectral primitive equations on a sphere (Tanaka and Watarai (1999)). It is important to note that the ratio of the magnitude of the linear term and nonlinear term in (5.17) determines the distinct behaviors of the mode. When σ_i is small (c_i is small), the nonlinear term dominates the linear term in (5.17), causing a turbulent behavior. Conversely, when σ_i is large (c_i is large), the linear term dominates the nonlinear term in (5.17), causing a simple normal mode behavior with the Laplace's tidal frequency σ_i. In this section, the ratio of the nonlinear term to the linear term is referred to as a spherical Rhines ratio R_i which characterizes the turbulence regime and the wave regime:

$$R_i = \frac{|\sum_{jk} r_{ijk} w_j w_k|}{|\sigma_i w_i|}. \tag{5.70}$$

The spherical Rhines ratio R_i is represented as a function of the scale parameter of the phase speed c_i.

5.5.3 Energy Spectrum for Geostrophic Turbulence

Figure 5.24 illustrates the result of the energy spectrum of E_i for the barotropic component as a function of the dimensionless phase speed of the Rossby modes c_i (only the magnitude is concerned). Energy levels are connected by dotted lines for the same zonal wavenumber k with different meridional mode numbers n. According to the result, the spectrum indicates two different regimes with distinct slopes for small c_i and large c_i. The energy injected at the synoptic scale (small c_i) cascades up to the larger scale (larger c_i) obeying a specific power law. The upscale energy cascade at the turbulence regime with small c_i would terminate at the wave regime with large c_i. The scale of the energy cascade arrest is referred to as Rhines scale on the β-plane (Rhines (1975)). Since the Rhines scale is defined on the sphere and measured by the phase speed instead of the wavenumber, we will refer to it as spherical Rhines speed c_R. The spectral peak at c_i=0.02 is clearly explained by the spherical Rhines speed c_R which separates the turbulence regime and wave regime. The line in the figure describes the spectral slope of c_i^2. Evidently, the spectrum obeys the power law at the turbulence regime, and the slope is close to c_i^2.

Figure 5.25 illustrates the spherical Rhines ratio R_i as a function of c_i. For small c_i, the ratio exceeds 30, indicating the dominant nonlinear term and negligible linear term. This range is characterized by the nonlinear turbulent behavior with

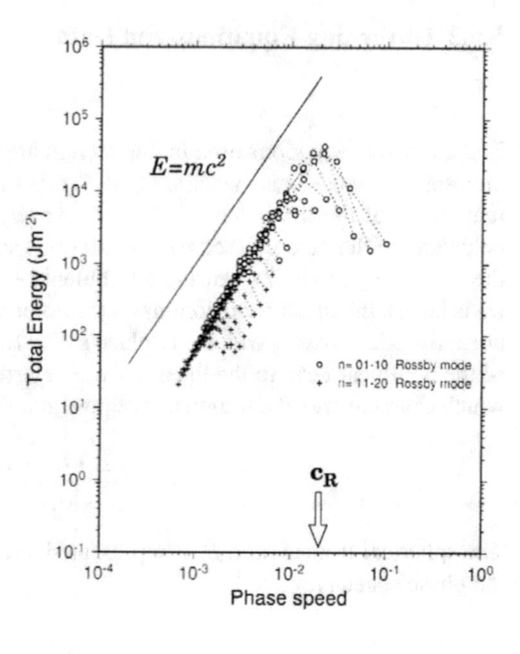

Fig. 5.24 Total energy spectrum E_i for the barotropic component as a function of the dimensionless phase speed of the Rossby mode $|c_i|$. Energy levels are connected by dotted lines for the same zonal wavenumber k with different meridional mode numbers n. The spherical Rhines speed c_R is marked by an arrow. The line of $E = mc^2$ represents the spectral slope derived by the condition of Rossby wave breaking, $\partial q/\partial y < 0$. From Tanaka and Kanda (2004), Fig. 2.

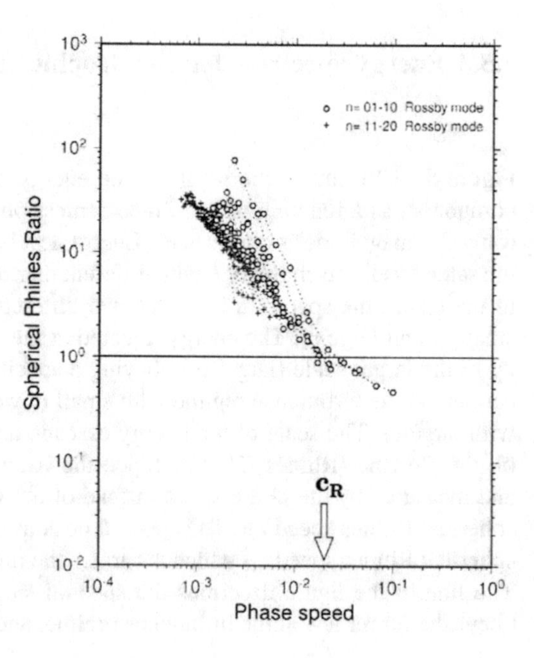

Fig. 5.25 Spherical Rhines ratio R_i as a function of $|c_i|$. The ratio represents the magnitude of nonlinear term divided by that of linear term in (5.70). The spherical Rhines speed c_R is defined by the phase speed with R_i=1. From Tanaka and Kanda (2004), Fig. 3.

strong coupling of different modes. The ratio becomes unity at c_i=0.02, which is the spherical Rhines speed c_R by definition. The upscale energy cascade must terminate at c_R because the nonlinear wave-wave interactions becomes negligible beyond this point. The spherical Rhines speed is also the speed where the westward phase speed of the Rossby wave balances with the eastward flow speed of the zonal motion. As a result, the Rossby wave becomes stationary, and appreciable amount of energy supply occurs at this range by the topographic forcing. For the large c_i beyond c_R, the ratio becomes less than unity, indicating the dominant linear term in (5.70). At this wave regime, we can confirm that the time behavior of the spectral coefficient w_i in the harmonic dial is characterized as a simple normal mode with the eigen frequency σ_i as inferred by (5.70).

Hence, the observed energy spectrum can be understood, at least for the spectral peak, from the upscale energy cascade in the turbulence regime and its arrest at the spherical Rhines speed. However, the spectral slope of c^2 has not been understood.

5.5.4 Energy Slope

In this section, a theoretical explanation is attempted for the characteristic spectral slope of c^2 based on the criterion of the Rossby wave breaking discussed by Garcia (1991). In the normal mode energetics, the barotropic component of the atmosphere describes the vertical mean of the state variables, which is governed formally by the shallow water system. Divergence of the shallow water system is contained mostly in the gravity modes with large c_i, but it is negligible for the Rossby modes with small c_i because divergence is proportional to σ_i. Therefore, the non-divergent quasi-geostrophic model is sufficient to represent the Rossby wave breaking for the barotropic component.

According to the Garcia's criterion, an amplifying Rossby wave breaks down if the local meridional gradient of the quasi-geostrophic potential vorticity q becomes negative somewhere in the domain:

$$\frac{\partial q}{\partial y} < 0, \quad q = \nabla^2\psi + f, \tag{5.71}$$

where ψ and f are stream function and the planetary vorticity, respectively. If the mid-latitude β plane is assumed, the criterion of the Rossby wave breaking provides the saturation point for the zonal wind speed u of the disturbance:

$$\frac{\partial}{\partial y}\left(\nabla^2\psi + f\right) = -\nabla^2 u + \beta < 0, \tag{5.72}$$

On the other hand, the saturation point for the zonal wind speed can be specified by the phase speed c of the Rossby waves:

$$u < \frac{-\beta}{k^2 + n^2} = c, \tag{5.73}$$

where k and n are zonal and meridional wavenumbers, respectively. The magnitude of u is apparently larger than v in the wave regime with large c, but they are comparable for the turbulence regime with small c as seen in Shepherd (1987). We can thus assume isotropy for the turbulence regime. Hence, we hypothesize that the energy spectrum E (Jm^{-2}) derived by Rossby wave saturation is proportional to c^2 and obtain the following relation:

$$E = \frac{1}{g}\int_0^{p_s} \frac{1}{2}\left(u^2 + v^2\right) dp = \frac{p_s}{g}c^2 = m\,c^2. \tag{5.74}$$

Here, $m = p_s/g$ represents a total mass of the atmosphere for unit area. Actual energy level must be smaller than that because u and v are defined by the peak values of a mode. Under the plane wave assumption, m is factored by 1/4 for the mean energy level. The theoretical slope then agrees quantitatively with the observation in Fig. 5.24.

For the spherical geometry, the saturation point of the zonal wind velocity may be represented by the dimensionless phase speed c_k of Haurwitz waves as in Wiin-Nielsen (1971):

$$\frac{u}{2\Omega a \cos\theta} < \frac{-1}{\hat{k}(\hat{k}+1)} = c_k, \tag{5.75}$$

where a is the radius of the Earth and $\hat{k}$ is the total wavenumber of the associated Legendre polynomial. The relation may be extended to the shallow water system on a sphere using the dimensionless phase speed of the normal mode c_i as:

$$\frac{u}{\sqrt{gh_0}} < \frac{\sigma_i}{k} = c_i. \tag{5.76}$$

The theoretical energy slope is compared with observation in Fig. 5.24 with the line designated as $E = mc^2$. The agreement is quite reasonable, supporting the hypothesis that the large-scale energy spectrum is determined by the Rossby wave saturation.

5.5.5 Rossby Wave Saturation Theory

In this section, energy spectrum of the large-scale atmospheric motions is examined in the framework of the 3D normal mode decomposition. Attention is concentrated to the barotropic component of the atmosphere where the low-frequency variabilities dominate. The representative horizontal scale of disturbance is measured by the phase speed of a Rossby mode $c_i = \sigma_i/k$ instead of the horizontal wavenumber $\hat{k}$, based on the wave dispersion relation for the normal mode on a sphere.

According to the result of the observational analysis, we obtain a characteristic energy spectrum with its peak at the spherical Rhines speed c_R. For the range of small c_i, the nonlinear term dominates the linear term, creating a turbulence regime. Energy injected at the synoptic eddies by the baroclinic instability cascades up to larger scale for the barotropic component of the atmosphere under the constraint of the 2D fluid mechanics. The upscale energy cascade terminates at the scale of the spherical Rhines speed c_R because the linear term dominates beyond that scale.

The observed spectral slope has been the major concern in this section because there is no theory to explain the characteristic slope in the turbulence regime. We put forward a hypothesis that the slope of c^2 can be derived from the criterion of Rossby wave breaking, $\partial q/\partial y < 0$, discussed by Garcia (1991). With a proportional constant m, describing a factored total mass of the atmosphere for unit area, we have shown that the barotropic energy spectrum of the general circulation can be represented as $E = mc^2$. The theoretical inference of the slope agrees quite well with the observation. A further study may be desired for the gravity modes and for the energy spectrum of the baroclinic component of the atmosphere.

5.5.6 Formation of Blocking in the Rossby Wave Saturation Theory

Low-frequency variabilities such as blocking have long been a major concern in the operational weather forecasting as well as the geophysical fluid dynamics. Although there is a wide consensus such that the nonlinearity is essential for the blocking formation, a simple linear dynamics of resonant stationary Rossby waves or the analysis of wave activity flux occasionally succeeds to explain the formation. Unified theory of blocking formation is difficult to be established despite the long history of observational and theoretical studies. Tung and Lindzen (1979); Shutts (1983); Holopainen and Fortelius (1987); Nakamura and Anderson (1997); Tanaka (1991, 1998).

When the blocking phenomenon is examined in the wavenumber domain, the importance of the strong nonlinearity reflects the key role of the scale interactions by the wave-wave interactions. Contrasted to the k^{-3} law over the sub-synoptic scales, there is no appropriate theory to describe the spectral characteristics at synoptic to planetary scales because of the existence of the energy source due to baroclinic instability.

Welch and Tung (1998) argued that the theory of nonlinear baroclinic adjustment Stone (1978) is responsible to determine the spectrum over the energy source range. They introduced a saturation criterion proposed by Garcia (1991) to determine the upper bound in meridional heat flux by the disturbances. According to the criterion, a Rossby wave breaks down when a local meridional gradient of potential vorticity becomes negative, i.e., $\partial q / \partial y < 0$, somewhere in the domain. The process of Rossby wave breaking was extensively studied as the theory of Rossby wave critical layers (e.g. Warn and Warn, 1978; Haynes, 1989). The excessive energy of the breaking Rossby wave is then transferred nonlinearly toward the so-called Rhines scale of cascade arrest Rhines (1975). According to the result of the normal mode energetics analysis by Tanaka (1985), the energy spectrum of the barotropic component of the atmosphere obeys characteristic slope over the eigenfrequency σ. Moreover, it is found by Tanaka and Kasahara (1992) that the energy spectrum is uniquely determined as a sole function of the phase speed of the Rossby mode c when the energy spectrum is plotted against c instead of σ.

Once the origin of the spectral peak and the power law is understood by the Rossby wave saturation theory, it is easily speculated that the barotropic energy would ultimately be accumulated at the spherical Rhines speed c_R. The excessively accumulated energy at c_R would stay for long time because the triad wave-wave interactions of turbulence no longer breaks down the amplified Rossby wave. The accumulated energy is subject to be transferred to the zonal motion by zonal-wave interactions, which is referred to zonalization (Williams (1978)). In this section, we propose a hypothesis such that the atmospheric blocking is a realization of accumulated energy at the spherical Rhines speed c_R exceeding the Rossby wave saturation level.

The purpose of this section is first to confirm the up-scale energy cascade from the synoptic eddies to the spherical Rhines speed c_R in the phase speed domain, as described by Tanaka and Terasaki (2006) (hereafter TT2006). Energetics analysis is conducted for the 51 years of NCEP/NCAR reanalysis during winter. Second, we confirm our speculation such that the blocking phenomenon is characterized as the excessive accumulation of barotropic energy at the spherical Rhines speed c_R. Finally, we confirm the intensification of the up-scale energy cascade to c_R in the phase speed domain during the blocking events in winter by the composite analysis.

5.5.7 Governing Equations and Data

By differentiating energy E_i with respect to time and substituting (5.17), we obtain the energy balance equation:

$$\frac{dE_i}{dt} = N_i + S_i\,, \tag{5.77}$$

where N_i and S_i designated the nonlinear interactions and the energy sources, respectively.

We then define energy flux in the phase speed domain F_i by the summation of the nonlinear interactions N_i with respect to c_i in descending order of magnitude:

$$F_i = \sum_{l=1}^{i} N_l\,. \tag{5.78}$$

The energy flux is further decomposed in contributions from zonal-wave interactions F_{Zi} and wave-wave interactions F_{Wi}, by the summation of the zonal-wave interactions N_{Zi} and wave-wave interactions N_{Wi}, respectively, with respect to c_i in descending order of magnitude in the phase speed domain:

$$F_{Zi} = \sum_{l=1}^{i} N_{Zl}\,, \tag{5.79}$$

$$F_{Wi} = \sum_{l=1}^{i} N_{Wl}\,. \tag{5.80}$$

Since the summation of N_{Wi} for all indices becomes zero, the positive and negative values of F_{Wi} represent up-scale and down-scale energy flux over the phase speed domain, respectively. Likewise, the summation of N_{Zi} for all indices becomes zero. The positive values of F_{Zi} represent the energy flux from eddy to zonal components.

5.5.8 Energy Flux and Spherical Rhines Speed

Figure 5.26 illustrates the result of the barotropic energy spectrum E_i as a function of c_i. According to the result, the spectrum indicates two distinct regimes with different slopes for small c_i and large c_i. The spectral peak at c_i=0.02 is clearly explained by the spherical Rhines speed c_R which separates the turbulence regime and wave regime. The spherical Rhines speed (18 m s^{-1}) is also the speed where the westward phase speed of the Rossby wave balances with the eastward flow speed by the zonal motion. For the NCEP/NCAR reanalysis used here, the energy levels at short waves are lower than that of JMA analysis by Tanaka and Kanda (2004) due to the apparent smoothing.

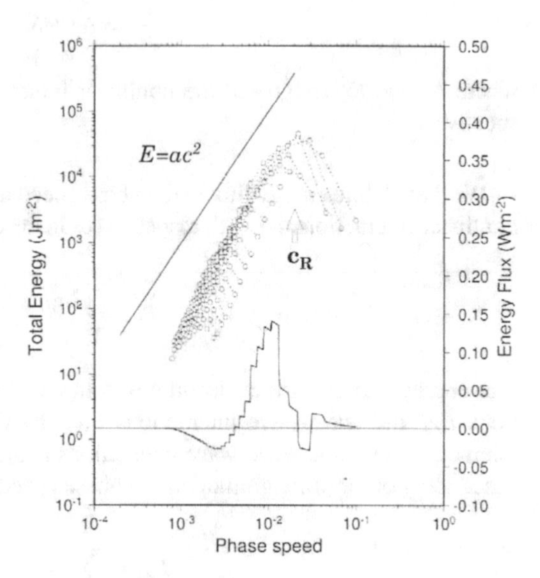

Fig. 5.26 Barotropic energy spectrum E_i and the energy flux for wave-wave interactions F_{Wi} as a function of the dimensionless phase speed of Rossby waves $|c_i|$. The spherical Rhines speed c_R is marked by an arrow. The line of $E = mc^2$ represents the spectral slope derived by the Rossby wave saturation theory. The lower graph of energy flux F_{Wi} for wave-wave interactions represents up-scale (down-scale) energy flux for positive (negative) values. From Tanaka and Terasaki (2006), Fig. 3.

Figure 5.26 illustrates also the energy flux F_{Wi} for the wave-wave interactions in the phase speed domain evaluated for 51 winters of the NCEP/NCAR reanalysis. The energy injected at the synoptic scale (small c_i) cascades up to the larger scale (larger c_i) obeying a specific power law. According to the result, the energy flux diverges at the synoptic scale around c_i=0.004 and cascades up toward c_R showing the peak value of 0.15 W m^{-2}. The up-scale energy flux converges at c_R as the cascade arrest discussed by Rhines (1975) in conjunction with a small down-scale energy flux from the wave regime beyond c_R. We notice also that a part of the energy injected at synoptic eddies cascades down to short waves as seen from the negative value of the flux at smallest c_i (see Tung and Orlando (2003)). It may be important to note that there is no apparent energy peak at the energy source range in synoptic scale around c_i=0.004. The result implies that the excessive energy produced by the Rossby wave breaking promptly cascades up toward c_R maintaining the spectral slope of c^2 bounded by the principle of Rossby wave saturation.

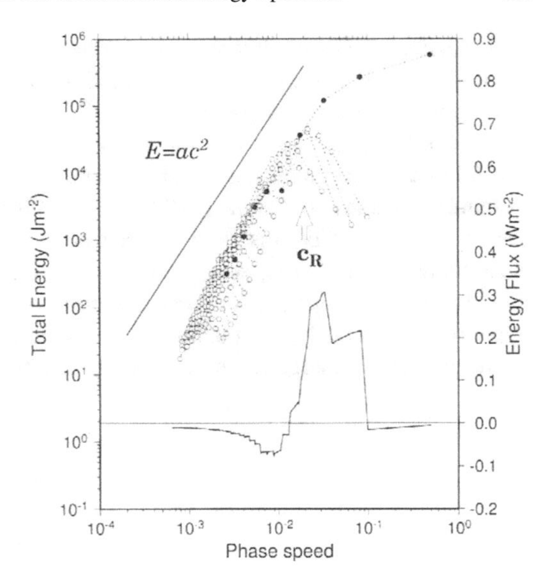

Fig. 5.27 As in Fig. 5.26, but with the energy flux F_{Zi} for the zonal-wave interactions in the phase speed domain. Energy spectrum for the zonal wavenumber k=0 is superimposed on the eddy energy spectrum by black dots after Tanaka (2003). The energy flux F_{Zi} for zonal-wave interactions represents up-scale (down-scale) energy flux for positive (negative) values. From Tanaka and Terasaki (2006), Fig. 4.

Figure 5.27 shows the same energy spectrum E_i, but with the energy flux F_{Zi} for the zonal-wave interactions in the phase speed domain. Energy spectrum for the zonal wavenumber k=0 is superimposed on the eddy energy spectrum by black dots after Tanaka (2003). According to the result, the energy flux diverges at c_R and converges at c_i=0.10 showing the peak value of 0.30 W m^{-2}. Evidently, the accumulated eddy energy at c_R is transferred to zonal motion, mostly at the meridional index k=3. This is a realization of the so-called zonalization by Williams (1978). It is found that a small down-scale energy flux is observed at synoptic to short waves around c_i=0.01 which is not documented in the previous studies. It may be interesting to note that time variation in the energy flux F_{Zi} is responsible for the formation of the Arctic Oscillation (see Tanaka and Terasaki (2005)).

5.5.9 Formation of Blocking by the Up-Scale Energy Cascade

The principle of Rossby wave saturation would lead to a number of new understanding of atmospheric phenomena; one of which may be the idea of blocking formation by the accumulation of barotropic energy at the spherical Rhines speed at c_R. A sequence of amplification and breaking of Rossby waves at the synoptic scale results in an expansion of the turbulence regime toward the larger c_i along the slope of c^2. At the cascade arrest of c_R, however, the accumulated excessive energy cannot break down by the triad wave-wave interaction because the nonlinearity no longer dominates the linear process. Accordingly, the amplified Rossby wave at c_R will persist long time showing a characteristic structure of blocking with $\partial q/\partial y < 0$. The direction of the overturning in potential vorticity can be either clockwise or anti-clockwise. It is definitely anti-clockwise for extra-tropical cyclones by the nature of baroclinic instability Tanaka and Watarai (1999). For blocking, however, clockwise overturning seems to dominate because anti-cyclone is breaking Nakamura (1994). When the accumulated energy at c_R is relaxed so as to extend the slope of c^2 beyond c_R, the blocking would propagate westward as often observed. In order to confirm those hypotheses, data analysis is conducted for blocking events in the same framework as the normal mode energetics in the phase speed domain.

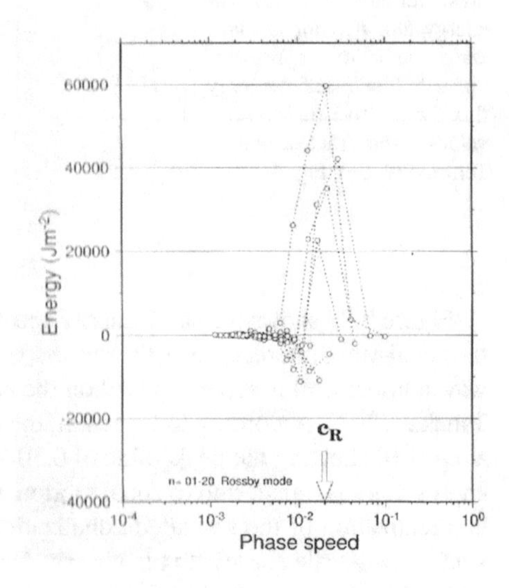

Fig. 5.28 Distribution of barotropic energy anomaly in the phase speed domain for 0000Z 2 February to 1800Z 6 February 1989 when the blocking occurred at Alaska. The spherical Rhines speed c_R is marked by an arrow. From Tanaka and Terasaki (2006), Fig. 7.

Figure 5.28 illustrates the barotropic energy anomaly in the phase speed domain for 0000Z 2 February to 1200Z 6 February 1989 when the large blocking event Tanaka and Milkovich (1990); Tan and Curry (1993) was in its mature stage. Energy levels are connected by dotted lines for the same zonal wavenumber k with different meridional mode numbers n. The spherical Rhines speed c_R is marked by an arrow. We can confirm that the excessive energy anomaly is concentrated just over c_R when the blocking is in its mature stage. The amount of the anomaly reaches 6.0×10^4 J m^{-2}, which corresponds to the double of the climate level.

Extensive analysis by TT2006 showed that blocking events are characterized by the excessive energy accumulation just over c_R. The up-scale energy flux from the synoptic-scale source range to c_R was enhanced during blocking events and the energy flux convergence increases evidently at c_R where the energy anomaly shows the peak. The anomaly of 0.5 W m^{-2} has a potential to produce energy anomaly of

2.0×10^5 J m^{-2} for 5 days, which is quantitatively sufficient to explain the energy anomaly during the blocking events.

TT2006 found that the accumulated excessive energy just over c_R results in a blocking. The energy level at c_R can exceed the saturation criterion of $E = mc^2$ because it no longer breaks down by the triad wave-wave interactions. Therefore, the accumulated excessive energy at c_R persists sufficiently longer time showing the specific structure with $\partial q/\partial y < 0$ which is characteristic to blocking. The amplified nonlinear Rossby wave at c_R stays at the same location for long time because c_R coincides with the scale of stationary Rossby waves.

Although the up-scale energy cascade may be the principal mechanism for the accumulation of energy at c_R, other mechanisms such as the resonant Rossby wave with topographic forcing Tung and Lindzen (1979), baroclinic instability of planetary waves Tanaka and Kung (1989), or accumulation of wave activity density flux by the quasi-stationary Rossby wave train Nakamura and Anderson (1997) may act as the excitation mechanism for the stationary Rossby wave at c_R. Those are recognized as different processes to produce the common large-scale configuration as blocking.

In conclusion, the essential feature of a blocking can be understood as the excessive accumulation of barotropic energy at the spherical Rhines speed c_R, where an amplified Rossby wave persists long time at the same location showing the breaking criterion $\partial q/\partial y < 0$. The conclusion of this section is based on the Rossby wave saturation theory in the spectral domain. Further descriptions may be desired to connect the special view of the blocking with that in the regional behavior of the blocking.

5.5.10 Excitation of Arctic Oscillation by the Up-Scale Energy Cascade

Tanaka and Terasaki (2006) postulated that the atmospheric blocking is formed when excessive energy is accumulated at the spherical Rhines speed c_R exceeding the Rossby wave saturation theory. Similar analogy of the energy flow in the phase speed domain will lead to a hypothesis such that the accumulated energy at the spherical Rhines speed c_R is transferred to the zonal flow by the zonal-wave interaction, creating the Arctic Oscillation.

In this section we first examine the up-scale energy cascade from the sperical Rhines speed to the zonal field in the phase speed domain. Second, we confirm our speculation such that the AO is characterized as the accumulation of barotropic energy at a specific meridional mode of the zonal field. Finally, we confirm the intensification of the zonal-wave interaction during the AO positive phase by the composite analysis.

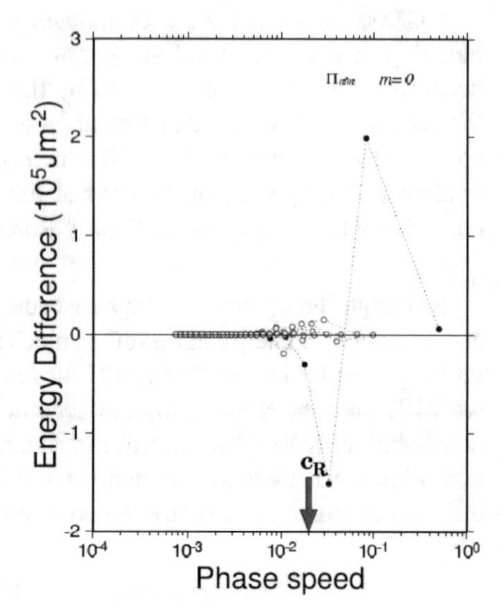

Fig. 5.29 Distribution of barotropic energy difference for the AO positive subtracted by AO negative in the phase speed domain. From Tanaka and Terasaki (2005), Fig. 3.

Figure 5.29 illustrates the distribution of barotropic energy difference for the AO positive subtracted by AO negative in the phase speed domain. The AO positive and AO negative are the composite of the AO time series for the standard deviation ±1.5 and above, respectively. It is found that the AO is characterized by the accumulation of energy at k=3 and reduction at k=5 of the zonal field. The positive and negative peaks reach 2.0 and -1.5 × 10^5 J m^{-2}, respectively.

Figure 5.30 shows the energy flux associated with the zonal-wave interactions F_{Zi} in the phase speed domain for the composite of the AO positive (solid line) and AO negative (dashed line) compared with the climate (dotted line). The up-scale energy flux from the synoptic-scale source range F_{Wi} converges at c_R during blocking events. It is noteworthy that the accumulated energy at c_R is further transferred to k=3 by the energy flux F_{Zi}. The excessive energy at k=3 is evidently resulted from the energy flux convergence at k=3. The reduction of energy at k=5 is also explained by the reduced flux convergence at k=5. It is confirmed that the up-scale energy flux F_{Zi} is clearly instrumental for the AO.

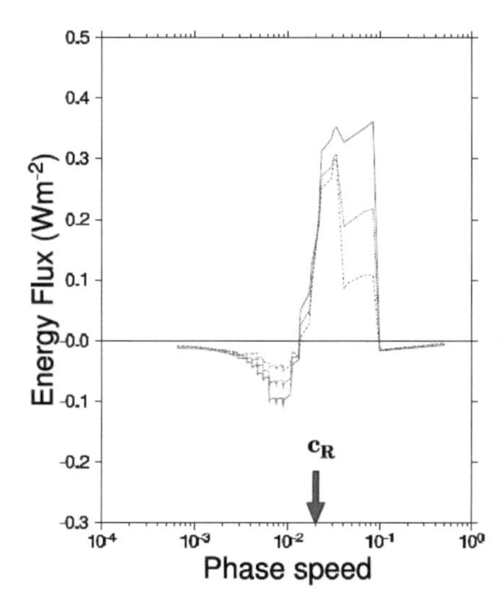

Fig. 5.30 Energy flux in the phase speed domain associated with the zonal-wave interactions evaluated for the AO positive (solid line), AO negative (dashed line) and climate (dotted line). From Tanaka and Terasaki (2005), fig. 4.

5.5.11 Summary

In this section, energy spectrum of the large-scale atmospheric motions was examined in the framework of the 3D normal mode energetics. Attention is concentrated to the barotropic component of the atmosphere where low-frequency variability dominates.

According to the presented studies, the AO in the phase speed domain is represented by the energy increase at $k = 3$ mode with simultaneous decrease at $k = 5$ of the zonal field. The energy accumulation at $k = 3$ is explained by the enhanced energy flux F_{Zi} associated with the zonal-wave interaction.

It was shown that the energy flux F_{Zi} comes from the spherical Rhines speed c_R where the planetary-scale Rossby waves are stationary. Kimoto and Yasutomi (2001) suggested that the AO is induced by the interactions with the forced steady planetary waves. In contrast, Tanaka (2003) suggested that the AO is induced by the interactions with the active synoptic eddies. It is shown in this section that the up-scale energy flux by transient eddies is once accumulated at c_R by F_{Wi}. The accumulated energy is then transferred to zonal field by F_{Zi} to cause the AO. The result shows that the low-frequency variability associated with the AO is maintained by energy flux from c_R, which is compensated by the up-scale cascade from synoptic eddies rather than the forced steady planetary waves.

References

Akahori, K. and Yoden, S. (1996). Zonal flow vacillation and bimodality of baroclinic eddy life cycles in a simple global circulation model. *J. Atmos. Sci.*, 54:2349–2361.

Balasubramanian, G. and Garner, S. (1997). The role of momentum fluxes in shaping the life cycle of a baroclinic wave. *J. Atmos. Sci.*, 54:510–533.

Boer, G. J. and Shepherd, T. G. (1983). Large-scale two-dimensional turbulence in the atmosphere. *J. Atmos. Sci.*, 40:164–1184.

Buizza, R. and Molteni, F. (1996). The role of finite-time barotropic instability during transition to blocking. *J. Atmos. Sci.*, 53:1675–1697.

Chang, E. K. M. and Fu, Y. (2002). Interdecadal variations in northern hemisphere winter storm track intensity. *J. Clim.*, 15:642–658.

Chang, E. K. M. and Fu, Y. (2003). Using mean flow change as a proxy to infer interdecadal storm track variability. *J. Clim.*, 16:2178–2196.

Charney, J. G. (1947). The dynamics of long waves in a baroclinic westely current. *J. Meteorol.*, 4:135–162.

Charney, J. G. and DeVore, J. G. (1979). Multiple flow equilibria in the atmosphere and blocking. *J. Atmos. Sci.*, 36:1205–1216.

Charney, J. G. and Eliassen, A. (1949). A numerical method for predicting the perturbations of the middle-latitude westerlies. *Tellus*, 1:38–54.

Charney, J. G. and Stern, M. E. (1962). On the stability of internal baroclinic jet in a rotating atmosphere. *J. Atmos. Sci.*, 19:159–172.

Chen, T.-C. and Trenberth, K. E. (1988). Forced planetary waves in the northern hemisphere winter: Wave coupled orographic and thermal forcings. *J. Atmos. Sci.*, 45:657–680.

Colucci, S. J. (1986). Explosive cyclogenesis and large-scale circulation change: Implications for atmospheric blocking. *J. Atmos. Sci.*, 42:2701–2717.

Colucci, S. J. and Alberta, T. L. (1996). Planetary-scale climatology of explosive cyclogenesis and blocking. *Mon. Wea. Rev.*, 124:2509–2520.

Edmon, H.J., H. B. and McIntyre, M. (1980). Eliassen-palm cross-sections for the troposphere. *J. Atmos. Sci.*, 37:2600–2616.

Eichelberger, S. J. and Hartmann, D. L. (2007). Zonal jet structure and the leading mode of variability. *J. Clim.*, 15:5149–5163.

Frederiksen, J. S. (1982). A unified three-dimensional instability theory of the onset of blocking and cyclogenesis. *J. Atmos. Sci.*, 39:969–982.

Frederiksen, J. S. and Bell, R. C. (1987). Teleconnection patterns and the roles of baroclinic, barotropic and topographic instability. *J. Atmos. Sci.*, 44:2200–2218.

Fritts, D. C. (1984). Gravity wave saturation in the middle atmosphere: A review of theory and observation. *Rev. Geophys. Space Phys.*, 22:275–308.

Gall, R. (1976). A comparison of linear baroclinic instability theory with the eddy statistics of a general circulation model. *J. Atmos. Sci.*, 33:349–373.

Garcia, R. R. (1991). Parameterization of planetary wave breaking in the middle atmosphere. *J. Atmos. Sci.*, 48:1405–1419.

Govindasamy, B. and Garner, S. (1997). The equilibration of short baroclinic waves. *J. Atmos. Sci.*, 54:2850–2871.

Green, J. S. A. (1960). A problem in baroclinic instability. *Quart. J. Roy. Meteor. Soc.*, 86:237–251.

Hansen, A. R. and Chen, T.-C. (1982). A spectral energetics analysis of atmospheric blocking. *Mon. Wea. Rev.*, 110:1146–1165.

Hartmann, D. and Zuercher, P. (1998). Response of baroclinic life cycle to barotropic shear. *J. Atmos. Sci.*, 42:865–883.

Hartmann, D. L. (1979). Baroclinic instability of realistic zonal-mean states to planetary waves. *J. Atmos. Sci.*, 36:2336–2349.

Haynes, P. (1989). The effect of barotropic instability on the nonlinear evolution of a Rossby wave critical layer. *J. Fluid Mech.*, 207:231–266.

Holopainen, E. and Fortelius, C. (1987). High-frequency transient eddies and blocking. *J. Atmos. Sci.*, 44:1632–1645.

Hoskins, B. J. (1983). *Modelling of the transient eddies and their feedback on the mean flow: Large-scale Dynamical Processes in the Atmosphere*. Academic Press.

Ioannou, P. and Lindzen, R. S. (1986). Baroclinic instability in the presence of barotropic jets. *J. Atmos. Sci.*, 43:2999–3014.

Jucks, M. (1989). A shallow water model of the winter stratosphere. *J. Atmos. Sci.*, 46:2934–2955.

Jucks, M. and McIntyre, M. E. (1987). A high-resolution one-layer model of breaking planetary waves in the stratosphere. *Nature*, 328:590–596.

Kalnay, E., Kanamitsu, M., Kistler, R., Collins, W., Deaven, D., Gandin, L., Iredell, M., Saha, S., White, G., Woollen, J., Zhu, Y., Chelliah, M., Ebisuzaki, W., Higgins, W., Janowiak, J., Mo, K., Ropelewski, C., Wang, J., Leetmaa, A., Reynolds, R., Jenne, R., , and Joseph, D. (1996). The NCEP/NCAR 40-year reanalysis project. *Bull. Amer. Meteor. Soc.*, 77:437–471.

Kasahara, A. (1977). Numerical integration of the global barotropic primitive equations with hough harmonic expansions. *J. Atmos. Sci.*, 34:687–701.

Kasahara, A. (1984). The linear response of a stratified global atmosphere to a tropical thermal forcing. *J. Atmos. Sci.*, 41:2217–2237.

Kasahara, A. and Puri, K. (1981). Spectral representation of three-dimensional global data by expansion in normal mode functions. *Mon. Wea. Rev.*, 109:37–51.

Kimoto, M., Jin, F. F. Watanabe, M. and Yasutomi, N. (2001). Zonal-eddy coupling and a neutral mode theory for the Arctic oscillation. *Geophys. Res. Lett.*, 28:737–740.

Kraichnan, R. H. (1967). Inertial ranges in two-dimensional turbulence. *Phys. fluids*, 10:1417–1423.

Kung, E. (1988). Spectral energetics of the general circulation and time spectra of transient waves during the FGGE year. . *J. Clim.*, 1:5–19.

Kung, E. C. and Baker, W. E. (1986). Spectral energetics of the observed and simulated northern hemisphere general circulation during blocking episodes. *J. Atmos. Sci.*, 43:2729–2812.

Leovy, C., Sun, C. R., Hitchman, M., Remsberg, E., Russell, J., Gordley, L., Gille, J., and Lyjak, L. (1985). Transport of ozone in the middle stratosphere: Evidence for planetary wave breaking. *J. Atmos. Terr. Phys.*, 42:230–244.

Limpasuvan, V. and Hartmann, D. L. (2000). Wave-maintained annular mode of climate variability. *J. Clim.*, 13:4414–4429.

Lindzen, R. S. (1981). Turbulence and stress owing to gravity wave and tidal break down. *J. Geophys. Res.*, 86:9707–9714.

Lorenz, D. J. and Hartmann, D. L. (2003). Eddy-zonal flow feedback in the northern hemisphere winter. *J. Clim.*, 16:1212–1227.

Lorenz, E. N. (1955). Available potential energy and the maintenance of the general circulation. *Tellus*, 7:157–167.

McIntyre, M. E. and Palmer, T. H. (1983). Breaking planetary waves in the stratosphere. *Nature*, 305:593–600.

McIntyre, M. E. and Palmer, T. H. (1984). The 'surf zone' in the stratosphere. *J. Atmos. Terr. Phys.*, 46:825–849.

McIntyre, M. E. and Palmer, T. H. (1985). A note on the general concept of wave breaking for Rossby and gravity waves. *Pure Appl. Geophys.*, 123:964–975.

Mullen, S. L. (1987). Transient eddy forcing of blocking flows. *J. Atmos. Sci.*, 44:3–22.

Nakamura, H. (1994). Rotational evolution of potential vorticity associated with a strong blocking flow configuration over europe. *Geophys. Res. Lett.*, 21:2003–2006.

Nakamura, H., Nakamura, M. and Anderson, J. L. (1997). The role of high- and low-frequency dynamics in blocking formation. *Mon. Wea. Rev.*, 125:2074–2093.

Nastrom, G. D., Gage, K. S., and Jasperson, W. H. (1984). Kinetic energy spectrum of large- and mesoscale atmospheric processes. *Nature*, 310(5972):36–38.

Ohashi, M. and Tanaka, H. L. (2009). Data analysis of recent warming pattern in the Arctic. *SOLA*, 6A:1–4.

Oort, A. (1964). On the energetics of the mean and eddy circulations in the lower stratosphere. *Tellus*, 16:309–327.

Rex, D. F. (1950). Blocking action in the middle troposphere and its effect upon regional climate. i. an aerological study of blocking action. *Tellus*, 2:196–211.

Rhines, P. (1975). Waves and turbulence on a beta-plane. *J. Fluid Mech.*, 69:417–443.

Robinson, W. (1988). Irreversible wave-mean flow interactions in a mechanistic model of the stratosphere. *J. Atmos. Sci.*, 45:3413–3430.

Salby, M. L., Garcia, R. R., O'Sullivan, D. and Tribbia, J. (1990). Global transport calculations with an equivalent barotropic system. *J. Atmos. Sci.*, 47:188–214.

Saltzman, B. (1970). Large-scale atmospheric energetics in the wavenumber domain. *Rev. of Geophys. and Space Phys.*, 8:289–302.

Sasaki, Y. K. and Chang, L. P. (1985). Numerical solution of the vertical structure equation in the normal mode method. *Mon. Wea. Rev.*, 113:782–793.

Schilling, H. D. (1986). On atmospheric blocking types and blocking numbers. anomalous atmospheric flows and blocking. Advances in Geophisics. 29, Academic Press, 71–99.

Seki, S., Tanaka, H. L., and Fujiwara, F. (2011). Modification of the baroclinic instability associated with positive and negative Arctic oscillation index: A theoretical proof of the positive feedback. *SOLA*, 7:53–56.

Shepherd, T. G. (1987). A spectral view of nonlinear fluxes and stationary-transient interaction in the atmosphere. *J. Atmos. Sci.*, 44:1166–1178.

Shutts, G. J. (1983). The propagation of eddies in difluent jet streams: Eddy vorticity forcing of blocking flow fields. *Quart. J. Roy. Meteor. Soc.*, 109:737–761.

Simmons, A. J. and Hoskins, B. J. (1976). Baroclinic instability on the sphere: Normal modes of the primitive and quasi-geostrophic equations. *J. Atmos. Sci.*, 33:1454–1477.

Simmons, A. J. and Hoskins, B. J. (1978). The life cycles of some nonlinear baroclinic waves. *J. Atmos. Sci.*, 35:414–432.

Simmons, A. J., Wallace, J. M., and Branstator, G. W. (1983). Barotropic wave propagation and instability, and atmospheric teleconnection patterns. *J. Atmos. Sci.*, 40:1363–1392.

Staniforth, A., Beland, M. and Coté, J. (1985). An analysis of the vertical structure equation in sigma coordinates. *Atmos.-Ocean*, 23:323–358.

Stone, P. H. (1978). Baroclinic adjustment. *J. Atmos. Sci.*, 35:561–571.

Swarztrauber, P. N. and Kasahara, A. (1985). The vector harmonic analysis of Laplace tidal equations. *SIAM J. Stat. Comput.*, 6:464–491.

Tan, Y.-C. and Curry, J. A. (1993). A diagnostic study of the evolution of an intense north american anticyclone during winter 1989. *Mon. Wea. Rev.*, 121:961–975.

Tanaka, H. (1985). Global energetics analysis by expansion into three-dimensional normal-mode functions during the FGGE winter. *J. Meteor. Soc. Japan*, 63:180–200.

Tanaka, H. L. (2003). Analysis and modeling of the Arctic oscillation using a simple barotropic model with baroclinic eddy forcing. *J. Atmos. Sci.*, 60:1359–1379.

Tanaka, H. L. and Kung, E. (1988). Normal-mode expansion of the general circulation during the FGGE year. *J. Atmos. Sci.*, 45:3723–3736.

Tanaka, H. L. and Kung, E. (1989). A study of low-frequency unstable planetary waves in realistic zonal and zonally varing basic states. *Tellus*, 41A:179–199.

Tanaka, H. L. and Terasaki, K. (2006). Blocking formation by an accumulation of barotropic energy exceeding the Rossby wave saturation level at the spherical rhines scale. *J. Meteor. Soc. Japan*, 84:319–332.

Tanaka, H. L. and Tokinaga, H. (2002). Baroclinic instability in high latitudes induced by polar vortex: A connection to the Arctic oscillation. *J. Atmos. Sci.*, 59:69–82.

Tanaka, H. L. (1991). A numerical simulation of amplification of low-frequency planetary waves and blocking formations by the upscale energy cascade. *Mon. Wea. Rev.*, 119:2919–2935.

Tanaka, H. L. (1993). Low-frequency unstable planetary waves in realistic basic states. *Science Report of theInstitute of Geoscience, University of Tsukuba, Section A*, 14:61–82.

Tanaka, H. L. (1995). A life-cycle of nonlinear baroclinic waves represented by 3-d spectral model. *Tellus*, 47A:697–704.

Tanaka, H. L. (1998). Numerical simulation of a life-cycle of atmospheric blocking and the analysis of potential vorticity using a simple barotropic model. *J. Meteor. Soc. Japan*, 76:983–1008.

Tanaka, H. L. and Kasahara, A. (1992). On the normal modes of Laplace's tidal equations for zonal wavenumber zero. *Tellus*, 44A:18–32.

Tanaka, H. L. and Matsueda, M. (2005). Arctic oscillation analyzed as a singular eigenmode of the global atmosphere. *J. Meteor. Soc. Japan*, 83:611–619.

Tanaka, H. L. and Milkovich, M. F. (1990). A heat budget analysis of the polar troposphere in and around Alaska during the abnormal winter of 1988/89. *Mon. Wea. Rev.*, 118:1628–1639.

Tanaka, H. L. and Seki, S. (2013). Development of a three-dimensional spectral linear baroclinic model and its application to the baroclinic instability associated with positive and negative Arctic oscillation indices. *J. Meteor. Soc. Japan*, 91:193–213.

Tanaka, H. L. and Terasaki, K. (2005). Energy spectrum and energy flow of the Arctic oscillation in the phase speed domain. *SOLA*, 1:65–68.

Tanaka, H. L. and Watarai, Y. (1999). A numerical experiment of breaking Rossby waves in the barotropic atmosphere withparameterized baroclinic instability. *Tellus*, 51A:552–573.

Tanaka, H. L., Watarai, Y. and Kanda, T. (2004). Energy spectrum proportional to the squared phase speed of Rossby modes in the general circulation of the atmosphere. *Geophys. Res. Lett.*, 31(13):13109.

Thompson, D. W. J. and Wallace, J. M. (1998). The Arctic oscillation signature in the wintertime geopotential height and temperature fields. *Geophy. Res. Lett.*, 25:1297–1300.

Thorncroft, C. D., Hoskins, B. J. and McIntyre, M. (1993). Two paradigms of baroclinic-wave life-cycle behavior. *Q. J. Roy. Meteor. Soc.*, 119:17–55.

Tung, K. K. and Lindzen, R. S. (1979). A theory of stationary long waves. part 1: A simple theory of blocking. *Mon. Wea. Rev.*, 107:714–734.

Tung, K. K. and Orlando, W. W. (2003). The k^{-3} and $k^{-5/3}$ energy spectrum of atmospheric turbulence: Quasigeostrophic two-level simulation. *J. Atmos. Sci.*, 60:824–835.

Valdes, P. J. and Hoskins, B. J. (1989). Linear stationary wave simulations for the time-mean climatological flow. *J. Atmos. Sci.*, 46:2509–2527.

Warn, T. and Warn, H. (1978). The effects of spherical geometry on the evolution of baroclinic waves. *J. Atmos. Sci.*, (24):597–612.

Welch, W. T. and Tung, K. K. (1998). On the equilibrium spectrum of transient waves in the atmosphere. *J. Atmos. Sci.*, 55(201-205):2833–2851.

Whitaker, J. and Snyder, C. (1993). The effects of spherical geometry on the evolution of baroclinic waves. *J. Atmos. Sci.*, 50:597–612.

Wiin-Nielsen, A. (1967). On the annual variation and spectral distribution of atmospheric energy. *Tellus*, 19:540–559.

Wiin-Nielsen, A. (1971). On the motion of various vertical modes of transient very long waves. *Tellus*, (19):207–217.

Williams, G. (1978). Planetary circulations: I. barotropic representation of jovian and terrestrial turbulence. *J. Atmos. Sci.*, 35:1399–1426.

Zhang, K. S. and Sasamori, T. (1985). A linear stability analysis of the stratospheric and mesospheric zonal mean state in winter and summer. *J. Atmos. Sci.*, 42:2728–2750.

Chapter 6

Applications to Predictions and Climate Studies

Hiroshi L. Tanaka

Abstract In this chapter, numerical simulations of atmospheric blocking and Arctic Oscillation (AO) are conducted using a simple barotropic model with the basis of the 3D-NMFs that considers the barotropic-baroclinic interactions as the external forcing. Using the perfect forcing evaluated as the residual of the model equation from the state variables, we can construct a best-fit forcing by solving an inverse problem. The model is referred to as a barotropic S-model. We integrated the model for 50 years under a perpetual January condition and analyzed the dominant EOF modes in the model atmosphere to obtain the AO as the EOF-1. The AO appears chaotically as an internal variability of the barotropic dynamics, induced by the up-scale energy cascade from stationary planetary waves. The result suggests that the AO can be understood as a dynamical eigenmode of the barotropic component of the atmosphere, and is not the statistical artifact as was argued by many researchers. Since the eigenmode is associated with zero eigenvalue of the dynamical system, the mechanism is called the singular eigenmode theory of the AO.

6.1 Construction of Barotropic P-model, B-model, and S-model

6.1.1 Introduction

A medium-range numerical weather prediction has been hampered by the prediction barrier caused by the chaotic nature of nonlinear fluid systems. According to the pioneer work by Lorenz (1963); Lorenz (1969), formally deterministic fluid systems having multiple scales of motion are observationally indistinguishable from

N. Žagar and J. Tribbia (eds.), *Modal View of Atmospheric Variability*,
Mathematics of Planet Earth 8, https://doi.org/10.1007/978-3-030-60963-4_6

indeterministic systems, due to the existence of chaos. The basic principle of chaos is a rapid growth of small initial errors, which deviates from the true trajectory of the solution. Lorenz showed that each scale of motion possesses an intrinsic finite range of predictability; cumulus-scale motions can be predicted by about one hour, synoptic-scale motions by a few days, and global-scale motions by a maximum of two weeks. It is our contention that deterministic medium-range forecasting is impossible beyond two weeks of the chaotic barrier, even if we can have a perfect prediction model. At present, the useful deterministic predictability with operational numerical weather prediction models is 7 to 10 days (Lorenz, 1985; Dalcher and Kalnay, 1987; Kalnay and Baker, 1990; Kalnay and McPherson, 1998).

The predictability might be extended by predicting averaged quantities. Anomaly correlation is better for planetary waves than for synoptic or short waves, when those waves are separated (e.g. Vallis, 1983; Shukla, 1985; Bengtsson, 1985). This implies that spatial averaging increases predictability. Miyakoda and Ploshay (1986), on the other hand, constructed a one-month prediction model which predicts 5-day or 10-day mean field by separating slowly moving low-frequency variability from unpredictable high-frequency eddies. Here, the contribution from transient eddies interacting with the low-frequency part appears to be the key issue.

An attempt was also made to construct a 2-D model that predicts the zonal mean state in the meridional height section (Satoh, 1994). The model needs to parameterize zonal eddies which accelerate zonal mean flows. The role of baroclinic wave activities induced by baroclinic instability was extensively investigated in the context of baroclinic adjustment (Stone, 1978). Transformed Eulerian Mean Equations were derived, and Eliassen-Palm flux divergence was extensively analyzed as a unique eddy contribution term which interacts with zonal motions (Edmon and McIntyre, 1980). However, the parameterization of such an eddy contribution is far from satisfaction. Recently, the main approach to medium-range weather forecasting is the ensemble prediction, in which a number of ordinary numerical model predictions are averaged to reduce the uncertainty caused by chaos (e.g. Toth and Kalnay, 1997; Molteni and Petroliagis, 1996). Although this method extended predictability to some extent, the chaotic nature of nonlinear fluid systems still limits its effectiveness.

One viable approach which has not pursued much to extending predictability by averaging is to predict the vertical mean of the state variables. In this method, primitive equations are averaged with respect to the vertical, and the resulting barotropic component of the atmosphere is predicted. The barotropic-baroclinic interaction appears to be the key issue to be parameterized. In general, the low-frequency variabilities, such as blocking phenomenon, PNA-like teleconnections, and Arctic Oscillation (Thompson and Wallace, 1998) are characterized by their barotropic structures. Atmospheric blocking has long been a major concern in medium-range forecasting due to its long life time beyond two weeks. Understanding and predicting atmospheric blocking is expected to improve the forecasting skill.

We review here a new type of barotropic general circulation model which we developed from a 3-D spectral primitive equation with a basis of vertical normal modes and Hough harmonics (Tanaka, 1991, 1998). The model predicts the vertical mean state (barotropic component) of the state variables. The barotropic component in this model is defined by the vertical transform weighted by the vertical normal mode function G_0. Since the vertical structure function for the barotropic mode G_0 is approximately constant with respect to the vertical, the method is equivalent to predict the vertical mean state of the primitive equations. In this model, baroclinic instability is parameterized as a major contribution to the baroclinic-barotropic interactions. It is demonstrated that the model can simulate realistic blocking at the right location and structure with the right behavior. According to the results by Tanaka (1998), the blocking in the model is caused by the inverse energy cascade from synoptic eddies to planetary waves in the wavenumber space, consistent with eddy straining hypothesis in the physical space (Shutts, 1986). Therefore, it is interesting to next investigate the predictability of the model which predicts the vertical mean state of the atmosphere.

Predictability of the quasi-geostrophic (QG) barotropic model was investigated by some researchers. Basdevant and Sadourny (1981) showed that the presence of Rossby wave regime, separated by the Rhines scale (Rhines (1975)), increases the predictability due to inhibitation of nonlinear error growth at the anisotropic part of the Rossby wave regime. Vallis (1983) showed that the predictability becomes 25 days with his QG barotropic model. He suggests that baroclinic instability induced by the two-layer QG model substantially reduces the predictability. Holloway (1983) suggests that inclusion of finite equivalent depth, as well as the β effect, increases the predictability of the QG barotropic model four time longer than previous estimates based on 2-D turbulence by Leith (1971) and Leith and Kraichnan (1972). The barotropic model in this section is based on vertically truncated primitive equations and is clearly different from the former (QG) barotropic models. If we can extend the limit of the deterministic predictability by the vertical averaging with a suitable additional forcing, the method would be useful for the medium-range weather forecasting.

The purpose of this section is to examine the predictability of the barotropic component of the atmosphere using the barotropic model by conducting perfect-twin model experiments as descrived by Tanaka and Nohara (2000) (hereafter TN2000). The model sensitivity to the initial error is examined for certain blocking events in the real and model atmospheres. First, we evaluate the barotropic-baroclinic interactions based on the observed data during the event of a pronounced blocking in the north Pacific. Given the barotropic-baroclinic interactions as the external forcing, the perfect-twin model experiments are conducted for the blocking life-cycle in the real atmosphere. Next, we integrate the barotropic model with the parameterized baroclinic instability to generate large-scale persistent blocking in the model atmosphere. When a realistic blocking appears, we superimpose a small error on the solution trajectory two weeks before the blocking formation. We can see whether the robust blocking life-cycle in the model atmosphere is predictable or not from two weeks in advance using the perfect-twin model with noise in the initial data. Finally, some examples of actual predictions of the real atmosphere are presented using the barotropic model with a suitable external forcing.

Note that any random observation error contained in the state variables can be reduced for the vertical mean in proportion to the square root of the number of sample data. Hence, predicting the vertical mean states can be a viable approach to medium-range weather prediction. The relation between the magnitude of the initial error and the predictability limit is investigated and discussed in the summary.

6.1.2 Model and Data

In the 3-D spectral representation, the vertical expansion basis functions may be divided into barotropic and baroclinic components. We may construct a simple spectral barotropic model, using only the barotropic components (m=0) of the Rossby modes, by truncating all the baroclinic modes and high-frequency gravity modes (see Kasahara (1977)). The governing equation for such a barotropic model has the same form as the baroclinic model equation of (5.17) described in Chapter 5, but for m =0.

$$\frac{dw_i}{d\tau} + i\sigma_i w_i = -i \sum_{jk} r_{ijk} w_j w_k + f_i, \quad i = 1, 2, 3, \ldots \quad (m = 0), \tag{6.1}$$

where the indices of the subscripts run only for the barotropic modes. The zonal and meridional wave truncation of the present model (6.1) is equivalent to rhomboidal 20 with an equatorial wall. The degree of freedom of the system is reduced enormously by these truncations, except for the fact that the barotropic-baroclinic interactions should be included formally in the external forcing f_i.

In the first part of this section, the external forcing f_i is evaluated as the residual of (6.1) using the observed state variables w_i. As described in Chapter 5, the dynamical part of (6.1) contains about 1 percent of error, since the parameters are evaluated analytically, except for the truncation in the domain integral. The dataset used in this section is the operational global analysis provided by the Japan Meteorological Agency for January to March, 1997, which is referred to as GANAL/JMA data. The dataset contains four-times daily (0000, 0600, 1200, and 1800Z) meteorological variables of horizontal wind vector (u, v) and geopotential ϕ at 1.25° longitude by

1.25° latitude grids on 17 mandatory vertical levels of 1000, 925, 850, 700, 600, 500, 400, 300, 250, 200, 150, 100, 70, 50, 30, 20, and 10 hPa.

Given the external forcing interpolated on one-hour intervals, the model equation (6.1) is then integrated with respect to time, starting from initial states of w_i on certain days during the analysis period. Since the perfect value of f_i is provided when a time τ is specified, the model is called a pseudo-perfect model ("P-model" in short) for the real atmosphere. It is, of course, not the true perfect model, but an imitation of it, since f_i is given from outside, rather than computed internally.

Figure 6.1 shows the time series of anomaly correlation for the P-model and the exponential growth of the error energy (J m^{-2}) for 200 days. The pseudo-perfect model experiments are repeated for 21 cases from the initial date of 1 January 1989 to 21 January 1989. Anomaly correlation above 0.6 is considerred as the range of predictability. Interestingly, the P-model predictions are parfect upto 150 days from the initial state, despite the fact that the model is highly nonlinear. The initial error is located at the last digit of the state variables in double precision. The error energy norm is at the order of 10^{-30} for the initial state. The initial error grows exponentially with a speed of 10 times per 5 days, and is saturated after 150 days. The model is certainly a chaotic system, but the Lyapunov exponent is very small for the barotropic model. The result suggests that the initial error is not important for the prediction, as long as the external forcing can be parameterized accurately. To confirm this property, we superimpose an initial error for the P-model run to conduct a pseudo-perfect-twin model experiment for the real atmosphere, as will be described in the next.

From the experiments with the P-model, we know both of f_i and w_i from the observation. Establishing a functional relation of $f_i(w_i, \tau)$ given w_i and τ is the next subject to close the system. In the second part of this section, the external forcing is parameterized considering the following physical processes as described in (5.56) and (5.57):

$$f_i = (BC)_i + (TF)_i + (DF)_i + (DZ)_i + (DE)_i, \tag{6.2}$$

where $(BC)_i$ represents the baroclinic instability, $(TF)_i$ the topographic forcing, $(DF)_i$ the biharmonic diffusion, $(DZ)_i$ the zonal surface stress, and $(DE)_i$ the friction by Ekman pumping. For simplicity, this model is called "B-model" which has been developed specifically to simulate a realistic blocking. When a pronounced blocking appears in the control run by the B-model, we conduct a perfect-twin model experiment for the blocking in the model atmosphere. In the experiment run, an initial noise is superimposed on the trajectory of the control run to investigate the growth of the initial noise.

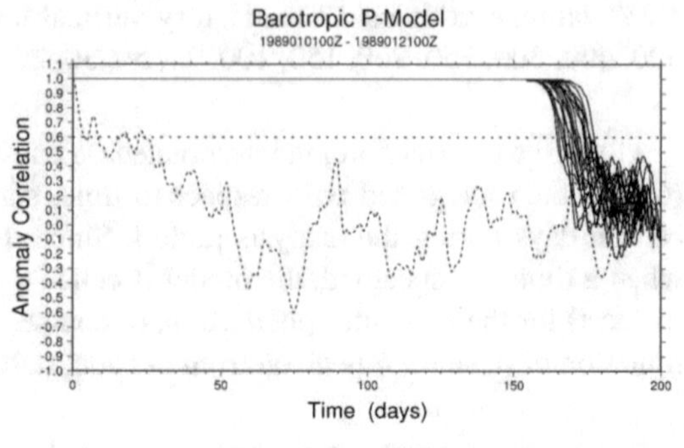

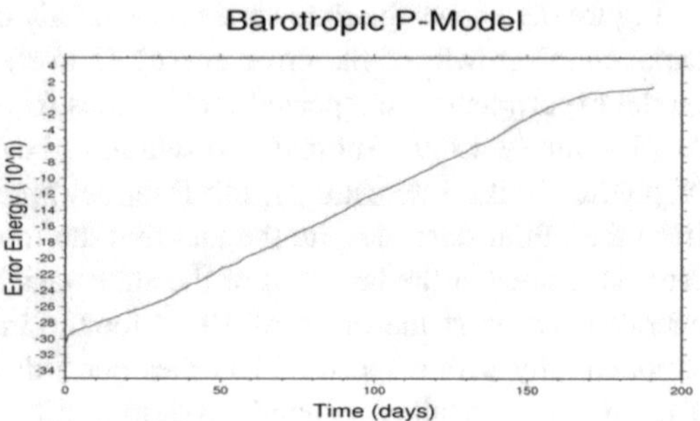

Fig. 6.1 (a) Time series of anomaly correlation for the P-model. Prediction by persistency is plotted by the dashed line. (b) The exponential growth of the error energy (J m^{-2}) in a logarithmic coordinate for the first 200 days.

The parameterization of (6.2) is fundamentally a linear approximation to the true forcing and is never perfect for the real atmosphere even if we know the perfect f_i from the observation. In the last part of this section, we attempt to obtain the best linear fit of the forcing f_i based on 17 years of the global reanalysis data provided by the NCEP/NCAR:

$$f_i = \tilde{f}_i + \mathbf{A}_{ij} w_j + \epsilon_i, \tag{6.3}$$

where $\tilde{f}_i$ is the climate of f_i with a seasonal change, the unknown linear matrix $\mathbf{A}_{ij}$ may be obtained by the standard method of the least square fitting to minimize the residual of ϵ_i:

$$\mathbf{A}_{ij} = \overline{f_i' w_j^+}. \tag{6.4}$$

Here, $f_i' = f_i - \tilde{f}_i$ is the anomaly of f_i, $\overline{()}$ is the time mean, and w_j^+ is referred to as pseudo-inverse of w_j defined by

$$w_j^+ = w_k^H \overline{(w_j w_k^H)}^{-1}, \tag{6.5}$$

where the superscript H denotes a complex conjugate transpause. The model with such a forcing is named as "S-model" since the forcing is obtained statistically from observation. The application of the S-model to the real forecasting is demonstrated later in this chapter. The linear matrix of (6.3) is computed from the NCEP/NCAR reanalysis for 1979 to 1995, and the real forecasting is carried out for the GANAL/JMA data in 1997 to avoid statistically dependent forecasting.

The S-model is further improved by adding a complex conjugate term of w_i as:

$$f_i = \tilde{f}_i + \mathbf{A}_{ij} w_j + \mathbf{B}_{ij} w_j^* + \delta_i, \tag{6.6}$$

where the matrix $\mathbf{B}_{ij}$ may be obtained by minimizing the reminder δ_i by regressing the former residual ϵ_i:

$$\mathbf{B}_{ij} = \overline{\epsilon_i w_j^{*+}}, \tag{6.7}$$

using the pseudo-inverse of w_j^*. Here, the terms with the matricies $\mathbf{A}_{ij}$ and $\mathbf{B}_{ij}$ represent a down scale and up sclae energy interaction terms. We call them as system matricies which are obtained by sovling the inverse problem using data.

In the spectral domain, the difference between the experiment run and the control run, which is referred to as error, is measured quantitatively by the error energy defined as

$$\Delta_i = \frac{1}{2} p_s h_m |w_i - \hat{w}_i|^2, \tag{6.8}$$

where w_i is the state variable of the experiment run and $\hat{w}_i$ is that of the control run. The total error energy is the sum of Δ_i over all indices.

The present barotropic model is different from a standard shallow water system in that all possible high-frequency gravity modes have been truncated out. Those gravity modes, if contained in the model, can be a great source of error energy, even though the amplitude itself may be kept small using an advanced technique. The baroclinic component of the atmosphere is also the major source of the error energy associated with the rapid growth of the local baroclinic instability. The present model is unique in that the baroclinic instability is parameterized as an important component of the baroclinic-barotropic interactions. Therefore, strong dynamical instabilities associated both with baroclinic instability and gravity waves have been ruled out from the dynamical core of the mode by truncations of gravity mode basis and baroclinic mode basis. It is therefore intriguing to find whether the vertical mean state of the atmosphere indicates longer predictability than a 3-D local quantity or not.

6.1.3 Pseudo-Perfect-Twin Model Experiments Using the P-model

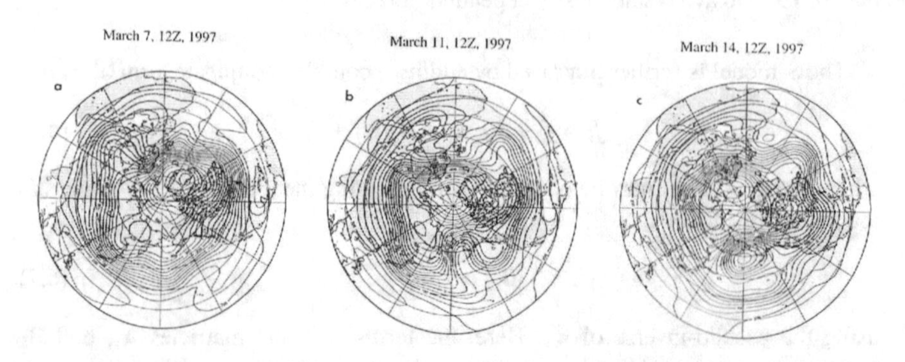

Fig. 6.2 Distributions of the barotropic component of geopotential height in the Northern Hemisphere at 1200Z on (a) March 7, (b) March 11, and (c) March 14 in 1997 evaluated from GANAL/JMA data. Contour interval is 50 gpm. Large-scale blocking is developing in the north Pacific. From Tanaka and Nohara (2000), Fig. 1.

Barotropic components of the state variables $(u, v, \phi)_0^T$ are evaluated using w_i for January 1 to March 31, 1997 using the GANAL/JMA datasets. Figure 6.2 illustrates the resulting geopotential height distributions in the Northern Hemisphere at 1200Z on March 7, March 11, and March 14 in 1997. The distribution is similar to 500 hPa height. It is found that the overall features of the westerly jet, planetary waves, and synoptic waves in the general circulation are contained in the barotropic component. The wind field (not shown) is very close to the geostrophic balance, because only the Rossby modes are retained in the state variables w_i in (6.1) after the spectral truncation. On March 7, a high-index circulation pattern dominates over the Pacific sector with a strong zonal jet. On March 11, a synoptic scale wave appears to develop over the Pacific sector, creating a strong meridional flow. On March 14, a pronounced blocking high establishes over the north Pacific. According to the analysis of the water vapor channel in the GMS-5 satellite, we can see a huge downburst of dry air spreading around the blocking high during this period. Compared with the 500 hPa height, we find that the characteristic features of the blocking high and polar vortex are almost completely retrieved by the barotropic component of the atmosphere. However, the cut-off lows, especially in the south of the blocking high, are projected relatively weak, since those contain a large fraction of baroclinic components. Since the barotropic component contains overall features of the low-frequency variability, predicting this component alone may be still meaningful for medium-range weather forecasting.

The mean linear term is approximately balanced with the nonlinear term, and the small residual results in the barotropic forcing F_Z. The mean distribution of F_Z

indicates positive values at the west coast of the American and Eurasia continents and negative values at the east coast of those continents. The pattern may be consistent with the vertical motions induced by topographic up-lift (e.g., Ringler and Cook (1999)).

Associated with the development of the Pacific blocking in Fig. 6.2, large positive forcing of F_Z is seen over 170°E for days 8 through 14 (not shown). The positive forcing shows retrogression to the west after the mature stage of the blocking. The similar Hovmöller diagrams for the linear and nonlinear terms indicate clear progression of synoptic-scale eddies advected by the westerly. It is interesting to note that such progression is not seen for the barotropic forcing.

Establishing the functional relation of $f_i(w_i, \tau)$ given w_i and τ is our ultimate goal for constructing the perfect model of the real atmosphere. Since it is extremely difficult, $f_i(w_i, \tau)$ is provided from observations in the P-model. Therefore, this model is not a true perfect model, but a pseudo-perfect-model run for the real atmosphere, because the perfect external forcing f_i is provided at every time step. Given the time series of the external forcing evaluated from observations, the model equation (6.1) is integrated from the initial data at 1200Z, March 7, 1997.

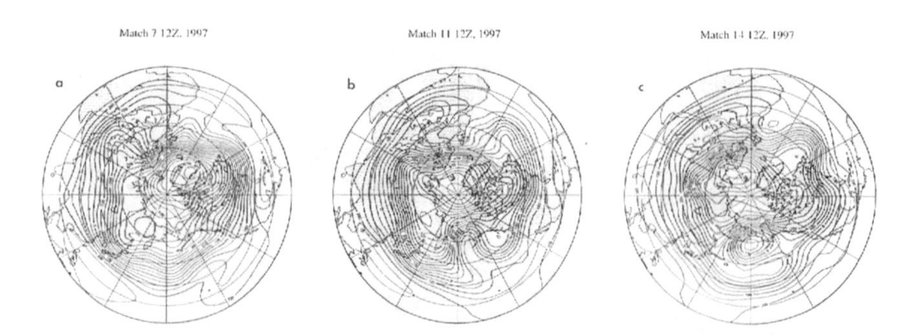

Fig. 6.3 Distributions of the geopotential height for the P-model run at 1200Z on (a) March 7, (b) March 11, and (c) March 14 in 1997. The model equation (6.1) is integrated from 1200Z on March 7 given the diagnostically evaluated external forcing f_i. From Tanaka and Nohara (2000), Fig. 4.

According to the result of the P-model run in Fig. 6.3, the life-cycle of the pronounced blocking described in Fig. 6.2 is perfectly reproduced by this model. In fact, we demonstrated that the time behavior of the real atmosphere can be reproduced, even for the entire three months of the analysis period, starting from the initial value on January 1, 1997. Further analysis shows that the model atmosphere behaves almost exactly as the real atmosphere up to a hundred days when the perfect external forcing is provided. The result may be non-trivial in reference to our current understanding of chaos. Since the model should contain more than an infinitesimal noise in the numerical procedures of the time integration, the noise ought to amplify rapidly to abandon the forecast, according to the theory of chaos. However, what we have found from this experiment is that the initial value problem perfectly follows the real atmosphere for a hundred days, as long as the correct external forcing is provided as demonstrated in Fig. 6.1.

In order to examine the growth of initial error, we integrate the P-model giving small white noise on the initial data. The run is referred to as a pseudo-perfect-twin model experiment. Hereafter, we refer to the white noise as 10% error compared to the total eddy energy. Evidently, the noise level is higher than the observed energy spectrum for short waves of $k \geq 15$, implying that those short waves have more than 100% initial error. The noise amounts to about 30 m in geopotential height, which may be greater than the (vertical mean) observational error.

Superimposing the random noise on the initial data, the model equation (6.1) is integrated again from the initial data at 1200Z, March 7, 1997. As in Fig. 6.2, zonal flow is replaced by meridional flow on March 11 and the pronounced blocking appears on March 14. The result demonstrates that the blocking may be predictable one week in advance, even if the initial data contain 100% error for the short waves, as long as the model error is sufficiently small.

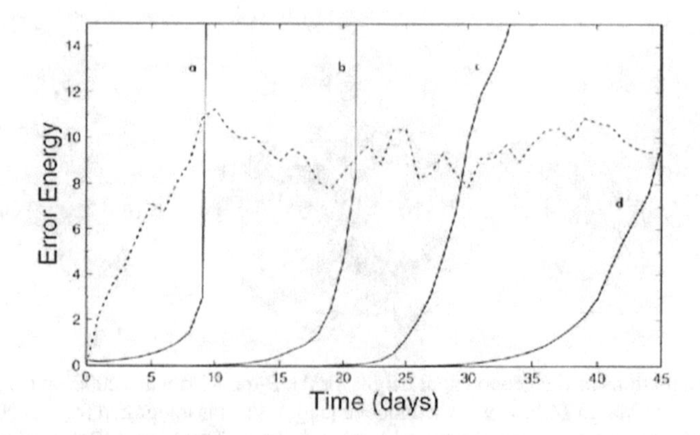

Fig. 6.4 Growth of error energy for various amount of initial error. The line (a) stands for the 10% white noise, (b) for 1%, (c) for 0.1%, and (d) for 0.01%, respectively. Also plotted by dashed line represents the error growth of the persistence forecast started from January 1, 1997. From Tanaka and Nohara (2000), Fig. 7.

The model sensitivity to the initial noise is further examined in Fig. 6.4 for various amounts of initial white noise superimposed on January 1, 1997. The lines stand for the growth of error energy in (6.8) with the units of 10^4 J m^{-2} for the case of (a) 10% white noise, (b) for 1%, (c) for 0.1%, and (d) for 0.01%, respectively. Also plotted by the dashed line is the error growth for the persistence forecast started from January 1, 1997. Since the model predicts the vertical mean rather than a local quantity in the 3-D atmosphere, observational error can be reduced in theory for a large number of the vertical observations. In those experiments, the initial error grows slowly at first, then it diverges suddenly when the error energy reaches 1/10 of the saturation level specified by the persistence forecast. The diverging error energy is reasonable since the external forcing is predetermined, no matter how the state variable develops. The result is an important clue to the relation between the initial noise and the predictability. It is found that the predictability increases 10 days when the initial error energy decreases to 1/10. Namely, the predictability limit is inversely proportional to the logarithm of the error energy level. The predictability limit extends up to a hundred days when no error is superimposed in the pseudo-perfect-twin model as mentioned in Fig. 6.1.

The same experiments as in Fig. 6.4 are repeated 50 times using NCEP/NCAR reanalysis to increase the statistical confidence. The result supports the hypothesis such that the predictability has no upper bound when the initial error approaches zero as pointed out by Lorenz (1963) for a 2-D turbulent flow.

6.1.4 Perfect-Twin Model Experiments Using the B-model

In this section, a long-term integration of the model equation (6.1) was first carried out for a few years as a control run, using the internally calculated external forcing (6.2). The forcing may be too simple compared with the complexity of the real atmosphere. Yet, we can improve each of the physical process of the B-model based on the knowledge of the true forcing f_i evaluated by the P-model in the previous section.

The initial state of the B-model is an axisymmetric flow. Disturbances are soon excited by topographic forcing, and the eddy is saturated at an equilibrium energy level about 20 days after the time integration. Some pronounced blockings occasionally appear in the model atmosphere, especially over the north Pacific and north Atlantic sectors. In the following, we examine a typical Pacific blocking that occurred around 955 model day, which was extensively analyzed in Tanaka (1998).

Then, in the experiment run of the perfect-twin model, we superimpose a small error on the initial data and integrate it in parallel with the control run. Since the mature stage of the blocking is around day 955, the initial data for the experiment run is set on day 941, i.e., two weeks before the mature stage of the blocking.

We confirm that the pronounced dipole blocking over Alaska is successfully simulated by the experiment run, despite the initial error superimposed. The location, configuration, and strength are all satisfactory for the dipole blocking, as well as those of troughs and ridges in the rest of the domain. The prediction error is still of the order of 50 m. It should be mentioned that no notable error growth occurs during the two weeks of the experiment run.

The perfect model experiment with the B-model is further extended to day 962, three weeks after the beginning of the experiment. The experiment run adequately simulates the whole sequence of the life-cycle of the Pacific blocking, as evidenced from the agreement of the two geopotential fields on day 962. When the blocking is fading out to an ordinary ridge over the West Coast of the United States, the error attains to 100 m in high latitudes and continues to increase, even three weeks after the beginning of the experiment. The deterministic prediction for three weeks is still meaningful within the model atmosphere for this specific blocking event.

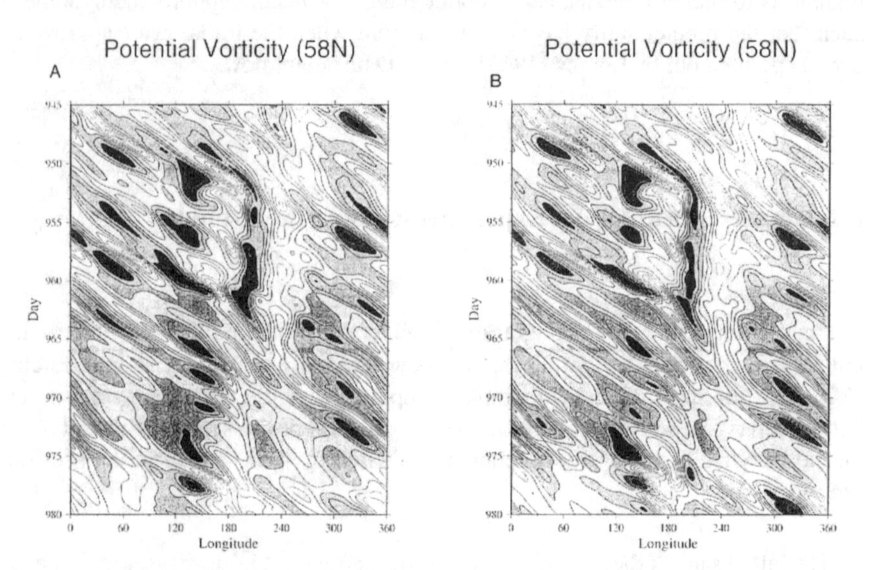

Fig. 6.5 Longitude-time section of potential vorticity q (PV units) along 58°N during days 945 to 970 for (a) the control run and (b) the experiment run with the B-model. From Tanaka and Nohara (2000), Fig. 12.

The sequence of the development, maintenance, and decay of the blocking event is expressed in Fig. 6.5a, using a Hovmöller diagram of barotropic potential vorticity q along 58°N for days 945 to 980. The blocking high is characterized by a low value of q (white area) and the troughs by a high value of q (black area). As described by Tanaka (1998) the progressive free waves excited by the baroclinic instability approach the stationary ridge induced by the topography near 180°E and break there. The ridge of the quasi-stationary planetary wave interacts with a progressive free wave and captures it at a fixed geographical location to initiate a blocking. Subsequent free waves are then blocked by the blocking and stretched in a north-south direction. An eddy straining mechanism supplies fresh low q into the blocking high during 950 to 965 days. When the lifetime of the block is over, the pair of high and low q is swept away downstream.

We can clearly understand the formation of blocking by a breaking Rossby wave and the maintenance by the eddy straining mechanism as discussed in Tanaka (1998). Figure 6.5b illustrates the same Hovmöller diagram of q, but for the experiment run with an initial error starting from day 941. Evidently, the entire sequence of development, maintenance, and decay of the blocking event is simulated rather accurately, despite the considerable amount of the initial error.

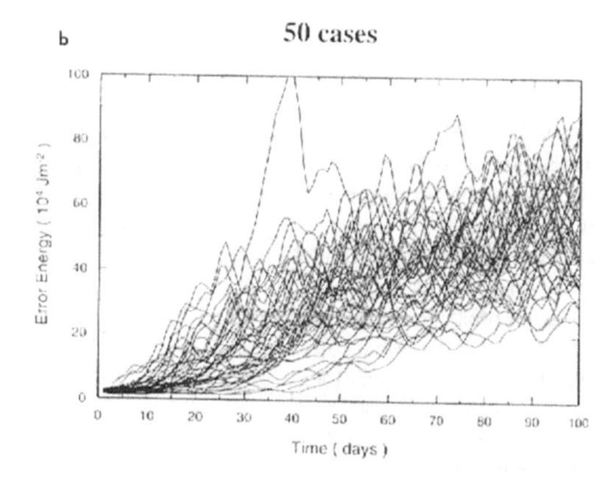

Fig. 6.6 (a): Mean growth of the error energy for the 50 samples of the perfect-twin model experiments with the B-model (solid line), its standard deviation (dashed lines), and the mean error energy of a persistence forecasting (dotted line) over the 100 days into the forecasts. (b): The 50 individual samples of growing error energy for the perfect-twin model experiments with the B-model. Units are $10^4 \mathrm{Jm}^{-2}$. From Tanaka and Nohara (2000), Fig. 13.

The perfect-model experiment using the B-model is repeated for 50 cases to increase the statistical confidence of this model's predictability. Initial random error is superposed on a certain day of the control run, and the experiment run is integrated with initial error for 100 days. Such a 100-day prediction test is repeated 50 times under the similar experimental setting. Figure 6.6a illustrates the mean growth of the error energy for the 50 randomly sampled forecasts. The range of one standard deviation is plotted by dashed lines. Also plotted by dotted line is the mean error growth by persistence forecasting evaluated for the analysis period. Compared with the saturation level by the persistence forecasting, the forecast error of the model grows very slowly and attains the saturation level about 100 days after the initial date. Figure 6.6b plots the result of all 50 individual experiments. Although there are cases that attain the saturation level before 100 days, some experiments indicate extremely slow growth of the error energy. Evidently, the slow error growth demonstrated for a blocking event in the previous section is not an accident for a specific case, but a real feature of the present model.

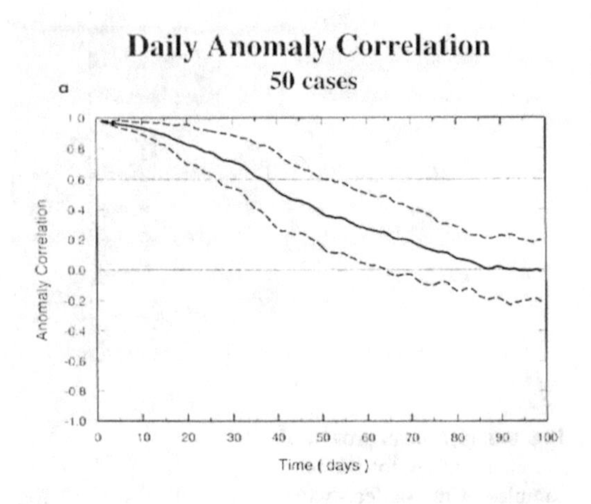

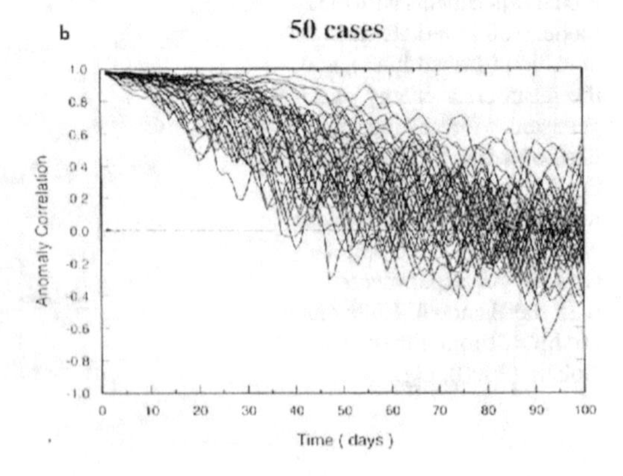

Fig. 6.7 (a): Mean anomaly correlation of geopotential heights for the 50 samples of the perfect-twin model experiments with the B-model (solid line) and its standard deviation (dashed lines) over the 100 days into the forecasts. (b): The 50 individual samples of the anomaly correlation for the perfect-twin model experiments with the B-model. From Tanaka and Nohara (2000), Fig. 14.

Since the predictability limit is often measured by an anomaly correlation instead of the MSE (error energy in this section), we computed the anomaly correlation of the geopotential height between the control run and the experiment run. Figure 6.7a illustrates the mean anomaly correlation of geopotential heights for the 50 randomly sampled forecasts. The range of one standard deviation is plotted by dashed lines. When the anomaly correlation drops to 0.6 and below, the forecasting is considered as meaningless. Therefore, the intersect of the dropping anomaly correlation and the 0.6 line is defined as the predictability limit. The result shows that the predictability limit of the B-model is 35 days. Figure 6.7b plots the results of all 50 individual experiments. The shortest predictability is about 15 days, whereas the longest predictability is about 70 days.

The same experiments are repeated using red noise initial error instead of white noise. Here, the red noise contains larger error in planetary waves in proportion to the observed energy level, while the total amount of noise energy is set to 10% as in the case of the white noise. We confirmed that the results of the very-long predictability of the model are unchanged for the red noise case (not shown). The relation between the predictability limit and the initial error energy levels is also investigated for the B-model. It is shown that the predictability increases 25 days when the initial error energy decreases to 1/10. The logarithmic relation discussed for the P-model certainly holds also for the B-model indicating longer predictability.

6.1.5 Forecast Experiments Using the S-model

It is noted that the parameterization of the baroclinic instability by the B-model generates right amount of synoptic eddies, which is essential in the model atmosphere to excite the realistic blocking by means of the inverse energy cascade. It appears, however, that the parameterization is not satisfactory with regards to the forecasting of the real atmosphere. Baroclinic instability occurs locally in response to the local baroclinicity. The spectral approach to excite local disturbance results in considerable error. A slight difference in phase speed of the growing eddies also causes another source of error.

Since the parameterization of (6.2) is basically a combination of linear matrices, one approach to the real forecasting is to find out the best fit of a single matrix $\mathbf{A}_{ij}$ without $\mathbf{B}_{ij}$ in (6.3) from observation. The anomaly of the forcing $f_i' = f_i - \tilde{f}_i$ is quite noisy as inferred from the large transient eddies. Approximately a half of the variance of f_i' is explained by the linear regression in (6.3). It is demonstrated (not shown) that the perfect-twin model experiments with the S-model result in the same long predictability as in B-model.

The forecasting experiment of the real atmosphere is repeated 23 times with the S-model for the period from December 1996 to March 1997 using the GANAL/JMA data. The result shows that the model has a predictability to some extent, but is not satisfactory compared with the present standard of operational forecasting models. The S-model still has a considerable climate bias. The energy spectrum indicates weaker synoptic eddies. Yet, the major difficulty is contained in the complexity of the external forcing f_i.

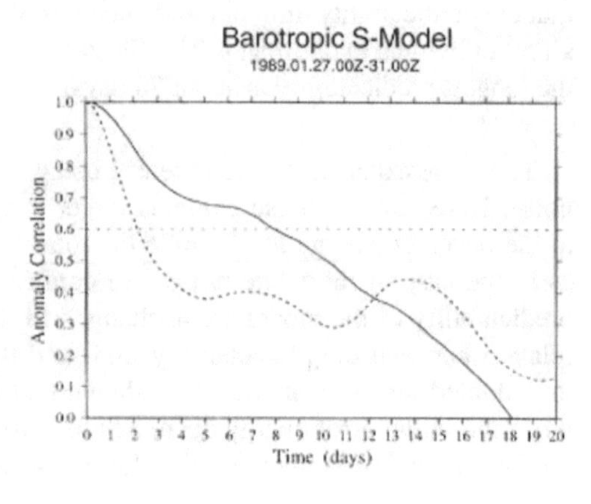

Fig. 6.8 The 20 individual samples of the anomaly correlation for the forecasting experiments with the improved S-model. The experiments are carried out for the period from 27 to 31 January 1989 using the NCEP/NCAR data. From Tanaka and Nohara (2000), Fig. 18.

Figure 6.8 illustrates the result of the anomaly correlation for prediction starting from 27 January 1989. The forecasting is repeated 20 times for different initial conditions for the 4 times daily data during 27 to 31 January. The result shows that the actual predictions have improved, exceeding the mean predictability of about 8 days. According to the analysis of the forecasting error, it is found that the erroneous baroclinic instability parameterization and the low-frequency anomaly in the forcing are the main cause of the forecasting error. Therefore, the S-model may be useful as the extended range prediction model by a further improvement in the parameterization of the low-frequency variability. The accurate parameterization of the external forcing is the key problem to improve the S-model for the successful forecasting.

6.1.6 Summary

It has long been stated that the deterministic atmospheric predictability may be of the order of two weeks due to the chaotic nature of nonlinear fluid systems. Even if we can have a perfect prediction model, a deterministic, medium-range weather forecast has been pessimistic beyond the chaotic barrier of two weeks (see Chen (1989)). However, the possibility still remains to predict some averaged quantities. In this section, the predictability of the vertical mean state (barotropic component) of the atmosphere is examined by conducting perfect-twin model experiments.

In the first part of the section, we evaluated the barotropic-baroclinic interactions from the observations during a pronounced blocking event in the north Pacific. It is demonstrated that the model accurately simulates the life-cycle of the blocking event in the real atmosphere. We found that the present model accurately follows the trajectory of the real atmosphere not just for a week, but for a hundred days after the initial date, as long as the correct external forcing is provided. The result may be non-trivial, since the model should contain a certain amount of noise during the computation which is bigger than the butterfly effect discussed by Lorenz.

Based on the result of the P-model, the pseudo-perfect-twin model experiments are then conducted to examine how the initial error grows. The result shows that the predictability increases 10 days when the initial error energy is reduced by 1/10. Predictability limit is inversely proportional to the logarithm of the error energy level. This implies that the predictability increases as long as the initial error decreases as noted by Lorenz (1969) for 2-D turbulent flows. This property of the barotropic model can be useful for the actual medium-range forecasting. Hence, the result suggests that the vertical mean state may be predictable beyond two weeks "if" we can have a perfect model and sufficient number of observations in the vertical. However, it must be a big "if" as a working hypothesis.

In the second part of the section, we integrated the B-model which computes the forcing internally within the model. We demonstrated that the robust blocking life-cycle in the model atmosphere is reproduced using the initial data two weeks in advance despite the abundant initial noise. It is concluded that predicting the vertical mean of the atmosphere is less sensitive to the initial noise. A nonlinear fluid system is chaotic only when the nonlinear system has strong dynamical instability which grows the initial error. The result implies that the nonlinearity is not the essential factor of chaos, but the dynamical instability contained in the system is essential for it. For this reason, the vertical mean of the atmosphere is much predictable than we thought even though it is a highly nonlinear system. The result of present section suggests that predicting the vertical mean of the atmosphere can be one of the viable approaches to the medium-range weather prediction, provided that the barotropic-baroclinic interactions are accurately parameterized.

Using a quasi-geostrophic barotropic model, similar long predictability is discussed by previous studies by Leith (1971), Leith and Kraichnan (1972), Basdevant and Sadourny (1981), Holloway (1983), and Vallis (1983). Among those, Vallis (1983) shows the longest predictability of 25 days. It is shown, however, that the predictability of the present barotropic model is much longer than those by the quasi-geostrophic barotropic model. The parameterized baroclinic instaibity in this section, even though it is linear and simple, causes the error to grow fast with the e-folding time of 2 to 3 days. Yet, the error growth is much slower than that. It is the nonlinearity in the 2-D flow that reduces the error growth. The inverse energy cascades seem to be important to understand the phenomenon. Holloway (1983) demonstrated that initial error in small scales also cascades up to planetary waves similar to the energy itself in the 2-D turbulence. The planetary waves are strongly forced by topography maintaining stationary Rossby waves. It seems that error energy cascading up to planetary waves collapses by the balance of the topographic forcing and Ekman damping.

In the last part of the section, we demonstrated an example of actual forecasting for the Pacific blocking using the S-model, in which a best linear fit of the external forcing is obtained from 17 years of the NCEP/NCAR reanalysis. As a result, we found that the predictability extends up to 8 days for the improved S-model. We can expect a longer predictability if the parametarization of the external forcing has higher accuracy. The expectation may be reserved in the future subject.

6.2 Long-Term Climatic Variations Induced by the Arctic Oscillation

6.2.1 Background

The Arctic Oscillation (AO) postulated by Thompson and Wallace (1998) has attracted more attention in climate studies. The AO is a north-south seesaw of the atmospheric mass between the Arctic region poleward of 60°N and a surrounding zonal ring in the mid-latitudes. It is defined as a primary mode of an empirical orthogonal function (EOF) for the sea-level pressure field in the Northern Hemisphere. The spatial pattern of the AO is characterized by its zonally symmetric or "annular" structure centered in the Arctic. The AO bears a superficial resemblance to the leading mode of low-frequency variability in the Southern Hemisphere, which is now referred to as a Southern Hemisphere annular mode (SAM). In this regard, a terminology of the AO which is defined at the sea-level pressure is distinguished from the the Northern Hemisphere annular mode (NAM) which is defined at each vertical level. In this section, we will refer to the AO as the EOF-1 of atmospheric variability with one sign over the Arctic and the opposite sign over the two poles at Pacific and Atlantic sectors, while the terminology of the annular mode is reserved essentially for the zonally symmetric variation at each vertical level.

The AO is successfully simulated by a number of realistic general circulation models with fixed forcing (Yamazaki and Shinya, 1999; Fyfe and Flato, 1989; Shindell and Pandolfo, 1989; Limpasuvan and Hartmann, 2000; Boer and Yu, 2001; Robertson, 2001). The basic features of the zonally symmetric structures are also simulated using simple quasi-geostrophic and primitive equation models with simple physics and no topography or seasonal cycle (James and James, 1992; Robinson, 1996; Akahori and Yoden, 1996; Frederiksen and Lee, 1998). Because the external forcing of the model is independent of time, the variability found in these simple models is the unforced variability associated with dynamical processes internal to the atmosphere. The dynamical examination of the annular modes shows that the baroclinic eddies in mid-latitudes played an important role for the internal variability of the zonal mean flow through the positive feedback in momentum budget between the eddies and the mean flow (Karoly, 1990; Shiotani, 1990; Hartman, 1995; Yamazaki and Shinya, 1999; Lorenz and Hartmann, 2001; Tanaka and Tokinaga, 2002).

Shiotani (1990) showed that the vacillation between the eddies and the mean flow is the essential mechanism for the SAM. Yamazaki and Shinya (1999) analyzed the EP-flux divergence and found that the wave-mean flow interactions by planetary waves provide the largest contribution to the positive feedback while the synoptic-scale waves contribute destructively to the NAM. Lorenz and Hartmann (2001) analyzed the zonal-mean zonal wind anomaly in the Southern Hemisphere and showed that an equivalent barotropic dipole with opposite anomalies at 40°S and 60°S indicates a positive feedback with low-frequency eddy forcing to maintain the persistent anomaly associated with the SAM. A similar conclusion was attained by Tanaka and Tokinaga (2002), who analyzed baroclinic instability associated with the polar jet. While the ordinary Charney mode excited by the subtropical jet brings eddy momentum southward in high latitudes to intensify the subtropical jet, the baroclinic instability excited by the polar jet brings the eddy momentum northward to intensify the polar jet. Since the polar baroclinic mode tends to dominate when the polar jet is strong, the positive feedback plays an essential role in maintaining the anomaly of the polar jet, which may in turn result in the AO.

Despite the argument with simple general circulation models, the dynamical interpretation of the AO is still an open question. An attempt to understand the essential part of the dynamics of the AO may be desired with further simplified models. According to the observational analysis, the AO has an equivalent barotropic structure from the surface to the lower stratosphere in that the positive or negative geopotential anomaly occurs consistently from the troposphere to the stratosphere (Thompson and Wallace, 1998). Based on this fact, even a simple barotropic model may be used to simulate a robust structure of the AO.

The purpose of this section is to demonstrate that the realistic AO can be simulated by a simple barotropic model with a suitable external forcing representing the barotropic-baroclinic interactions as demonstrated by Tanaka (2003) (hereafter T2003). Such a barotropic model has been constructed in our previous studies from a 3-D spectral primitive equation model using only the component of the vertical wavenumber zero of the model. According to the observational analysis, the barotropic-baroclinic interactions are achieved mostly by synoptic disturbances (Tanaka, 1985; Tanaka and Kung, 1988). With the knowledge of the linear solutions of baroclinic instability on a sphere (Tanaka and Kung, 1989; Tanaka and Sun, 1990), the barotropic-baroclinic interactions have been parameterized by (Tanaka, 1991, 1998) to close the barotropic model. With the parameterized baroclinic instability as the main energy source, the barotropic model was integrated to simulate realistic blocking. The model has been applied to the actual forecasting of the barotropic component of the atmosphere (Tanaka and Nohara, 2000). We demonstrated that the model has a meaningful deterministic predictability of 8 days, which is comparable to the operational weather forecasting models. In this section, the model is integrated for 51 years under a perpetual January condition to simulate the realistic pattern of the AO. The result of the EOF analysis of the model atmosphere is compared with the observed EOF patterns using the NCEP/NCAR reanalysis for 50 years from 1950 to 1999 (Kalnay et al., 1996).

6.2.2 Method and Data

6.2.2.1 Construction of an Advanced Barotropic S-model

The barotropic components capture the essential features of the low-frequency variability of planetary-scale motions. The governing equation to be solved for the barotropic model is written as (6.1). The main problem of the model is to formulate the external forcing f_i as a function of w_i. Since we know the perfect f_i evaluated by the P-model, we first calculate the climete $\tilde{f}_i$ and the anomaly $f_i' = f_i - \tilde{f}_i$ based on the 50 years of the NCEP/NCAR reanalysis from 1950 to 1999. First four harmonics of the seasonal change are synthesized to get a smooth seasonal march of $\tilde{f}_i$.

We then attempted to obtain the best linear fit of the forcing anomaly f_i' by

$$f_i' = \mathbf{A}_{ij} w_j + \mathbf{B}_{ij} w_j^* + \epsilon_i, \tag{6.9}$$

by minimizing the norm of the residual ϵ_i. In general, w_i and w_i^* are linearly independent, and both are required in the complex-valued regression since only the positive (and zero) zonal wavenumbers are considered for the state variables of the model.

In Tanaka and Nohara (2000), the unknown matrix $\mathbf{A}_{ij}$ was first evaluated by (6.4) and (6.5), namely:

$$\mathbf{A}_{ij} = \overline{f_i' w_k^H}\, \overline{(w_j w_k^H)}^{-1}, \tag{6.10}$$

and the second matrix $\mathbf{B}_{ij}$ was evaluated sequentially by minimizing the residual.

In this section, the unknown system matrices $\mathbf{A}_{ij}$ and $\mathbf{B}_{ij}$ are obtained simultaneously by the real-valued least square fitting for f_i' to minimize the regression residual ϵ_i in (6.9). Let the subscripts R and I be real and imaginary parts of the vector and matrix. Then we can obtain following relation:

$$\begin{pmatrix} A_R + B_R & -A_I + B_I \\ A_I + B_I & A_R - B_R \end{pmatrix} = \overline{\begin{pmatrix} f_R' \\ f_I' \end{pmatrix} \begin{pmatrix} w_R \\ w_I \end{pmatrix}^T} \; \overline{\begin{pmatrix} w_R \\ w_I \end{pmatrix} \begin{pmatrix} w_R \\ w_I \end{pmatrix}^T}^{-1} \tag{6.11}$$

Here the overbar denotes time mean, and the superscript T denotes a transpose. Once we obtian the left hand side, we can solve for $\mathbf{A}_{ij}$ and $\mathbf{B}_{ij}$. It is found from the residual that the simultaneous method produces slightly a better solution than the sequential method.

Although the barotropic S-model demonstrates considerable predictability, the synoptic eddies appear to be weak by this formulation. The system matrices $\mathbf{A}_{ij}$ and $\mathbf{B}_{ij}$ mostly represent topographic forcing (TF). For this reason, we have added the energy source in terms of the parameterized baroclinic instability (BC) described by Tanaka (1998). The increased energy source at the synoptic scale requires an additional energy sink, so the biharmonic diffusion (DF), the zonal surface stress (DZ), and the Ekman pumping (DE) are also included in reference to the resulting energy spectrum of the model climate. The final form of the external forcing f_i is expressed as:

$$f_i = \tilde{f}_i + \mathbf{A}_{ij} w_j + \mathbf{B}_{ij} w_j^* + (BC)_i + (DF)_i + (DZ)_i + (DE)_i. \tag{6.12}$$

Note that the topographic forcing is contained in $\tilde{f}_i$ as wall as the system matrices.

Since the steady climatological forcing $\tilde{f}_i$ is introduced, the zonal suface friction is revised to simply relax the anomaly of the state variable w_i to the climate $\tilde{w}_i$ by a Rayleigh friction in the barotropic S-model.

$$(DZ)_i \; = -K_Z(w_i - \tilde{w}_i) \quad \text{for } n = 0, \tag{6.13}$$

where the dimensionless damping coefficient is set as $K_Z = 3.0 \times 10^{-7} s^{-1}/(2\Omega)$, which corresponds to the damping time of 38.6 days.

Although the barotropic-baroclinic interaction is implicitly considered in the model, the forcing has been modeled entirely by the barotropic component of the state variables w_i. Hence, any variability generated within the model may be considered as a natural variability governed by the barotropic dynamics internal to the model atmosphere.

6.2.2.2 Data and Procedures

The data used in this section are four-times daily NCEP/NCAR reanalysis for 51 years from 1950 to 2000 (see Kalnay et al. 1996). The data contain horizontal winds $\mathbf{V} = (u, v)$ and geopotential deviation from the mean ϕ, defined at every 2.5° longitude by 2.5° latitude grid point over 17 mandatory vertical levels from 1000 to 10 hPa.

The expansion coefficients w_i are obtained by the inverse Fourier transform from the dataset of $\mathbf{U} = (u, v, \phi)$. Although the data is four-times daily, we refer to it as daily data in subsequent sections since the diurnal variation of w_i is removed by a time filter. The external forcing f_i is then diagnostically calculated by (6.1) as the residual of the equation, which provides the climate $\tilde{f}_i$ and the anomaly f_i' in (6.9). The EOF analysis is applied to the time series of w_i and f_i for the 50-year data from 1950 to 1999 to find the dominant mode in the real atmosphere. With the known w_i and f_i, the system matrices $\mathbf{A}_{ij}$ and $\mathbf{B}_{ij}$ in (6.9) are computed based on the data for December to February (DJF) during the 50 years. Finally, the barotropic S-model (6.1) is integrated for 51 years under a perpetual January condition, starting from the initial data of w_i at 0000Z on 1 January 1950 with the value of $\tilde{f}_i$ fixed on 1 January.

6.2.3 Dominant Modes of the Observed Atmosphere

6.2.3.1 Barotropic Component of the Atmosphere and AO

In general, the low-frequency variabilities such as AO or blocking are characterized by the equivalent barotropic structure. In this section, the 3-D structures of atmospheric variables are projected onto the barotropic component of the vertical normal mode functions as in (6.14).

$$(u, v, \phi')_0^T = \frac{1}{p_s} \int_0^{p_s} (u, v, \phi')^T G_0 dp. \tag{6.14}$$

Here, G_0 is the vertical normal mode of m=0, and p_s is the surface pressure of the reference state. It is a part of the inverse Fourier transform associated with the vertical transform. The vertical transform (6.14) is followed by the Fourier-Hough transforms to obtain the expansion coefficient w_i in (6.1). Since the structure of G_0 is approximately constant with no node in the vertical, the vertical transform for m=0 may be regarded as the vertical mean of the state variables. In this section, the left hand side of (6.14) is referred to as a barotropic component of the atmosphere that depends only on longitude, latitude, and time.

It may be important to note that the AO is defined as the EOF-1 of the time variation in sea-level pressure p_s, which is dynamically related to the time variation of the barotropic component of the atmosphere as follows:

$$\frac{\partial p_s}{\partial t} \simeq -\int_0^{p_s} \nabla \cdot \mathbf{V} dp \simeq -p_s \nabla \cdot \mathbf{V}_0 \simeq \frac{p_s}{g h_0} \frac{\partial \phi_0}{\partial t}. \tag{6.15}$$

The time variation of p_s is caused by the vertical integral of mass flux convergence $\nabla \cdot \mathbf{V}$. Here, the vertical integral coincides with the barotropic component of the atmosphere (noted by the subscript 0) which controls the time variation of the barotropic geopotential ϕ_0 in a shallow water system with the depth h_0 and gravity

g. Hence, the AO represented by the variation in p_s is dynamically equivalent to the variation in ϕ_0. For this reason, the essential features of the AO are contained in the barotropic component of the atmosphere governed by the 2D fluid mechanics which characterizes the low-frequency variability. This property clearly separates the AO from the NAM: the AO represents the barotropic dynamics of the low-frequency variability, but NMA is only a statistical product without a dynamical basis.

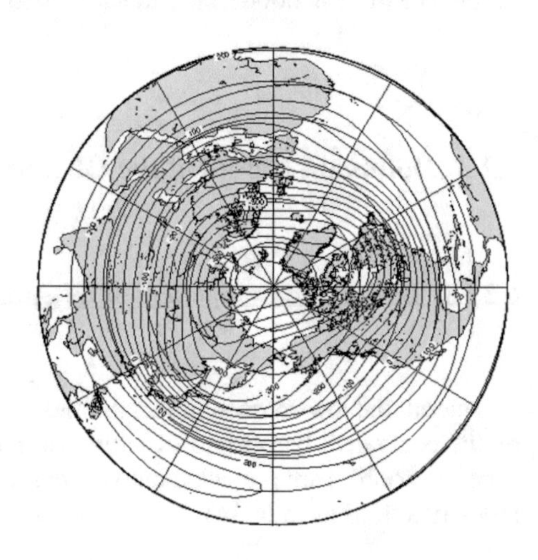

Fig. 6.9 Barotropic height (m) for the winter mean (DJF) of the NCEP/NCAR reanalysis during 1950 to 1999 used for the climate basic state. Values are the deviation from the global mean. The conter interval is 50 m. From Tanaka (2003), Fig. 1. ©American Meteorological Society. Used with permission.

Figure 6.9 illustrates the barotropic component of the geopotential height averaged for the winter (DJF) from 1950 to 1999. In order to obtain this figure, the expansion coefficients w_i are averaged for 50 years in the spectral domain, then the mean w_i is converted to the geopotential field. Since geopotential height is defined by the deviation from the global mean, the value is negative at the polar region and is positive in the tropics. We will refer to it as a barotropic height.

In the spectral domain, total energy E is simply the sum of the energy elements E_i defined by the expansion coefficient w_i. The energy spectrum of E_i as a function of the scale index $|c_i|$ was analyzed by Tanaka and Kasahara (1992), showing that the spectrum obeys a unique power law even for the zonal wavenumber zero. All eigenvalues of σ_i are zero for k=0. The difficulty was overcome by the use of Shigehisa modes (Shigehisa 1983) where mathematical limits of $c_i = \sigma_i/k$ converge to finite values. It was demonstrated in Tanaka (1991) that the phase speed of the geostrophic mode can be approximated by that of the Haurwitz wave on a sphere. Using this definition of the scale index $|c_i|$ for k=0, we can analyze the energy spectrum for all zonal waves, including k=0.

Figure 6.10 plots the barotropic (m=0) energy spectrum of $E_i = E_{knm}$ as a function of the scale index $|c_i| = |c_{knm}|$ evaluated for the 50 years of the NCEP/NCAR reanal-

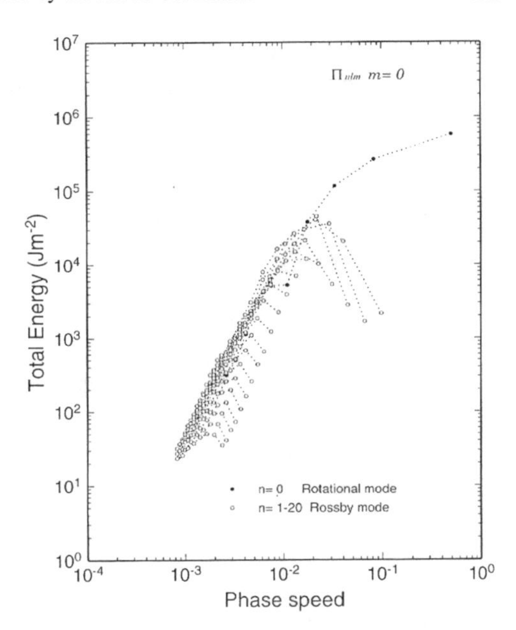

Fig. 6.10 Barotropic energy spectrum of E_{knm} as a function of the dimensionless scale index $|c_{knm}|$ evaluated for the 50 years of the NCEP/NCAR reanalysis during winter. (k is noted as n in the figure.) The black and white dots denote k=0 and $k \neq 0$, respectively. The unit is in J m^{-2}. From Tanaka (2003), Fig. 2. ©American Meteorological Society. Used with permission.

ysis during winter. The dotted lines connect spectra for the same zonal wavenumber k over different meridional mode numbers n. The spectrum obeys approximately the 2nd power of $|c_i|$, except for the largest-scale meridional modes, which indicate less energy levels. Refer to Chapter 5.5 of the Rossby wave saturation theory for the mechanism of the spectrum. The peak of the energy spectrum for the zonal eddy $k \neq 0$ is basically determined by the Rhines' scale, where the inverse energy cascade from short waves to planetary waves terminates (Rhines 1975). Due to the low model resolution of NCEP/NCAR reanalysis, the tail of the spectrum is elliminated from the plot. It is interesting to note that the spectrum for k=0 coincides with that of $k \neq 0$ for a smaller $|c_i|$, and it continues to increase as the meridional index n decreases. We can confirm that the scale index of k=0 for geostrophic mode is a good approximation to the energy spectrum of the Sigehisa mode presented in Tanaka and Kasahara (1992).

6.2.3.2 EOF Analysis of the Observed Atmosphere

The climate of w_i with its seasonal march is computed from 50 years of the time-series data. Once the climate is constructed, the anomaly of w_i is computed for the analysis period. The EOF analysis is then conducted for the anomaly time series of w_i in the spectral domain based on the daily data. The complex numbers of w_i are split in real and imaginary parts to compute the covariance matrix for the EOF analysis.

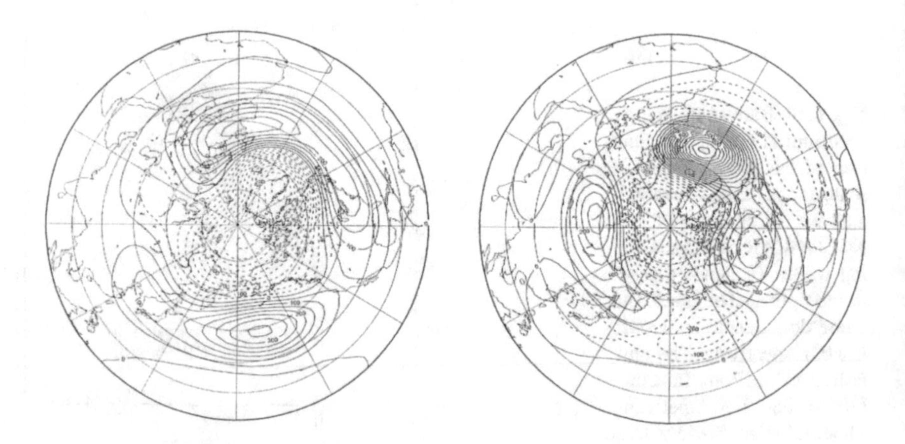

Fig. 6.11 First two EOFs components of the barotropic component of the observed atmosphere, analyzed by the state variable w_i with the daily NCEP/NCAR reanalysis from 1950 to 1999. From Tanaka (2003), Fig. 5. ©American Meteorological Society. Used with permission.

Figure 6.11 illustrates the first two EOF components of the atmosphere represented by the barotropic height. The structure of the EOF-1 indicates a negative region in high latitudes surrounded by a positive region in mid latitudes. The positive region shows two peaks over the Pacific and Atlantic sectors. The zero crossing is located around 60°N. The structure compares quite well with the AO analyzed by Thompson and Wallace (1998), so the EOF-1 of this section is easily identified as the AO. The original definition of the AO is based on sea-level pressure, which is a single variable at a single level. It should be noted that our result is based on the whole column of vertical data using all state variables (u, v, ϕ) during 50 years of the daily data. Although our EOF analysis is performed for the barotropic component, the EOF-1 should be recognized as the most dominant low-frequency variability in the 3-D atmosphere since most of the low-frequency variabilities are contained in the barotropic component. Since the daily data contain active synoptic-scale eddies, it may be worth noting that the AO remains the most dominant mode even for the analysis with the daily data.

The EOF-1 explains 5.7% of the total variance of the daily data. The score time series (365-day running mean) of the EOFs are plotted in Fig. 6.12. The time series of the EOF-1 is referred to as AO index, which represents considerable interannual variabilities. The positive value corresponds to a strong polar vortex while the negative value indicates a weak polar vortex. The AO index fluctuates between

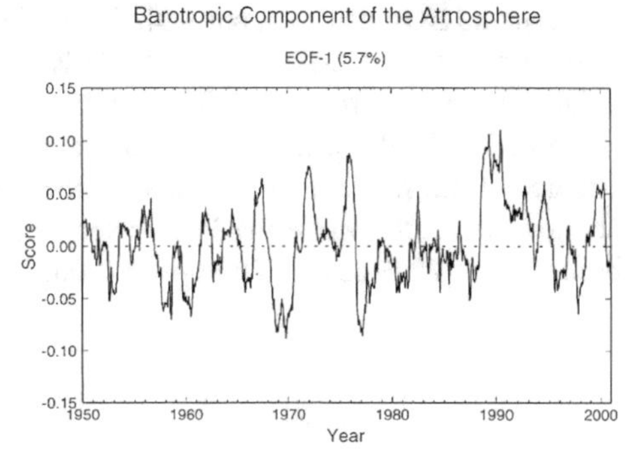

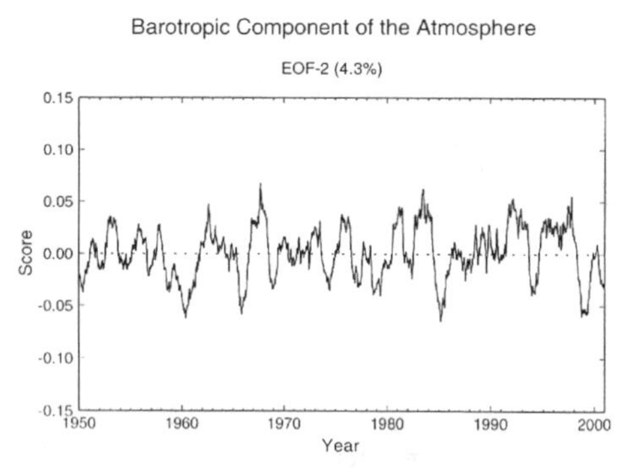

Fig. 6.12 The score time series of the first two EOFs components. The value is a 365-day running mean. From Tanaka (2003), Fig. 6. ©American Meteorological Society. Used with permission.

positive and negative values for 1950 to 1970, and it remains negative from 1976 to 1989. A notable shift from negative to positive occurs in 1989. The positive AO index then gradually decreases to negative in the late 1990s, but it returns to positive in 2000. Clear decadal variability is contained in the AO index, but no significant trend is detected for the barotropic component of the atmosphere. According to the original definition of the AO index by Thompson and Wallace (1998), the decadal variability is superimposed on a recent trend related to global warming. Such a trend may be characterized in the baroclinic component of the atmosphere.

The EOF-2 explains 4.3% of the total variance of the daily data. A negative area over the Arctic is surrounded by positive values in the mid-latitudes, with three peaks over the Atlantic Ocean, Siberia, and Canada. The structure is similar to the AO, except for the opposite sign over the Atlantic and the Pacific sectors. The score time series of the EOF-2 indicates interannual variability with a two-to-three year period. The apparent decadal variability seen for the EOF-1 is not obvious for the EOF-2.

The same EOF analysis, but with the winter mean (DJF) data from 1950 to 1999, was examined. The structure of the AO is nearly identical to that in Fig. 6.11. The variance increases from 5.7% for the daily data to 13.9% for the monthly mean data and to 21.0% for the winter mean (DJF) data. The EOF-2 explains 10.3% of the total variance of the winter mean data. The structure indicates an NAO-like pattern over the Atlantic sector while the Pacific sector clearly indicates a PNA-like pattern as discussed by Wallace and Thompson (2002).

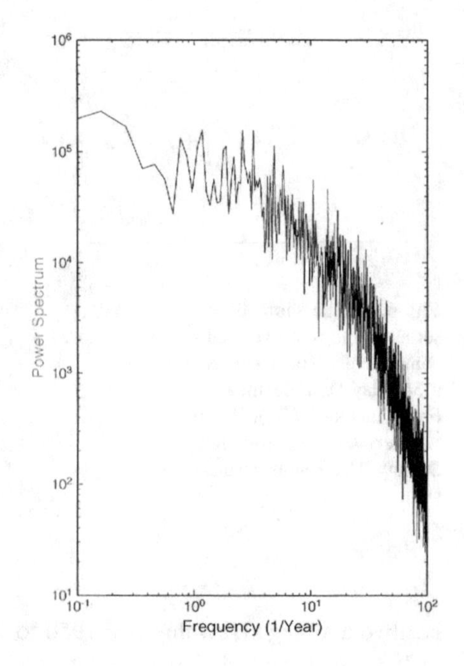

Fig. 6.13 The time spectrum of the daily AO index of the EOF-1. From Tanaka (2003), Fig. 8. ©American Meteorological Society. Used with permission.

The time spectrum of the daily AO index for the EOF-1 is plotted in Fig. 6.13. The result is characterized by a red noise spectrum over the higher frequency range. The red noise turns to white noise for the period longer than four months. It is interesting to note that the AO is the most dominant mode from the daily time scale to the decadal time scale, but there is no obvious peak in the time spectrum.

With the same procedure as to obtain the AO in Fig. 6.11 from the state variable w_i, we can examine the most dominant mode for the external forcing f_i. The structure clearly shows the characteristics of synoptic-scale disturbances of zonal wavenumber 6 in the mid-latitudes, located along the Pacific storm track. The EOF-2 has a similar structure as EOF-1, but with a phase shift by 90-degree for the zonal wavenumber 6. The combination of these two modes represents the progressive zonal wavenumber 6 along the Pacific storm track. A similar synoptic-scale forcing is seen along the Atlantic storm track in the higher EOF modes. These forcing represent the barotropic-baroclinic interactions f_i in (6.1) associated with baroclinic instability as parameterized by Tanaka (1998).

It is important to note that the major energy source of the barotropic component of the atmosphere appears in the synoptic-scale baroclinic eddies. The major response to the forcing, however, appears as the zonally symmetric annular mode. The nonlinear response to the synoptic-scale forcing may be connected by the up-scale energy cascade governed by the internal barotropic dynamics.

6.2.4 Dominant Modes of the Model Atmosphere

6.2.4.1 Perpetual January Run of the Barotropic S-model

In this section, the barotropic S-model of (6.1) is integrated for 51 years starting from the initial condition of 0000Z 01 January 1950 under the perpetual January condition. The time series of zonal energy (k=0) and eddy energy ($k \neq 0$) are investigated. Since the initial condition is set on 01 January 1950, the time is expressed, for convenience, from 1950 to 2000. The mean energy levels are approximately 12 and 4×10^5J m^{-2} for zonal and eddy energies, respectively. Those energy levels are consistent with winter observations. The energy levels fluctuate with respect to time within 10 to 14×10^5J m^{-2} for zonal energy and 3 to 6×10^5J m^{-2} for eddy energy, respectively. The fluctuation may be regarded totally as the natural variability of the model atmosphere since all model parameters and forcing parameters are fixed. Although the details are different, the overall features of the winter climate in Fig. 6.9 is simulated. The locations of storm tracks analyzed for the model atmosphere (not shown) agree quite well with observations. Therefore, both mean and transient fields are well described by the barotropic S-model.

6.2.4.2 EOF Analysis of the Model Atmosphere

The EOF analysis is conducted for the time series of the model atmosphere with the same method used for the observed atmosphere. Figure 6.14 illustrates the first two EOF components of the model atmosphere analyzed by the time series of w_i. The structure of the EOF-1 indicates a negative region in high latitudes surrounded by a positive region in mid latitudes. The positive region shows two peaks over the

Pacific and Atlantic sectors. The zero crossing is located around 60°N. The structure is almost identical with the AO analyzed by the observed atmosphere in Fig. 6.11. Hence, the EOF-1 in the model atmosphere may be identified as the AO. The positive peak over the Pacific sector is elongated in the zonal direction, which disagrees to some extent with observation. The elongated positive area in the observed Atlantic sector is now disconnected at the eastern Atlantic in the model. Although there are some differences, the overall features agree quite well with the observation in Fig. 6.11.

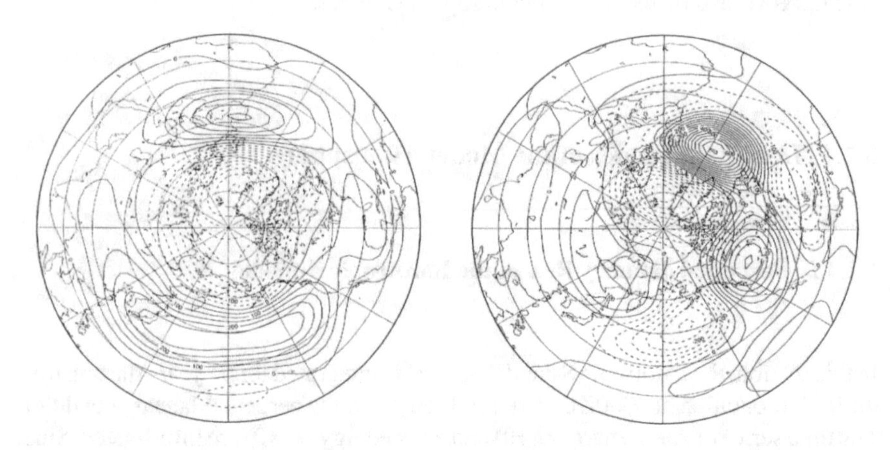

Fig. 6.14 As in Fig. 6.11, but for the perpetual January run of the barotropic S-model for 51 years. From Tanaka (2003), Fig. 12. ©American Meteorological Society. Used with permission.

The score time series (365-day running mean) of the EOFs are plotted in Fig. 6.15. The time series of the EOF-1 show a large interannual variability with two-to-three year periods, similar to the observation in Fig. 6.12. The decadal internal variability is dettectable for the model. The EOF-1 explains 15.6% of the total variance of the daily data. The large contribution of the AO may be a result from the relatively weak transient eddies in the synoptic and short waves in the model atmosphere. It is evident from the result that time series are totally diffrent from observation since it is not a forced change but a natural variability. The natural variability associated with the AO can be produced by the internal dynamics of the barotropic atmosphere.

The EOF-2 explains 7.9% of the total variance of the daily data. There is a positive peak over the Atlantic sector that extends over North America. The Arctic is negative and the Pacific is also negative. The structure of the EOF-2 agrees well with the observed EOF-2 in Fig. 6.11. The main discrepancy may be seen in the location and strength of the positive peak over Russia. Yet, the overall agreement with observation suggests that the barotropic S-model captures not only the essential features of the EOF-1 but also that of EOF-2 of the observed atmosphere. The

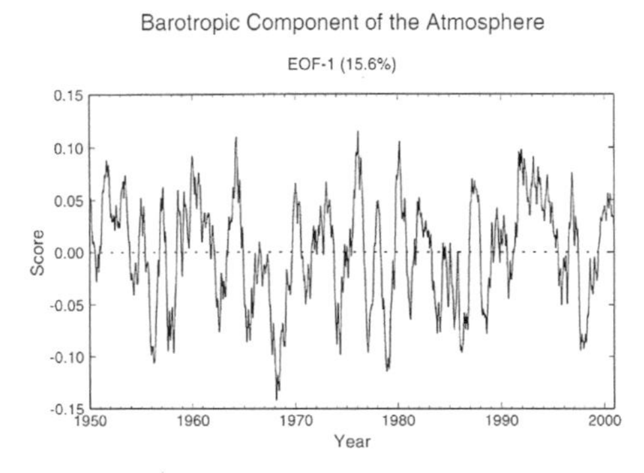

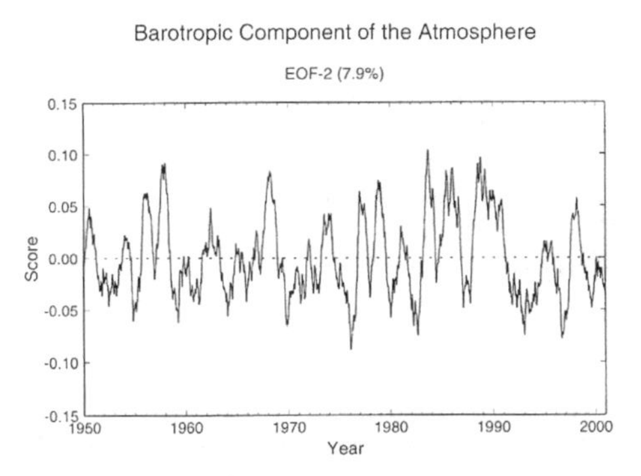

Fig. 6.15 As in Fig. 6.12, but for the perpetual January run of the barotropic S-model for 51 years. From Tanaka (2003), Fig. 13. ©American Meteorological Society. Used with permission.

score time series of the EOF-2 indicates a larger interannual variability than that of the observation. The dominant periodicity of two-to-three year agrees with the observation in Fig. 6.12.

We have compared the structures of EOF-3 and EOF-4 with that of the observation (not shown). The result shows not only the EOF-1 but also the EOF-2 to EOF-4 agree fairly well with the observed EOFs. Therefore, it is confirmed in this section that the major characteristics of the atmospheric low-frequency variability have been captured satisfactory by the barotropic S-model.

The time spectrum of the daily AO index for the model atmosphere is plotted in Fig. 6.16. The red noise spectrum at the high frequency range extends up to the period of four months, then it shifts to the white noise spectrum at the low frequency range. There is a minor spectral peak at two-year period, and the decadal variability indicates sufficient power in the result. According to the result of the 500-year perpetual January run (not shown here), we confirmed that the white noise extends all the way to the low-frequency range. It is interesting to note that even the

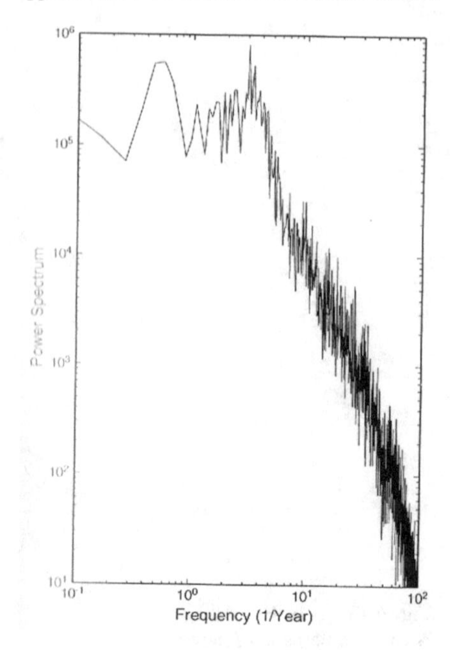

Fig. 6.16 As in Fig. 6.13, but for the perpetual January run of the barotropic S-model for 51 years. From Tanaka (2003), Fig. 14. ©American Meteorological Society. Used with permission.

multi-decadal oscillation has same magnitude as 4 month period. In the observed atmosphere in Fig. 6.13, the shift from red noise to white noise occurs gradually around the same period as model atmophsere. The spectrum is similar to that argued by James and James (1992). Since the white noise is characterized by the variability that has no memory in the past, the clear shift from the red noise to white noise suggests that the barotropic model has a memory up to four months as long as the AO index is concerned. The result suggests that the AO index has no specific periodicity, and the variation may be considered as a natural variability of stochastic random noise.

6.2.5 Energetics of the Model Atmosphere

6.2.5.1 Energy Spectrum and Conversions

In this section, the energy spectrum and conversions of the model atmosphere is compared with observations. Figure 6.17 plots the barotropic (m=0) energy spectrum of $E_i = E_{knm}$ as a function of the scale index $|c_i| = |c_{knm}|$ evaluated for the model atmosphere for one year. In order to avoid the influence of the initial condition, the second year of the time integration was analyzed. The energy distribution compares well with the observations in Fig. 6.10. The peak of the energy spectrum for the zonal eddy $k \neq 0$ is determined by the Rhines' scale, where the inverse energy cascade terminates (Rhines (1979)). The spectrum for k=0 coincides with that of $k \neq 0$ for a smaller $|c_i|$, but it continues to increase for larger $|c_i|$ beyond the Rhines' scale. The inverse energy cascade continues to even larger scale for k=0. The energy level is lower than observations for short waves with smaller $|c_i|$ because there is no energy source.

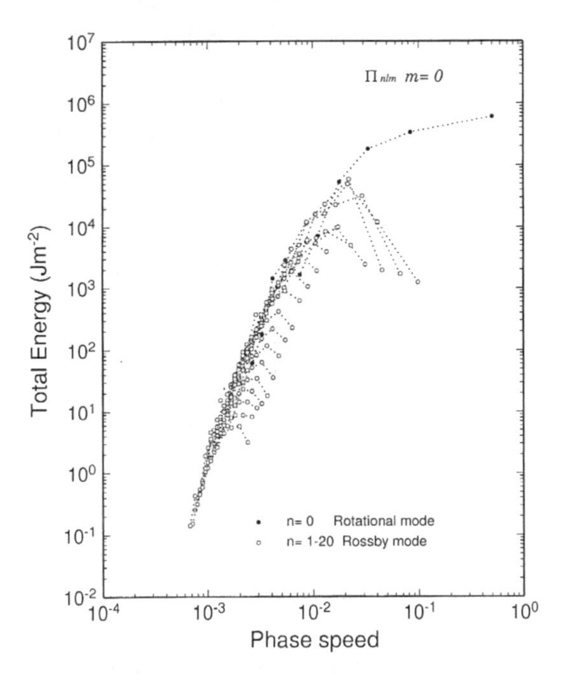

Fig. 6.17 As in Fig. 6.10, but for the barotropic S-model for the perpetual January run. The black and white dots denote k=0 and $k \neq 0$, respectively. The unit is in J m^{-2}. From Tanaka (2003), Fig. 15. ©American Meteorological Society. Used with permission.

The AO can be characterized by variations in zonally symmetric component of k=0. It is interesting to analyze the energy source and sink for the zonal flows in the model atmosphere. The mean (one year) energy levels and energy conversions are computed based on Tanaka (1998), and the result is listed in Table 6.1 as functions of zonal wavenumber k. Both EN and FN can be assessed directly from the observation. The contributions from the climate forcing $\tilde{f}_i$ are added, for convenience, to the system forcing (AB).

The mean zonal energy is 12.1×10^{5}J m^{-2}, which is comparable with observations. Two spectral peaks are seen for (EN) at the wavenumbers k=1 and 3, and the energy level decreases for short waves. The synoptic energy source is transformed mostly to large-scale motions and little comes to short waves under the restriction of the barotropic dynamics.

Table 6.1 Time mean (one year) energy (10^5J m^{-2}) and energy conversions (10^{-3} W m^{-2}) for the barotropic S-model as functions of the zonal wavenumber k. The symbols are EN represents total energy, AB system forcing, BC baroclinic instability, NL nonlinear wave-wave interaction, DS surface friction, DF diffusion, and FN total external forcing. From Tanaka (2003), Table 1. ©American Meteorological Society. Used with permission.

k	EN	AB	BC	NL	DS	DF	FN
0	1208261	-114	0	235	-86	-36	-236
1	115707	16	0	76	-60	-32	-75
2	70069	169	1	-109	-39	-22	109
3	108319	141	0	-50	-56	-33	53
4	24646	30	7	0	-14	-16	8
5	29521	5	54	-11	-15	-18	25
6	24069	34	56	-37	-11	-18	62
7	19117	16	77	-26	-10	-20	64
8	11316	17	55	-17	-6	-17	49
9	4794	6	23	0	-2	-10	16
10	2989	5	12	3	-1	-9	7
11	1376	2	4	5	-1	-6	0
12	794	0	1	6	0	-4	-3
13	436	1	0	4	0	-3	-2
14	235	0	0	3	0	-2	-2
15	153	0	0	2	0	-2	-1
16	99	1	0	1	0	-1	-1
17	66	0	0	1	0	-1	-1
18	45	0	0	1	0	-1	-1
19	28	0	0	1	0	-1	-1
20	19	0	0	0	0	-1	0

The system forcing (AB) excites planetary waves representing the topographic forcing, and a minor energy supply is seen at the synoptic-scale waves. This term works as an energy sink for k=0. The weak energy supply at the synoptic-scale waves is compensated by the baroclinic instability (BC). The nonlinear wave-wave interactions (NL) redistribute the energy from synoptic and planetary waves to mostly zonal motions by the nature of the barotropic dynamics. The wavenumber one and short waves also obtain energy by (NL). The energy is dissipated by the surface friction (DS) and diffusion (DF). The sum of these conversion terms is listed in the total external forcing (FN). The results in this section are quite reasonable compared with observations (Tanaka and Kung (1988); Tanaka (1998)).

The result of the energetics analysis indicates two major energy sources at synoptic-scale eddies and planetary waves. The energy is redistributed mostly to zonal motion by the nonlinear up-scale energy cascade within a framework of the barotropic dynamics (see Fig. 6.10). The nonlinear up-scale energy cascade by (NL) is the sole energy supply for the zonal motions where the AO is contained.

6.2.5.2 Energetics of the Anomaly Field

The AO can be analyzed as the variation in the anomaly field, especially for zonally symmetric component of k=0. For this reason, a similar energetics analysis is conducted for the anomaly field of the model atmosphere in this subsection. The energy spectrum is evaluated for the anomaly w_i' and is plotted in Fig. 6.18 as functions of $|c_i|$. We can notice that the zonal energy (black dots) is reduced substantially compared with that in Fig. 6.17. The result suggests that the large fraction of zonal energy is contained in steady motions. The spectral peak in the zonal eddy (white dots) is also reduced, indicating that a large fraction of planetary waves are explained by the steady component induced by topographic forcing. However, the energy levels for the synoptic and short waves are comparable with that in Fig. 6.17, showing the dominant transient motions for these eddies. The energy peak for the anomaly field appears in the intermediate scale of zonal field. The energy peak level for k=0 is as large as that of zonal eddy $k \neq 0$. It is in this range where the characteristics of the AO are contained.

The mean (one year) energy levels and energy conversions for the anomaly field are listed in Table 6.2 as functions of zonal wavenumber k. All variables including the nonlinear term are separated in the mean and anomaly, and the energy budget is computed with those anomalies. The energy levels (EN) are comparable for k=0 to 5, and the peak is seen at k=1. The baroclinic eddies (BC) are the main energy source of the anomaly for the synoptic-scale transients. The nonlinear term (NL) acts to reduce the anomaly for synoptic eddies. On the contrary, the nonlinear term appears to be the major energy source of the anomaly in planetary waves and the zonal motions. The surface friction (DS) and diffusion (DF) act to reduce the anomaly for all scale of motions. It is interesting to note that the system forcing (AB) is not the energy source but a sink of the anomaly field in planetary waves. The system forcing represents mostly the topographic forcing. The large energy source of (AB) in Table 6.1 is thus explained by a steady forcing induced by the topography. This term rather acts to reduce the anomaly in planetary waves.

A large time variation is a notable feature of the term (NL). Therefore, a large fraction of the time variation associated with the AO can be explained by the chaotic fluctuation of (NL) which is the sole energy source of the transient anomalies in planetary waves and zonal motions.

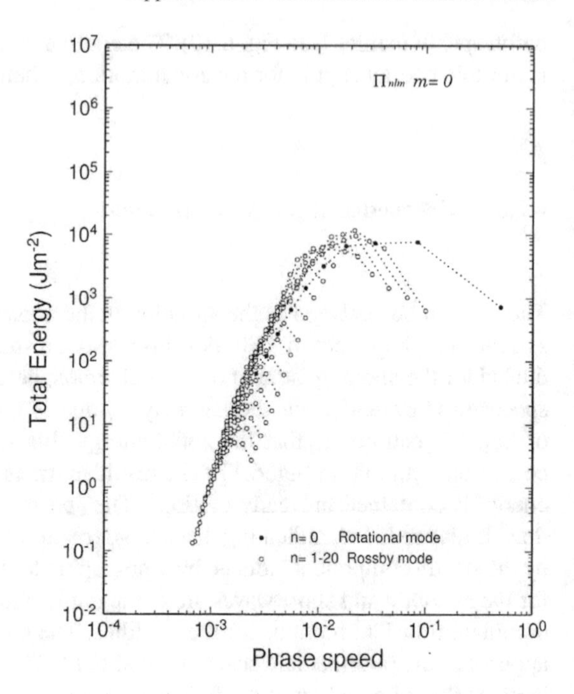

Fig. 6.18 As in Fig. 6.17, but for the anomaly of the model atmosphere w_i'. The black and white dots denote k=0 and $k \neq 0$, respectively. The unit is in J m^{-2}. From Tanaka (2003), Fig. 16. ©American Meteorological Society. Used with permission.

6.2.6 Summary

In this section, a numerical simulation of the Arctic Oscillation (AO) is conducted using a simple barotropic model that considers the barotropic-baroclinic interactions as the external forcing. The model is referred to as a barotropic S-model since the external forcing is obtained statistically from the long-term historical data solving an inverse problem. We have integrated the S-model for 51 years under the perpetual January condition and analyzed the dominant EOF modes in the model. The model results are compared with the EOF analysis of the barotropic component of the real atmosphere based on the daily NCEP/NCAR reanalysis for 50 years from 1950 to 1999.

According to the results, the first EOF of the model atmosphere appears to be the AO similar to the observation. The annular structure of the AO and two centers of action at Pacific and Atlantic sectors are simulated nicely by the barotropic S-model. The result suggests that the AO may be understood as the natural variability of the barotropic component of the atmosphere induced by the inherent barotropic dynamics, which is forced by the barotropic-baroclinic interactions. The EOF analysis of the model atmosphere shows not only the EOF-1 but also the EOF-2 to EOF-4 agree

Table 6.2 Time mean (1000 days) energy (10^5J m^{-2}) and energy conversions (10^{-3} W m^{-2}) for the anomaly of the S-model as functions of the zonal wavenumber k. The symbol EN represents total energy, AB system forcing, BC baroclinic instability, NL nonlinear wave-wave interaction, DS surface friction, DF diffusion, and FN total external forcing. From Tanaka (2003), Table 2.

n	EN	AB	BC	NL	DS	DF	FN
0	31639	-17	0	44	-19	-7	-43
1	36634	-19	0	56	-20	-14	-53
2	31125	-5	1	37	-17	-13	-34
3	22917	-3	0	32	-13	-12	-28
4	21333	5	6	23	-12	-13	-14
5	23153	8	50	-15	-12	-16	30
6	19586	15	59	-22	-10	-17	47
7	15806	21	69	-32	-8	-17	66
8	8716	14	43	-14	-4	-13	39
9	4367	5	20	0	-2	-9	13
10	2380	2	9	5	-1	-7	3
11	1169	0	3	6	-1	-5	-2
12	633	1	1	4	0	-3	-2
13	362	1	0	3	0	-2	-2
14	198	0	0	2	0	-2	-2
15	118	0	0	2	0	-1	-1
16	78	0	0	1	0	-1	-1
17	47	0	0	1	0	-1	-1
18	33	0	0	1	0	-1	-1
19	21	0	0	1	0	0	0
20	14	0	0	0	0	0	0

fairly well with the observed EOFs. Therefore, the characteristics of the atmospheric low-frequency variability have been captured satisfactory by the barotropic S-model.

The EOF analysis is further applied for the external forcing of the barotropic component of the observed atmosphere. According to the result, the most dominant mode of the forcing is the barotropic-baroclinic interactions associated with baroclinic eddies of zonal wavenumber 6 located along the Pacific storm track. A similar synoptic-scale forcing is seen along the Atlantic storm track in the higher EOF modes. The energy source at the synoptic-scale eddies is transformed to zonally symmetric motions by the nonlinear up-scale energy cascade, which is the sole energy source of the zonal wavenumber k=0. Hence, the annular structure of the AO or NAM has been excited by the nonlinear up-scale energy cascade from active transient eddies along the storm tracks. The positive feedback between transient eddies and zonal motions can be one of the possible excitation mechanism of the AO (Lorenz and Hartmann (2001); Tanaka and Tokinaga (2002)).

There are two active storm tracks over the Pacific and Atlantic sectors. These two storm tracks are known to behave independently, indicating insignificant correlation with each other (Ambaum and Stephenson (2001)). When the Atlantic storm track produces a positive and negative height anomalies over the Atlantic and the Arctic, respectively, the annular structure appears excited by the energy supply to the zonal motion. Similarly, when the Pacific storm track produces a positive and negative height anomalies over the Pacific and the Arctic, respectively, the same annular structure appears excited by the energy supply to the zonal motion. Although the activities of the two storm tracks are independent and not correlated with each other, the statistical EOF analysis reveals a single negative height anomaly over the Arctic and two positive height anomalies over the Pacific and Atlantic sectors, as demonstrated by Deser (2000), Ambaum and Stephenson (2001), and Wallace and Thompson (2002). The EOF-2 indicates opposite signs over the Pacific and Atlantic sectors in order to describe the independent hehavior between the two sectors as demonstrated in this section with the barotropic S-model.

The zonally symmetric annular mode would be excited when the storm track is distributed uniformly as seen for the SAM. In the Northern Hemisphere, the AO or NAM would be excited both by transient synoptic eddies along the storm tracks and standing planetary waves. The up-scale energy cascade from those eddies or waves to zonal motions would produce the largest-scale annular mode of k=0, which can exist in both hemispheres. The zonally symmetric annular mode can be excited by various arbitrary forcing such as a volcanic effect, global warming, or the Milankovitch cycle because energy of the impact is transformed eventually to the zonal flow by the characteristics of barotropic dynamics. In conclusion, the AO can be dynamically understood as the internal mode of the barotropic atmosphere where the largest-scale annular mode is excited by the nonlinear up-scale energy cascade from various energy sources.

6.3 Singular Eigenmode Theory of the Arctic Oscillation

6.3.1 Scientific Background

The AO is a north-south seesaw of the atmospheric mass between the Arctic region poleward of 60°N and a surrounding zonal ring in the mid-latitudes. It is defined as a leading mode of an empirical orthogonal function (EOF) of the sea-level pressure field and has an equivalent barotropic structure extending well into the stratosphere. The AO fluctuations dominate over the weekly and longer time scale, including decadal to long term trends associated with the global warming (e.g., Shindell and Pandolfo (1989); Fyfe and Flato (1989)).

It is generally agreed that the AO is essentially a mode internal to the atmosphere. However, the realization as to whether the AO is a physical mode of a linearized dynamical system or a simple statistical illusion of independent multiple teleconnections is an open question under the active debate, and no general consensus has yet been reached (Wallace and Gutzler, 1981; Deser, 2000; Ambaum and Stephenson, 2001; Itoh, 2002; Wallace and Thompson, 2002).

Beside the controversy based on the statistical arguments, a dynamical approach has been pursued by solving a singular mode with the smallest singular value of the linearized dynamical system, which is now referred to as the neutral mode theory (e.g., Navarra, 1993; Marshall and Molteni, 1993; Kimoto and Yasutomi, 2001; Watanabe and Jin, 2004). It is found that the principal singular mode for a forced steady state shows a considerable resemblance with the observed AO, suggesting that the AO may be a physical mode of the dynamical system for the global atmosphere. Kimoto and Yasutomi (2001) solved highly truncated baroclinic model linearized about a zonal basic state. In their computation, the wave-wave interactions are ignored because the matrix size to be solved is too large. Despite the introduced additional assumptions, the essential features of the AO are captured by the simplified model. The full matrix was solved by Watanabe and Jin (2004) with 11 vertical levels. The 3D structure of the AO-like singular mode was compared with the observed 300 hPa geopotential height. The barotropic structure of the singular mode with anti-correlated zonal-mean winds between 30°N and 60°N agrees quite well with observation. However, the similarity seems to have lost at the sea-level pressure where the essence of the AO fluctuations are contained by the following reason.

The AO is defined as the EOF-1 of the time variation in sea-level pressure p_s, which is dynamically related to the time variation of the barotropic component of the atmosphere as shown by (6.15). For this reason, the essential features of the AO are contained in the barotropic component of the atmosphere governed by the 2D fluid mechanics which characterizes the low-frequency variability.

Using a simple barotropic model derived from the 3D normal mode expansion, Tanaka (2003) conducted a numerical simulation of the AO and obtained the same structure as observed in the atmosphere. The low-frequency variabilities such as the AO and PNA are evidently contained in the internal dynamics of the barotropic component of the atmosphere. It was concluded that the inverse energy cascade from synoptic disturbances to the so-called Rhines scale is responsible for the low-frequency variabilities governed by the 2D fluid mechanics (Rhines, 1979; Tanaka, 2003; Tanaka and Kanda, 2004). However, the characteristic structure of the AO is yet to be explained.

We anticipate, therefore, that the AO may be understood as the eigenmodes or singular modes as demonstrated by Watanabe and Jin (2004), but for the linearized dynamical system of the barotropic atmosphere. Since the matrix size is small enough to be solved by the standard numerical packages, we may identify the AO mode in the eigenmodes or singular modes. In this section, the eigenmodes and singular modes are analyzed in the framework of the barotropic atmosphere in order to understand the dynamics of the Arctic Oscillation, as described by Tanaka and Matsueda (2005) (hereafter TM2005). If we can identify the AO in the dynamical system, we would find whether the AO is a physical mode or a simple statistical illusion of multiple teleconnections.

6.3.2 Governing Equations and Data

6.3.2.1 EVP and SVD for the Linearized System

The spectral barotropic equation (6.1) is linearized about a prescribed basic state in order to construct a linear system. The equation for the first-order term of perturbations becomes

$$\frac{dw_i}{d\tau} = -i\sigma_i w_i - i\sum_{j=1}^{K}(\sum_{k=1}^{K}(r_{ijk} + r_{ikj})\overline{w}_k)w_j + d_i\, w_i + f_i, \quad i = 1, 2, 3, \ldots, K \quad (6.16)$$

In this section, the frictional force is incorporated as a part of the dynamical system, and the external forcing f_i, which mainly represents baroclinic-barotropic interactions, is considered as a stochastic random process. For the barotropic component, the heat related source terms in f_i vanishes, and only the frictional forcing remains to be considered.

The model in this section is different from the traditional non-divergent barotropic quasi-geostrophic (QG) model in that the present model is basically a shallow water system (see Simmons et al. (1983); Navarra (1993)). It seems that the assumption of vanishing divergence in the QG theory is inaccurate for the global scale phenomenon such as AO.

In this section, the frictional forcing d_i is parameterized by the following hyper diffusion represented by c_i and the linear damping representing Rayleigh friction:

$$d_i\, w_i \;=\; -k_D\, c_i^{-4}\, w_i - \nu_S\, w_i, \quad (6.17)$$

where k_D is a diffusion coefficient with the value of $k_D(2\Omega\, a^8) = 2.7\times 10^{40}\ \mathrm{m^8 s^{-1}}$, and a is the radius of the Earth. The linear damping coefficient ν_S is first set to zero and will be added later to shift the eigenvalues so that the system becomes singular.

The complex-valued linear system (6.16) can be represented by a real-valued linear system, putting $\mathbf{x} = (w_{Ri}, w_{Ii})$ and $\mathbf{f} = (f_{Ri}, f_{Ii})$, where the subscripts R and I represent real and imaginary parts, respectively:

$$\frac{d\mathbf{x}}{d\tau} = \mathbf{A}\mathbf{x} + \mathbf{f}. \tag{6.18}$$

The real matrix $\mathbf{A}$ is determined by (6.16) using the basic state $\overline{w}_i$.

First, an eigenvalue problem (EVP) may be solved by assuming a wave-type solution for $\mathbf{x}(\tau) = \xi \exp(\nu\tau)$, disregarding the external forcing:

$$\nu\mathbf{x} = \mathbf{A}\mathbf{x}. \tag{6.19}$$

In general, the eigenvalues are complex due to the barotropic instability of the non-zonal basic state. The solutions have a life-cycle in its structure multiplied by the exponential growth or decay. As a special case, standing eigenmodes appear associated with real-valued eigenvalues with arbitrary signs in its fixed structure. It is those standing eigenmodes (ν_I=0) that may be important for the analysis of the Arctic Oscillation because eigenvalues with small magnitudes are searched in present section. If a structure similar to the AO is obtained as the standing eigenmode, we can support that the observed AO is a dynamical eigenmode.

Second, singular vectors of the linear system may be solved for neutral modes with respect to a stochastic random forcing $\mathbf{f}$ under the steady state (see Kimoto and Yasutomi (2001); Watanabe and Jin (2004)). In this case, the singular value decomposition (SVD) of the matrix $\mathbf{A}$ is conducted as

$$\mathbf{A} = \mathbf{U}\Sigma\mathbf{V}^{\mathbf{T}}, \tag{6.20}$$

where Σ is a diagonal matrix containing singular values s_i, $\mathbf{U}$ and $\mathbf{V}$ are orthonormal matrices containing left and right singular vectors, respectively (see Golub and van Loan, 1983). Then, the steady solution may be represented as

$$\mathbf{x} = -\mathbf{A}^{-1}\mathbf{f} = -\mathbf{V}\Sigma^{-1}\mathbf{U}^T\mathbf{f}. \tag{6.21}$$

Here, responding to the stochastic random forcing $\mathbf{f}$, the steady solution $\mathbf{x}$ is obtained as a summation of the right singular vectors $\mathbf{v}_i$, among those $\mathbf{v}_1$ associated with the smallest magnitude of the singular value (referred to as SVD-1) is considered as most important by the resonant behavior. If a structure similar to the AO is obtained as the SVD-1, we can support that the AO is a dynamical neutral mode.

The singular mode SVD-1 is a steady solution considered as a forced mode, while EVP-1 can be either a transient or steady solution considered as an intrinsic free mode derived with less assumptions. The resonant growth is allowed even for transient modes by transient forcing. We should note here that the matrix $\mathbf{A}$ is singular if the frictional force is not considered because all of the rotational Hough modes have zero eigenfrequency and the interaction terms are all zero for $k=0$. Hence, incorporation of the frictional force is essential for the SVD analysis which is not the case for the EVP.

6.3.3 Result for EVP and SVD Analyses

The basic state used in this section is illustrated in Fig. 6.9 for the barotropic component of the geopotential height in the Northern Hemisphere (referred to as barotropic height). A basic state of 300 hPa stream function is used in most of the QG theory as an equivalent barotropic level (e.g., Simmons et al., 1983; Navarra, 1993). However, the intensity of the westerly jet is rather different from our vertical mean basic state.

Observed EOF-1 of the Arctic Oscillation was shown in Fig. 6.11. The EOF-1 explains 21% of the total variance when computed for the DJF mean anomaly w_i' of the NCEP/NCAR reanalysis. The negative anomaly in the Arctic is surrounded by positive anomaly in mid-latitudes with two pronounced positive peaks at the North Pacific and the North Atlantic.

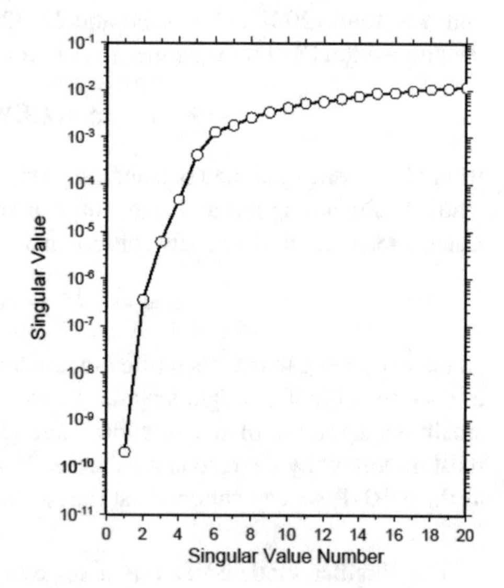

Fig. 6.19 The first 20 singular values s_i in ascending order for the matrix without the shift. The values are normalized by 2Ω. From Tanaka and Matsueda (2005), Fig. 3.

Using the basic state in Fig. 6.9, the SVD analysis (6.21) is performed first with the diffusion but without the Rayleigh friction. The singular values are plotted in Fig. 6.19 in ascending order. The result shows that the SVD-1 is well separated from the rest of the higher order modes. In fact, the SVD-1 is 1700 times smaller than that of the SVD-2.

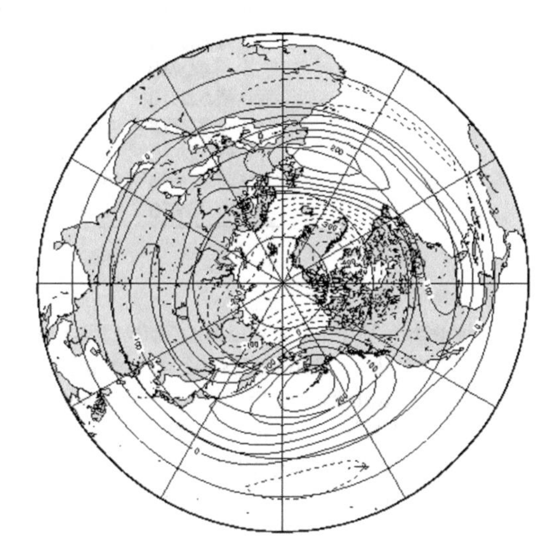

Fig. 6.20 Barotropic height of the singular mode SVD-1 without the shift. The structure is identical with EVP-3 in Table 6.3. The contour interval is 50 m with dashed lines for the negative area. From Tanaka and Matsueda (2005), Fig. 4.

The horizontal pattern of the SVD-1 is illustrated in Fig. 6.20. There is a negative anomaly in the Arctic centered at Greenland. The negative anomaly is surrounded by a ring of positive anomaly in mid-latitudes showing two positive poles at North Pacific and North Atlantic. There are other negative regions at low latitudes. Although the center of the negative anomaly in the Arctic is slightly shallower than the observation in Fig. 6.11, we can safely identify the SVD-1 as the AO pattern.

Next, the eigenvalue problem is solved for the system matrix **A** in (6.19). Table 6.3 lists the eigenvalues of the standing modes ν_R in descending order. The values (ν_R, ν_I) are normalized by 2Ω. It is found that the EVP-1 and EVP-2 are unstable modes and EVP-3 is a damping mode having the smallest magnitude of the eigenvalue. We have confirmed that the EVP-3 has the same structure as the SVD-1 shown in Fig. 6.20. One may anticipate that the results of the EVP are sensitive to the configuration of the basic state. We repeated the same analysis using NDJ (Nov. to Jan.) mean and JFM (Jan. to Mar.) mean and found the same structure as shown here. Hence, the results presented in this section are quite robust against the basic state.

Table 6.3 List of eigenvalues (EVP) for standing modes before the shift ν_R, after the shift $\nu_R - \nu_S$, and singular values (SVD) after the shift s_i with the 10 smallest magnitudes. The values are normalized by 2Ω. From Tanaka and Matsueda (2005), Table 1.

No.	ν_R	$\nu_R - \nu_S$	s_i
1	1.52×10^{-3}	0.0	0.0
2	7.78×10^{-4}	-7.44×10^{-4}	6.32×10^{-4}
3	-1.09×10^{-9}	-1.52×10^{-3}	1.27×10^{-3}
4	-1.41×10^{-6}	-1.52×10^{-3}	1.45×10^{-3}
5	-3.66×10^{-6}	-1.53×10^{-3}	1.53×10^{-3}
6	-5.52×10^{-5}	-1.58×10^{-3}	2.16×10^{-3}
7	-6.71×10^{-4}	-2.19×10^{-3}	2.53×10^{-3}
8	-2.52×10^{-3}	-4.04×10^{-3}	3.44×10^{-3}
9	-4.47×10^{-3}	-6.00×10^{-3}	4.19×10^{-3}
10	-9.43×10^{-3}	-1.09×10^{-2}	5.04×10^{-3}

As noted by Watanabe and Jin (2004) the neutral mode theory in a forced problem requires that the eigenmodes of **A** are all stable. Otherwise, the unstable mode emerges and dominates in the steady response on monthly or longer time scale (Branstator (1985)). For this reason, we have introduced the linear damping ν_S in (6.17) with the same magnitude as the EVP-1 in Table 6.3 to eliminate all unstable modes. The value of $\nu_S = 1.52 \times 10^{-3}$ corresponds to the e-folding time of 52 days. This is a technique called a shift of eigenvalues. The eigenvectors remain the same by this shift. The result of the new eigenvalues after the shift is listed as $\nu_R - \nu_S$ in Table 6.3. It is apparent that all eigenmodes are stable and the EVP-1 has the smallest (zero) magnitude of the eigenvalue, i.e., singular. Physically, the Rayleigh friction implies a combination of surface friction and transient eddy forcing. It is known that the eddy forcing has a positive feedback with the AO to relax the linear damping.

Figure 6.21 illustrates the horizontal pattern of the singular eigenmode of the EVP-1. The structure of the EVP-1 is identical before and after the shift and is identical also to that of SVD-1 for the shifted matrix. The structure of EVP-1 may be identified as the AO. It is rather similar to that of EVP-3 in Fig. 6.21, but the negative anomaly in the Arctic is as deep as the observation. We should note that EVP-2 also shows a structure similar to the AO.

The singular values after the shift are listed as s_i in Table 6.3. The first singular mode SVD-1 is the neutral mode after the shift. However, EVP-3 was the neutral mode recognized as the SVD-1 before the shift. The result has an important implication such that there are multiple eigenmodes similar to the AO, and any of them can be excited by the resonant response to the random forcing by a slight change in the frictional forcing or transient eddy forcing as discussed by Watanabe and Jin (2004). When the eigenvalue becomes zero, the AO mode is amplified resonantly by arbitrary forcing. This resonant behavior is the most important dynamical property to undersatand the AO. We refer to it as the singular eigenmode theoty of the AO.

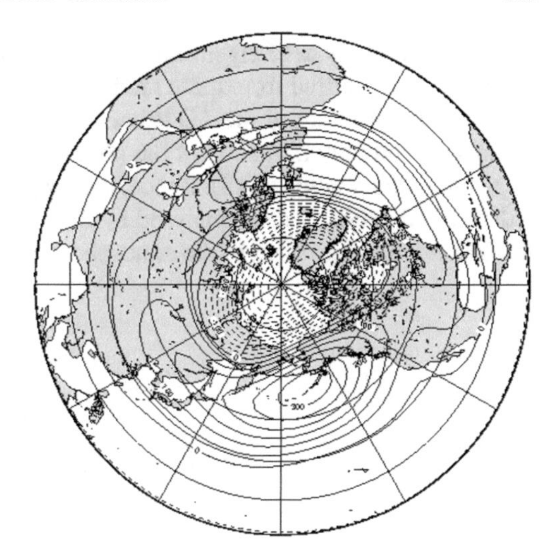

Fig. 6.21 Barotropic height of the standing eigenmode EVP-1 in Table 6.3 after the shift. The structure is identical with the SVD-1 in Table 6.3. The contour interval is 50 m with dashed lines for the negative area. From Tanaka and Matsueda (2005), Fig. 5.

6.3.4 Standing Rossby Waves and Teleconnections

Atmospheric teleconnections are prominent recurrent patterns in the general circulation, which is measured by one point correlation with surrounding remote areas. North Atlantic Oscillation (NAO) and the Pacific/North American (PNA) patterns are the most notable teleconnections appearing in the sea level pressure and 500 hPa height, respectively. The problem is whether the Arctic Oscillation is a physical mode with true dynamical basis or a statistical artifact by a combination of multiple teleconnections of the NAO and PNA.

According to Deser (2000), the NAO is clearly qualified as a teleconnection showing the correlation of −0.83 between the Arctic and Atlantic sectors for the winter-mean data. Yet, the AO is not qualified as a teleconnection because the correlation between the Atlantic and Pacific sectors is only −0.07, where the statistical significance at the 5% confidence level is 0.22 and more. Itoh (2002) explained the statistical trick of the AO using a three-point correlation model followed by a detailed analysis of teleconnectivity between the Atlantic and Pacific sectors by removing the data in the Arctic region. They concluded that the AO is not a dynamical mode but a statistical artifact by the multiple teleconnections.

Using a simple barotropic model derived from the 3D normal mode expansion, Tanaka (2003) conducted a numerical simulation of the AO and obtained the same structure as observed in the atmosphere. Tanaka and Matsueda (2005) identified that the characteristics of the AO are originated from the eigenmode of the dynamical system with nearly zero eigenvalue, i.e., singular eigenmode, for the global atmosphere. It was inferred from these model experiments and theoretical deduction that the essential property of the AO is contained in the barotropic dynamics of the atmosphere, and only a fraction of the property is analyzed through a single variable of p_s under the relation (6.15). Hence, the insignificant correlation in p_s between the Atlantic and Pacific sectors analyzed by Deser (2000) may become significant when the barotropic component of the atmosphere is analyzed.

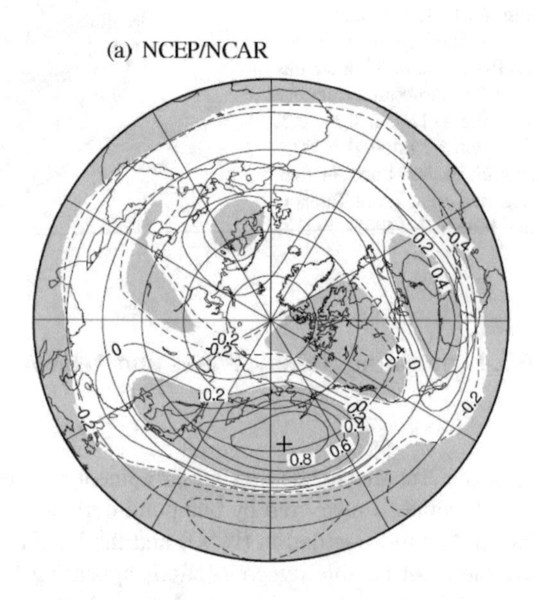

Fig. 6.22 Distribution of the one point correlation centered at the grid point (49°N, 170°W) for the NCEP/NCAR reanalysis. From Suzuki and Tanaka (2007), Fig. 4a.

In this section, one point correlations are examined for the PNA and NAO or AO. Figure 6.22 illustrates the one point correlation maps for the PNA using the barotropic height. The target grid points are chosen from the North Pacific peaks of the teleconnectivity. Monthly mean data for DJF during 51 years are used for the correlation (the number of data is 153). The computed one-month lag correlations are 0.26 and 0.31 for observation and model. The degree of freedoms with these lag correlations are estimated as 89 and 80, respectively. The statistical significance at the 5% confidence level becomes 0.21 and 0.22 for each data. The result shows clear wavetrain from the North Pacific via Canada to the southwestern Atlantic. The agreement for the PNA patterns is almost perfect for the model and observation as seen in Fig. 6.23. However, the model clearly shows higher correlations at Canada and southwestern Atlantic since the low-frequency variability is simulated in pure form with less transient noise.

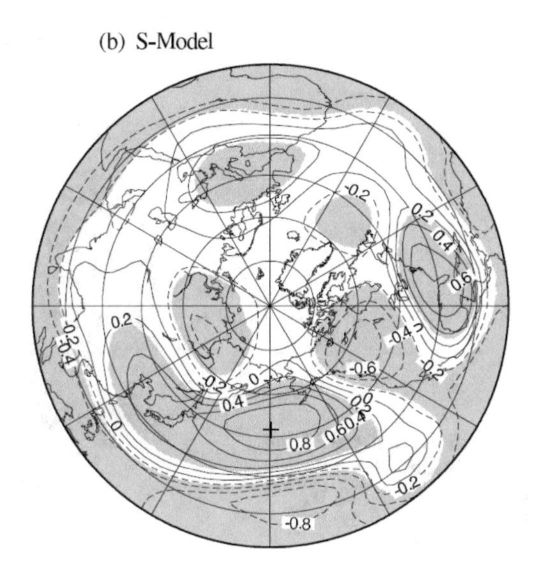

Fig. 6.23 As in Fig. 6.22, but for the barotropic S-model centered at (49°N, 180°W). From Suzuki and Tanaka (2007), Fig. 4b.

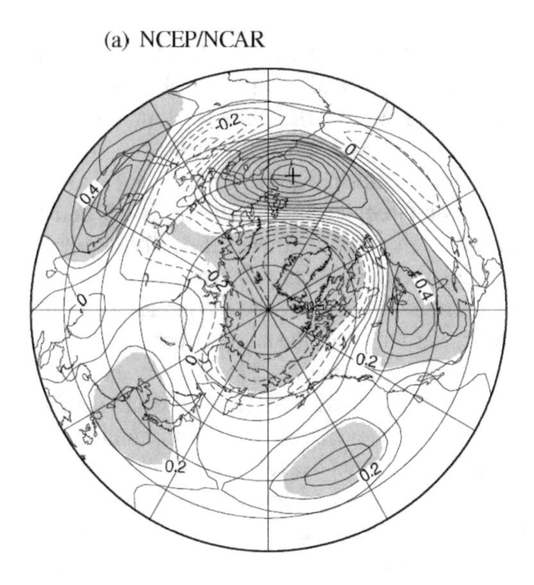

Fig. 6.24 Distribution of the one point correlation centered at the grid point (43°N, 10°W) for the NCEP/NCAR reanalysis. From Suzuki and Tanaka (2007), Fig. 5a.

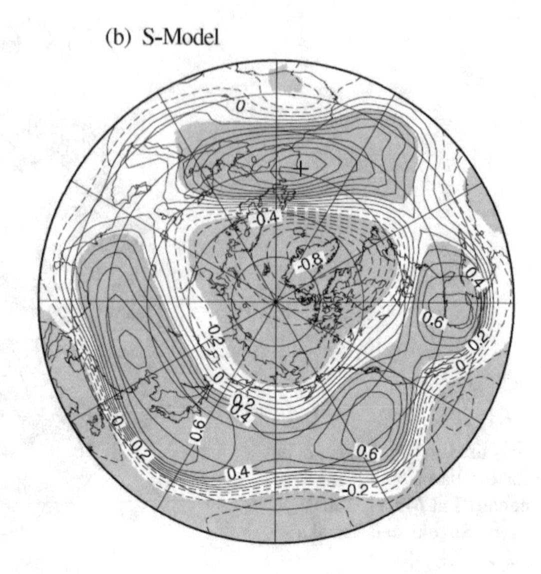

Fig. 6.25 As in Fig. 6.24, but for the barotropic S-model centerd at (43°N, 10°W). From Suzuki and Tanaka (2007), Fig. 5b.

Figure 6.24 illustrates the similar one point correlation maps for the NAO or AO. The target grid points are chosen from the positive center of action of the barotropic height at the eastern Atlantic because we are interested in the AO rather than NAO. It is noteworthy from the result of the model in Fig. 6.25 that the correlation pattern is quite similar to the AO pattern in Fig. 6.11, showing a negative pole over Greenland representing NAO and three positive poles at the Far East, eastern Pacific and southern USA. The result suggests the possibility that the EOF-1 contains the dynamical AO mode in part at least for the model atmosphere. When the correlation map for the observation in Fig. 6.24 is compared with that for the model, we can find the similar pattern with the model, although the correlations are substantially smaller. There is a negative pole over Greenland and three positive poles at the Far East, eastern Pacific and southern USA, with the correlations of 0.3, 0.3, and 0.5, respectively. It is important to note that these are statistically significant at the 5% confidence level. The less correlation compared with the model is probably due to the larger transient noise in the observation as mentioned before. The present model is not perfect for simulating the real atmosphere, but is sufficient to simulate the essential features of the low-frequency variability.

6.3.5 Summary

In this section, eigenmodes and singular modes are analyzed for a linearized dynamical system of the global atmosphere in order to understand the dynamics of the Arctic Oscillation (AO). Since the AO is defined by the fluctuation in the sea-level pressure, which is equivalent to the fluctuation in the barotropic component of the atmosphere, the AO is investigated in the framework of the shallow water system derived from the 3D normal mode decomposition. The neutral mode theory proposed by Kimoto and Yasutomi (2001) is examined using our barotropic model, which has demonstrated the capability of simulating the realistic AO in the nonlinear framework (see Tanaka (2003)).

As a result of the EVP and SVD analyses for the winter climate basic state, we find multiple eigenmodes which are similar to the AO with a negative pole in the Arctic and positive poles in the Pacific and Atlantic sectors. Some of them appear to be unstable. A linear drag representing Rayleigh friction is then introduced to shift the eigenvalue in order to pick up different eigenmodes as the neutral mode obtained as the SVD-1. As a result, these singular eigenmodes appear to be all similar to the AO.

It is concluded from the result that the singular eigenmode of the dynamical system emerges resonantly as the AO in response to the arbitrary forcing. The resonant growth is allowed for multiple eigenmodes including the unstable modes. The AO described as the neutral mode under the strong friction (see Kimoto and Yasutomi (2001); Watanabe and Jin (2004)) is therefore recognized as the least damping mode excited by the tail of the resonant response curve. In reference to the successful nonlinear simulation of the AO using the same barotropic model, we may conclude that the observed AO can be understood as a dynamical eigenmode of the global atmosphere with zero eigenvalue, which is excited by the resonant response to the stochastic forcing due mostly to the barotropic-baroclinic interactions.

Whether the AO is a physical mode or a statistical artifact by the multiple teleconnections of NAO and PNA has been a great concern for the study of understanding the AO. In this respect, the observational fact such that the correlation analyzed for p_s between Atlantic and Pacific sectors is insignificant (Deser (2000)) is the key factor to reject the dynamical theory in explaining the AO. In this section, similar analysis as Deser (2000) is performed for the barotropic height Z_0 instead of p_s. The essential feature of the low-frequency variability is contained in the barotropic component of the atmosphere which is governed by the 2D fluid mechanics with the inverse energy cascade. The variation of p_s is one of the measures of the barotropic dynamics of the global atmosphere, but contains considerable influence of the baroclinic component of the circulation. The question is focused whether the one point correlation for the NAO is restricted within the Atlantic sector or spread to the Pacific sector surrounding the Arctic.

According to the result, it is found that the one point correlation is not restricted within the Atlantic sector but spread to entire mid-latitudes surrounding the Arctic with a positive poles at the Far East, eastern Pacific, and southern USA. The positive correlations are all statistically significant at the 5% confidence level. The variances explained by the EOF-1 are 21.0% and 32.0% for the observation and model, respectively, when the DJF data are analyzed. From this fact, the reduced one point correlation in the observation may be the influence of the transient noise forced by the baroclinic eddies.

When the behavior in the barotropic component of the atmosphere is compared with that in p_s, the difference can be explained by the contribution from the baroclinic component. For example, p_s associated with the Southern Oscillation shows evident teleconnection between Tahiti and Darwin induced by the Walker circulation, which is a typical baroclinic phenomenon and is not detected in the barotropic component. Therefore, part of p_s analyzed by Deser (2000) may be controlled by baroclinic component of the atmosphere. It was inferred from these model experiments and theoretical deduction that the essential property of the AO is contained in the barotropic dynamics of the atmosphere. From the result, we may conclude that the dynamical AO mode is embedded in the statistical EOF-1 for the observed atmosphere.

References

Akahori, K. and Yoden, S. (1996). Zonal flow vacillation and bimodality of baroclinic eddy life cycles in a simple global circulation model. *J. Atmos. Sci.*, 54:2349–2361.

Ambaum, M. H. P., Hoskins, B. J. and Stephenson, D. B. (2001). Arctic oscillation or north atlantic oscillation? *J. Clim.*, 14:3495–3507.

Basdevant, C., Legras, B. and Sadourny, R. (1981). A study of barotropic model flows: Intermittency, waves and predictability. *J. Atmos. Sci.*, 38:2305–2326.

Bengtsson, L. K. (1985). Medium-range forecasting at the ECMWF. issues in atmospheric and oceanic modeling. *Advances in Geophysics*, 28:3–54.

Boer, G. J., Fourest, S. and Yu, B. (2001). The signature of the annular modes in the moisture budget. *J. Clim.*, 14:3655–3665.

Branstator, G. (1985). Analysis of general circulation model sea surface temperature anomaly simulations using a linear model. *J. Atmos. Sci.*, 42:2242–2254.

Chen, W. Y. (1989). Estimate of dynamical predictability from NMC DERF experiments. *Mon. Wea. Rev.*, 117:1227–1236.

Dalcher, A. and Kalnay, E. (1987). Error growth and predictability in operational ECMWF forecasts. *Tellus A*, 39A:474–491.

Deser, C. (2000). On the teleconnectivity of the Arctic oscillation. *Geophys. Res. Lett.*, 27:779–782.

Edmon, H. J., Hoskins, B. and McIntyre, M. (1980). Eliassen-Palm cross-sections for the troposphere. *J. Atmos. Sci.*, 37:2600–2616.

Frederiksen, J. S. and Lee, S. (1998). Is the atmospheric zonal index driven by an eddy feedback? *J. Atmos. Sci.*, 55:3077–3086.

Fyfe, J. C., Boer, G. J. and Flato, G. M. (1989). The Arctic and Antarctic oscillations and their projected changes under global warming. *Geophys. Res. Lett.*, 26:1601–1604.

Golub, G. and van Loan, C. (1983). *Matrix Computations*. Johns Hopkins.

Hartman, D. L. (1995). A PV view of zonal flow vacillation. *J. Atmos. Sci.*, 52:2561–267.

Holloway, G. (1983). Effects of planetary wave propagation and finite depth on the predictability of atmosphere. *J. Atmos. Sci.*, 40:314–327.

Itoh, H. (2002). True versus apparent Arctic oscillation. *Geophys. Res. Lett.*, 29:1268.

James, I. N. and James, P. M. (1992). Spatial structure of ultra-low frequency variability of the flow in a simple atmospheric circulation model. *Quart. J. Roy. Meteor. Soc.*, 118:1211–1233.

Kalnay, E., Kanamitsu, M., Kistler, R., Collins, W., Deaven, D., Gandin, L., Iredell, M., Saha, S., White, G., Woollen, J., Zhu, Y., Chelliah, M., Ebisuzaki, W., Higgins, W., Janowiak, J., Mo, K., Ropelewski, C., Wang, J., Leetmaa, A., Reynolds, R., Jenne, R., , and Joseph, D. (1996). The NCEP/NCAR 40-year reanalysis project. *Bull. Amer. Meteor. Soc.*, 77:437–471.

Kalnay, E., Kanamitsu, M. and Baker, W. E. (1990). Global numerical weather prediction at the national meteorological center. *Bull. Amer. Meteor. Soc*, 71:1410–1428.

Kalnay, E., Lord, S. J. and McPherson, R. D. (1998). Maturity of operational numerical weather prediction: Medium range. *Bull. Amer. Meteor. Soc*, 79:2753–2769.

Karoly, D. J. (1990). The role of transient eddies in the low-frequency zonal variations in the southern hemisphere circulation. *Tellus*, 42A:41–50.

Kasahara, A. (1977). Numerical integration of the global barotropic primitive equations with hough harmonic expansions. *J. Atmos. Sci.*, 34:687–701.

Kimoto, M., Jin, F.-F. Watanabe, M. and Yasutomi, N. (2001). Zonal-eddy coupling and a neutral mode theory for the Arctic oscillation. *Geophys. Res. Lett.*, 28:737–740.

Leith, C. E. (1971). Atmospheric predictability and two-dimensional turbulence. *J. Atmos. Sci.*, 28:145–161.

Leith, C. E. and Kraichnan, R. H. (1972). Predictability of turbulent flows. *J. Atmos. Sci.*, 29:1041–1058.

Limpasuvan, V. and Hartmann, D. L. (2000). Wave-maintained annular mode of climate variability. *J. Clim.*, 13:4414–4429.

Lorenz, D. J. and Hartmann, D. L. (2001). Eddy-zonal flow feedback in the southern hemisphere. *J. Clim.*, 13:4414–4429.

Lorenz, E. L. (1963). Deterministic nonperiodic flow. *J. Atmos. Sci.*, 20:130–141.

Lorenz, E. N. (1969). The predictability of a flow which possess many scales of motion. *Tellus*, XXI(3):289–307.

Lorenz, E. N. (1985). The growth of errors in prediction. Turbulence and Predictability in Geophysical Fluid Dynamics and Climate Dynamics.

Marshall, J. and Molteni, F. (1993). Toward a dynamical understanding of planetary-scale flow regimes. *J. Atmos. Sci.*, 50:1792–1818.

Miyakoda, K., Sirutis, J. and Ploshay, J. (1986). One-month forecast experiments-without anomaly boundary forcings. *Mon. Wea. Rev.*, 114:2363–2401.

Molteni, F., Buizza, R., Palmer, T. N. and Petroliagis, T. (1996). The ECMWF ensemble prediction system: Method and validation. *Quart. J. Roy. Meteor. Soc.*, 122:73–119.

Navarra, A. (1993). A new set of orthonormal modes for linearized meteorological problems. *J. Atmos. Sci.*, 50:2569–2583.

Rhines, P. (1975). Waves and turbulence on a beta-plane. *J. Fluid Mech.*, 69:417–443.

Rhines, P. (1979). Geostrophic turbulence. *Ann. Rev. Fluid Mech.*, 11:401–441.

Ringler, T. and Cook, K. H. (1999). Understanding the seasonality of orographically forced stationary waves: Interaction between mechanical and thermal forcing. *J. Atmos. Sci.*, 56:1154–1174.

Robertson, A. W. (2001). Influence of ocean-atmosphere interaction on the Arctic oscillation in two general circulation models. *J. Clim.*, 14:3240–3254.

Robinson, W. (1996). Does eddy feedback sustain variability in the zonal index? *J. Atmos. Sci.*, 53:3556–3569.

Satoh, M. (1994). Hadley circulations in radiative-convective equilibrium in an axially symmetric atmosphere. *J. Atmos. Sci.*, 51:1947 –1968.

Shigehisa, Y. (1983). Normal modes of the shallow water equations for zonal wavenumber zero. *J. Meteor. Soc. Japan*, 61:479–493.

Shindell, D. T., Miller, R. L., Schmidt, G. A. and Pandolfo, L. (1989). Simulation of recent northern winter climate trends by greenhouse-gas forcing. *Nature*, 399:452–455.

Shiotani, M. (1990). Low-frequency variations of the zonal mean state of the southern hemisphere troposphere. *J. Meteor. Soc. Japan*, 68:461–471.

Shukla, J. (1985). Predictability. issues in atmospheric and oceanic modeling. *Advances in Geophysics*, 28:87–122.

Shutts, G. J. (1986). A case study of eddy forcing during an Atlantic blocking episode. *Adv. in Geophys.*, 29:135–162.

Simmons, A. J., Wallace, J. M., and Branstator, G. W. (1983). Barotropic wave propagation and instability, and atmospheric teleconnection patterns. *J. Atmos. Sci.*, 40:1363–1392.

Stone, P. H. (1978). Baroclinic adjustment. *J. Atmos. Sci.*, 35:561–571.

Suzuki, I. and Tanaka, H. L. (2007). Teleconnections and the Arctic oscillation analyzed in the barotropic component of the model and observedatmosphere. *J. Meteor. Soc. Japan*, 85:933–941.

Tanaka, H. (1985). Global energetics analysis by expansion into three-dimensional normal-mode functions during the FGGE winter. *J. Meteor. Soc. Japan*, 63:180–200.

Tanaka, H. L. (2003). Analysis and modeling of the Arctic oscillation using a simple barotropic model with baroclinic eddy forcing. *J. Atmos. Sci.*, 60:1359–1379.

Tanaka, H. L. and Kung, E. (1988). Normal-mode expansion of the general circulation during the FGGE year. *J. Atmos. Sci.*, 45:3723–3736.

Tanaka, H. L and Kung, E. (1989). A study of low-frequency unstable planetary waves in realistic zonal and zonally varing basic states. *Tellus*, 41A:179–199.

Tanaka, H. L. and Tokinaga, H. (2002). Baroclinic instability in high latitudes induced by polar vortex: A connection to the Arctic oscillation. *J. Atmos. Sci.*, 59:69–82.

Tanaka, H. L. (1991). A numerical simulation of amplification of low-frequency planetary waves and blocking formations by the upscale energy cascade. *Mon. Wea. Rev.*, 119:2919–2935.

Tanaka, H. L. (1998). Numerical simulation of a life-cycle of atmospheric blocking and the analysis of potential vorticity using a simple barotropic model. *J. Meteor. Soc. Japan*, 76:983–1008.

Tanaka, H. L. and Kasahara, A. (1992). On the normal modes of Laplace's tidal equations for zonal wavenumber zero. *Tellus*, 44A:18–32.

Tanaka, H. L. and Matsueda, M. (2005). Arctic oscillation analyzed as a singular eigenmode of the global atmosphere. *J. Meteor. Soc. Japan*, 83:611–619.

Tanaka, H. L. and Nohara, D. (2000). A study of deterministic predictability for the barotropic component of the atmosphere. *Science Report, Institute of Geoscience, University of Tsukuba*, 21:1–21.

Tanaka, H. L. and Sun, S. (1990). A study of baroclinic energy source for large-scale atmospheric normal modes. *J. Atmos. Sci.*, 47:2674–2695.

Tanaka, H. L., Watarai, Y. and Kanda, T. (2004). Energy spectrum proportional to the squared phase speed of Rossby modes in the general circulation of the atmosphere. *Geophys. Res. Lett.*, 31(13):13109.

Thompson, D. W. J. and Wallace, J. M. (1998). The Arctic oscillation signature in the wintertime geopotential height and temperature fields. *Geophy. Res. Lett.*, 25:1297–1300.

Toth, Z. and Kalnay, E. (1997). Ensemble forecasting at NMC: The generation of perturbations. *Mon. Wea. Rev.*, 125:3297–3319.

Vallis, G. K. (1983). On the predictability of quasi-geostrophic flow: The effect of beta and baroclinicity. *J. Atmos. Sci.*, 40:10–27.

Wallace, J. M. and Gutzler, D. S. (1981). Teleconnections in the geopotential height field during the northern hemisphere winter. *Mon. Wea. Rev.*, 109:784–812.

Wallace, J. M. and Thompson, D. W. J. (2002). The pacific center of action of the northern hemisphere annular mode: Real or artifact? *J. Clim.*, 15:1987–1991.

Watanabe, M. and Jin, F.-F. (2004). Dynamical prototype of the Arctic oscillation as revealed by a neutral singular vector. *J. Clim.*, 17:2119–2138.

Yamazaki, K. and Shinya, Y. (1999). Analysis of the Arctic oscillation simulated by agcm. *J. Meteor. Soc. Japan*, 77:1287–1298.

Printed in the United States
by Baker & Taylor Publisher Services